METHODS IN MOLECULAR BIOLOGY

Series Editor
John M. Walker
School of Life and Medical Sciences
University of Hertfordshire
Hatfield, Hertfordshire, AL10 9AB, UK

For further volumes:
http://www.springer.com/series/7651

Coronaviruses

Methods and Protocols

Edited by

Helena Jane Maier

The Pirbright Institute, Compton, Newbury, Berkshire, UK

Erica Bickerton

The Pirbright Institute, Compton, Newbury, Berkshire, UK

Paul Britton

The Pirbright Institute, Compton, Newbury, Berkshire, UK

 Humana Press

Editors
Helena Jane Maier
The Pirbright Institute, Compton
Newbury, Berkshire, UK

Erica Bickerton
The Pirbright Institute, Compton
Newbury, Berkshire, UK

Paul Britton
The Pirbright Institute, Compton
Newbury, Berkshire, UK

ISSN 1064-3745   ISSN 1940-6029 (electronic)
Methods in Molecular Biology
ISBN 978-1-4939-2437-0   ISBN 978-1-4939-2438-7 (eBook)
DOI 10.1007/978-1-4939-2438-7

Library of Congress Control Number: 2015932490

Springer New York Heidelberg Dordrecht London

Printed on acid-free paper

Humana Press is a brand of Springer
Springer Science+Business Media LLC New York is part of Springer Science+Business Media (www.springer.com)

Preface

In this book we aimed to describe a variety of techniques that reflect the wide range of research currently being performed in the field of coronavirology. However, most of the techniques described are also applicable to a wide variety of other virology fields, so we hope that this book will have wider appeal. As such, we have started this book with an overview chapter of current understanding of coronavirus replication and pathogenesis to introduce nonspecialist readers to the field.

Since the emergence of SARS-Coronavirus in 2003, numerous new coronaviruses have been identified. The emergence of MERS-Coronavirus in 2012 and the continued occurrence of human cases highlight the importance of techniques to verify the presence of coronaviruses in a sample as well as identify new coronaviruses that may pose a potential threat to the health of both humans and livestock. As such, chapters have been chosen to describe identification, diagnosis, and study of evolution of coronaviruses.

To allow the study of viruses, propagation and quantification of virus is essential. Therefore, we have included chapters describing preparation of cells and organ cultures useful in propagating coronaviruses and titration techniques. In addition, several techniques for analyzing virus function require purification of virus, so purification protocols suitable for different downstream techniques have been included.

The ability to reverse engineer virus genomes and recover recombinant viruses with defined mutations is invaluable in the progression of understanding the mechanisms for virus pathogenicity, viral protein and RNA function and understanding virus-host interactions. Therefore, chapters describing two commonly used reverse genetics techniques for coronaviruses are included.

A key step in virus replication is attachment to and entry into the host cell. Techniques detailing identification of cellular receptors, binding profiles of viral attachment proteins, and virus-cell fusion are described.

Finally, a major area of coronavirus research currently is the interaction between the virus and the host cell to gain insight into requirements of the virus to enable replication but also how the host cell responds to virus infection. Understanding these processes is vital in enabling future control of virus replication with antiviral therapeutics or prevention through vaccination. Therefore, several chapters have been included covering a broad spectrum of techniques to identify virus-host protein-protein interactions, confirm the functional role of these proteins in virus replication, study host cell responses through genome-wide or pathway-specific approaches, and visualise virus replication complexes.

We would like to thank the authors who have contributed to this book for the time they have taken to prepare detailed methods as well as provide practical hints and tips that are often essential to get a new working protocol.

Compton, UK   *Helena Jane Maier*
Erica Bickerton
Paul Britton

Contents

Contributors

FERNANDO ALMAZÁN • *Department of Molecular and Cell Biology, Centro Nacional de Biotecnología (CNB-CSIC), Campus Universidad Autónoma de Madrid, Cantoblanco, Madrid, Spain*

IRESHA N. AMBEPITIYA WICKRAMASINGHE • *Department of Pathobiology, Faculty of Veterinary Medicine, Utrecht University, Utrecht, The Netherlands*

ERICA BICKERTON • *The Pirbright Institute, Compton, Newbury, Berkshire, UK*

BEREND-JAN BOSCH • *Virology Division, Department of Infectious Diseases & Immunology, Faculty of Veterinary Medicine, Utrecht University, Utrecht, The Netherlands*

PAUL BRITTON • *The Pirbright Institute, Compton, Newbury, Berkshire, UK*

MARTÍ CORTEY • *The Pirbright Institute, Pirbright, Woking, Surrey, UK*

DEIRDRE A. COSTELLO • *School of Chemical and Biomolecular Engineering, Cornell University, Ithaca, NY, USA*

SUSAN DANIEL • *School of Chemical and Biomolecular Engineering, Cornell University, Ithaca, NY, USA*

CORNELIS A.M. DE HAAN • *Virology Division, Department of Infectious Diseases & Immunology, Faculty of Veterinary Medicine, Utrecht University, Utrecht, The Netherlands*

JEROEN A.A. DEMMERS • *Proteomics Department, Erasmus Medical Center, Rotterdam, The Netherlands*

STUART DENT • *School of Biological Sciences, University of Reading, Reading, UK*

RONALD DIJKMAN • *Federal Department of Home Affairs, Institute of Virology and Immunology, Berne and Mittelhäusern, Switzerland*

MARIETTE F. DUCATEZ • *INRA, UMR1225, IHAP, Toulouse, France; Université de Toulouse, INP, ENVT, IHAP, Toulouse, France*

LUIS ENJUANES • *Department of Molecular and Cell Biology, Centro Nacional de Biotecnología (CNB-CSIC), Campus Universidad Autónoma de Madrid, Cantoblanco, Madrid, Spain*

ANTHONY R. FEHR • *Department of Microbiology, University of Iowa Carver College of Medicine, Iowa City, IA, USA*

MARIA FORLENZA • *Cell Biology and Immunology Group, Wageningen Institute of Animal Sciences, Wageningen University, Wageningen, The Netherlands*

JEAN-LUC GUÉRIN • *INRA, UMR1225, IHAP, Toulouse, France; Université de Toulouse, INP, ENVT, UMR1225, IHAP, Toulouse, France*

JAMES S. GUY • *Department of Population Health and Pathobiology, College of Veterinary Medicine, North Carolina State University, Raleigh, NC, USA*

BART L. HAAGMANS • *Department of Viroscience, Erasmus Medical Center, Rotterdam, The Netherlands*

MARNE C. HAGEMEIJER • *Virology Division, Department of Infectious Diseases & Immunology, Faculty of Veterinary Medicine, Utrecht University, Utrecht, The Netherlands; Laboratory of Host-Pathogen Dynamics, Cell Biology and Physiology Center (CBPC), National Heart Lung and Blood Institute (NHLBI), National Institutes of Health, Bethesda, MD, USA*

MOHAMMAD SYAMSUL REZA HARUN • *Institute of Bioscience, Faculty of Veterinary Medicine, Universiti Putra Malaysia, Serdang, Selangor, Malaysia*

PHILIPPA C. HAWES • *The Pirbright Institute, Pirbright, Guildford, Surrey, UK*

RUTH M. HENNION • *The Pirbright Institute, Compton, Newbury, Berkshire, UK*

GILLIAN HILL • *The Pirbright Institute, Compton, Newbury, Berkshire, UK*

TSUTOMU HOHDATSU • *Laboratory of Veterinary Infectious Disease, School of Veterinary Medicine, Kitasato University, Towada, Aomori, Japan*

CHENG-HUEI HUNG • *Department of Laboratory Medicine and Biotechnology, Tzu Chi University, Hualien, Taiwan*

YVES JACOB • *CNRS, UMR3569, Paris, France; Unité de Génétique Moléculaire des Virus à ARN, Département de Virologie, Institut Pasteur, Paris, France; Sorbonne Paris Cité, Unité de Génétique Moléculaire des Virus à ARN, Université Paris Diderot, Paris, France*

ERIK JAGT • *MSD Animal Health, Bioprocess Technology and Support, Boxmeer, The Netherlands*

LOUIS M. JONES • *Centre d'Informatique pour la Biologie, Institut Pasteur, Paris, France*

HULDA R. JONSDOTTIR • *Federal Department of Home Affairs, Institute of Virology and Immunology, Berne and Mittelhäusern, Switzerland*

SARAH M. KEEP • *The Pirbright Institute, Compton, Newbury, Berkshire, UK*

JOERI KINT • *Cell Biology and Immunology Group, Wageningen Institute of Animal Sciences, Wageningen University, Wageningen, The Netherlands; MSD Animal Health, Bioprocess Technology and Support, Boxmeer, The Netherlands*

MART M. LAMERS • *Department of Viroscience, Erasmus Medical Center, Rotterdam, The Netherlands*

HUI-CHUN LI • *Department of Biochemistry, Tzu Chi University, Hualien, Taiwan*

SHIH-YEN LO • *Institute of Medical Sciences, Tzu Chi University, Hualien, Taiwan; Department of Laboratory Medicine and Biotechnology, Tzu Chi University, Hualien, Taiwan; Department of Laboratory Medicine, Buddhist Tzu Chi General Hospital, Hualien, Taiwan*

HELENA JANE MAIER • *The Pirbright Institute, Compton, Newbury, Berkshire, UK*

SILVIA MÁRQUEZ-JURADO • *Department of Molecular and Cell Biology, Centro Nacional de Biotecnología (CNB-CSIC), Campus Universidad Autónoma de Madrid, Cantoblanco, Madrid, Spain*

PARVANEH MEHRBOD • *Institute of Bioscience, Faculty of Veterinary Medicine, Universiti Putra Malaysia, Serdang, Selangor, Malaysia*

JEAN KAORU MILLET • *Department of Microbiology and Immunology, Cornell University, Ithaca, NY, USA*

HUIHUI MOU • *Virology Division, Department of Infectious Diseases & Immunology, Faculty of Veterinary Medicine, Utrecht University, Utrecht, The Netherlands*

MUHAMMAD MUNIR • *The Pirbright Institute, Pirbright, Woking, Surrey, UK*

BÉATRICE NAL • *Department of Life Sciences, College of Health and Life Sciences, Brunel University, London, UK*

BENJAMIN W. NEUMAN • *School of Biological Sciences, University of Reading, Reading, UK*

GRÉGORY NEVEU • *Institut Pasteur, Unité de Génomique Virale et Vaccination, Paris, France; CNRS, UMR3569, Paris, France; Division of Infectious Diseases and Geographic Medicine, Department of Medicine, Stanford University School of Medicine, Stanford, CA, USA; Department of Microbiology and Immunology, Stanford University School of Medicine, Stanford, CA, USA*

AITOR NOGALES • *Department of Microbiology and Immunology, University of Rochester, Rochester, NY, USA*

ABDUL RAHMAN OMAR • *Institute of Bioscience, Faculty of Veterinary Medicine, Universiti Putra Malaysia, Serdang, Selangor, Malaysia*

STANLEY PERLMAN • *Department of Microbiology, University of Iowa Carver College of Medicine, Iowa City, IA, USA*

V. STALIN RAJ • *Department of Viroscience, Erasmus Medical Center, Rotterdam, The Netherlands*

PETER J.M. ROTTIER • *Virology Division, Department of Infectious Diseases & Immunology, Faculty of Veterinary Medicine, Utrecht University, Utrecht, The Netherlands*

JEAN-PIERRE ROUSSARIE • *Institut Pasteur, Unité de Génomique Virale et Vaccination, Paris, France; Laboratory of Molecular and Cellular Neuroscience, The Rockefeller University, New York, NY, USA*

AHMAD NAQIB SHUID • *Institute of Bioscience, Faculty of Veterinary Medicine, Universiti Putra Malaysia, Serdang, Selangor, Malaysia*

SASKIA L. SMITS • *Department of Viroscience, Erasmus Medical Center, Rotterdam, The Netherlands*

TOMOMI TAKANO • *Laboratory of Veterinary Infectious Disease, School of Veterinary Medicine, Kitasato University, Towada, Aomori, Japan*

FRÉDÉRIC TANGY • *Institut Pasteur, Unité de Génomique Virale et Vaccination, Paris, France; CNRS, UMR3569, Paris, France*

M. HÉLÈNE VERHEIJE • *Department of Pathobiology, Faculty of Veterinary Medicine, Utrecht University, Utrecht, The Netherlands*

PIERRE-OLIVIER VIDALAIN • *Institut Pasteur, Unité de Génomique Virale et Vaccination, Paris, France; CNRS, UMR3569, Paris, France*

CHEE-HING YANG • *Institute of Medical Sciences, Tzu Chi University, Hualien, Taiwan*

Chapter 1

Coronaviruses: An Overview of Their Replication and Pathogenesis

Anthony R. Fehr and Stanley Perlman

Abstract

Coronaviruses (CoVs), enveloped positive-sense RNA viruses, are characterized by club-like spikes that project from their surface, an unusually large RNA genome, and a unique replication strategy. Coronaviruses cause a variety of diseases in mammals and birds ranging from enteritis in cows and pigs and upper respiratory disease in chickens to potentially lethal human respiratory infections. Here we provide a brief introduction to coronaviruses discussing their replication and pathogenicity, and current prevention and treatment strategies. We also discuss the outbreaks of the highly pathogenic Severe Acute Respiratory Syndrome Coronavirus (SARS-CoV) and the recently identified Middle Eastern Respiratory Syndrome Coronavirus (MERS-CoV).

Key words Nidovirales, Coronavirus, Positive-sense RNA viruses, SARS-CoV, MERS-CoV

1 Classification

Coronaviruses (CoVs) are the largest group of viruses belonging to the *Nidovirales* order, which includes *Coronaviridae*, *Arteriviridae*, *Mesoniviridae*, and *Roniviridae* families. The *Coronavirinae* comprise one of two subfamilies in the *Coronaviridae* family, with the other being the *Torovirinae*. The *Coronavirinae* are further subdivided into four genera, the alpha, beta, gamma, and delta coronaviruses. The viruses were initially sorted into these genera based on serology but are now divided by phylogenetic clustering.

All viruses in the *Nidovirales* order are enveloped, non-segmented positive-sense RNA viruses. They all contain very large genomes for RNA viruses, with some viruses having the largest identified RNA genomes, containing up to 33.5 kilobase (kb) genomes. Other common features within the *Nidovirales* order include: (1) a highly conserved genomic organization, with a large replicase gene preceding structural and accessory genes; (2) expression of many non-structural genes by ribosomal

Helena Jane Maier et al. (eds.), *Coronaviruses: Methods and Protocols*, Methods in Molecular Biology, vol. 1282,
DOI 10.1007/978-1-4939-2438-7_1, © Springer Science+Business Media New York 2015

frameshifting; (3) several unique or unusual enzymatic activities encoded within the large replicase–transcriptase polyprotein; and (4) expression of downstream genes by synthesis of 3' nested subgenomic mRNAs. In fact, the *Nidovirales* order name is derived from these nested 3' mRNAs as *nido* is Latin for "nest." The major differences within the Nidovirus families are in the number, type, and sizes of the structural proteins. These differences cause significant alterations in the structure and morphology of the nucleocapsids and virions.

2   Genomic Organization

Coronaviruses contain a non-segmented, positive-sense RNA genome of ~30 kb. The genome contains a 5' cap structure along with a 3' poly (A) tail, allowing it to act as an mRNA for translation of the replicase polyproteins. The replicase gene encoding the nonstructural proteins (nsps) occupies two-thirds of the genome, about 20 kb, as opposed to the structural and accessory proteins, which make up only about 10 kb of the viral genome. The 5' end of the genome contains a leader sequence and untranslated region (UTR) that contains multiple stem loop structures required for RNA replication and transcription. Additionally, at the beginning of each structural or accessory gene are transcriptional regulatory sequences (TRSs) that are required for expression of each of these genes (*see* Subheading 4.3 on RNA replication). The 3' UTR also contains RNA structures required for replication and synthesis of viral RNA. The organization of the coronavirus genome is 5'-leader-UTR-replicase-S (Spike)-E (Envelope)-M (Membrane)-N (Nucleocapsid)-3' UTR-poly (A) tail with accessory genes interspersed within the structural genes at the 3' end of the genome (Fig. 1). The accessory proteins are almost exclusively nonessential for replication in tissue culture; however, some have been shown to have important roles in viral pathogenesis [1].

3   Virion Structure

Coronavirus virions are spherical with diameters of approximately 125 nm as depicted in recent studies by cryo-electron tomography and cryo-electron microscopy [2, 3]. The most prominent feature of coronaviruses is the club-shaped spike projections emanating from the surface of the virion. These spikes are a defining feature of the virion and give them the appearance of a solar corona, prompting the name, coronaviruses. Within the envelope of the virion is the nucleocapsid. Coronaviruses have helically symmetrical nucleocapsids, which is uncommon among positive-sense RNA viruses, but far more common for negative-sense RNA viruses.

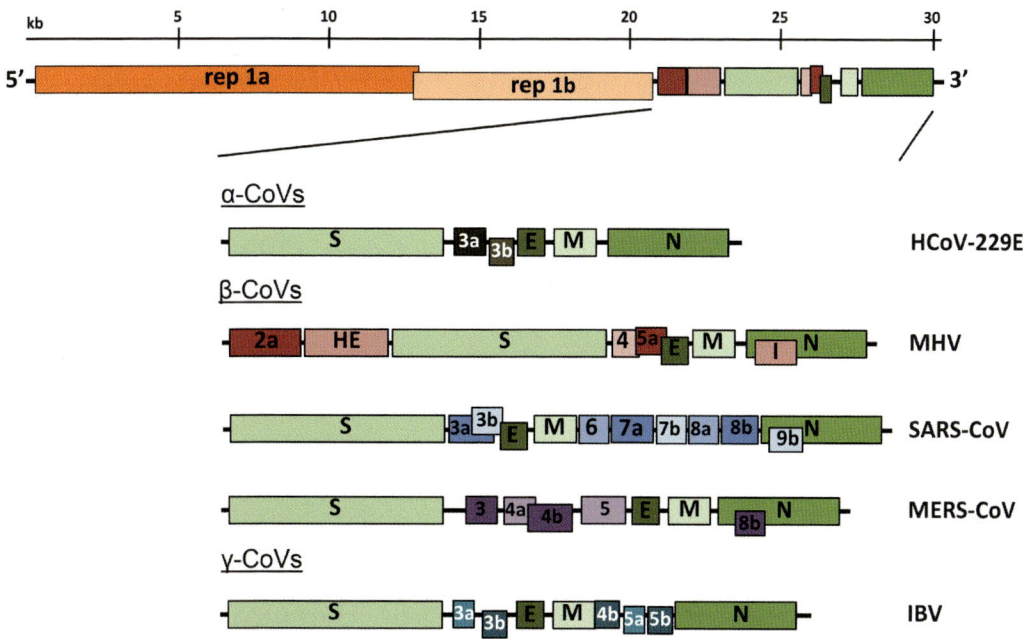

Fig. 1 Genomic organization of representative α, β, and γ CoVs. An illustration of the MHV genome is depicted at the *top*. The expanded regions below show the structural and accessory proteins in the 3′ regions of the HCoV-229E, MHV, SARS-CoV, MERS-CoV and IBV. Size of the genome and individual genes are approximated using the legend at the *top* of the diagram but are not drawn to scale. *HCoV-229E* human coronavirus 229E, *MHV* mouse hepatitis virus, *SARS-CoV* severe acute respiratory syndrome coronavirus, *MERS-CoV* Middle East respiratory syndrome coronavirus, *IBV* infectious bronchitis virus

Coronavirus particles contain four main structural proteins. These are the spike (S), membrane (M), envelope (E), and nucleocapsid (N) proteins, all of which are encoded within the 3′ end of the viral genome. The S protein (~150 kDa), utilizes an N-terminal signal sequence to gain access to the ER, and is heavily N-linked glycosylated. Homotrimers of the virus encoded S protein make up the distinctive spike structure on the surface of the virus [4, 5]. The trimeric S glycoprotein is a class I fusion protein [6] and mediates attachment to the host receptor [7]. In most, coronaviruses, S is cleaved by a host cell furin-like protease into two separate polypeptides noted S1 and S2 [8, 9]. S1 makes up the large receptor-binding domain of the S protein, while S2 forms the stalk of the spike molecule [10].

The M protein is the most abundant structural protein in the virion. It is a small (~25–30 kDa) protein with three transmembrane domains [11] and is thought to give the virion its shape. It has a small N-terminal glycosylated ectodomain and a much larger C-terminal endodomain that extends 6–8 nm into the viral particle [12]. Despite being co-translationally inserted in the ER membrane, most M proteins do not contain a signal sequence. Recent studies suggest the M protein exists as a dimer in the virion, and may adopt two different conformations, allowing it to promote membrane curvature as well as to bind to the nucleocapsid [13].

The E protein (~8–12 kDa) is found in small quantities within the virion. The coronavirus E proteins are highly divergent but have a common architecture [14]. The membrane topology of E protein is not completely resolved but most data suggest that it is a transmembrane protein. The E protein has an N-terminal ectodomain and a C-terminal endodomain and has ion channel activity. As opposed to other structural proteins, recombinant viruses lacking the E protein are not always lethal, although this is virus type dependent [15]. The E protein facilitates assembly and release of the virus (*see* Subheading 4.4), but also has other functions. For instance, the ion channel activity in SARS-CoV E protein is not required for viral replication but is required for pathogenesis [16].

The N protein constitutes the only protein present in the nucleocapsid. It is composed of two separate domains, an N-terminal domain (NTD) and a C-terminal domain (CTD), both capable of binding RNA in vitro, but each domain uses different mechanisms to bind RNA. It has been suggested that optimal RNA binding requires contributions from both domains [17, 18]. N protein is also heavily phosphorylated [19], and phosphorylation has been suggested to trigger a structural change enhancing the affinity for viral versus nonviral RNA. N protein binds the viral genome in a beads-on-a-string type conformation. Two specific RNA substrates have been identified for N protein; the TRSs [20] and the genomic packaging signal [21]. The genomic packaging signal has been found to bind specifically to the second, or C-terminal RNA binding domain [22]. N protein also binds nsp3 [18, 23], a key component of the replicase complex, and the M protein [24]. These protein interactions likely help tether the viral genome to the replicase–transcriptase complex (RTC), and subsequently package the encapsidated genome into viral particles.

A fifth structural protein, the hemagglutinin-esterase (HE), is present in a subset of β-coronaviruses. The protein acts as a hemagglutinin, binds sialic acids on surface glycoproteins, and contains acetyl-esterase activity [25]. These activities are thought to enhance S protein-mediated cell entry and virus spread through the mucosa [26]. Interestingly, HE enhances murine hepatitis virus (MHV) neurovirulence [27]; however, it is selected against in tissue culture for unknown reasons [28].

4   Coronavirus Life Cycle

4.1   Attachment and Entry

The initial attachment of the virion to the host cell is initiated by interactions between the S protein and its receptor. The sites of receptor binding domains (RBD) within the S1 region of a

Table 1
Coronavirus receptors

Virus	Receptor	References
Alphacoronaviruses		
HCoV-229E	APN	[115]
HCoV-NL63	ACE2	[116]
TGEV	APN	[117]
PEDV	APN	[118]
FIPV	APN	[119]
CCoV	APN	[120]
Betacoronaviruses		
MHV	mCEACAM	[121, 122]
BCoV	*N*-acetyl-9-*O*-acetylneuraminic acid	[123]
SARS-CoV	ACE2	[124]
MERS-CoV	DPP4	[100]

APN aminopeptidase N, *ACE2* angiotensin-converting enzyme 2, *mCEACAM* murine carcinoembryonic antigen-related adhesion molecule 1, *DPP4* dipeptidyl peptidase 4, *HCoV* human coronavirus, *TGEV* transmissible gastroenteritis virus, *PEDV* porcine epidemic diarrhea virus, *FIPV* feline infectious peritonitis virus, *CCoV* canine coronavirus, *MHV* murine hepatitis virus, *BCoV* bovine coronavirus, *SARS-CoV* severe acute respiratory syndrome coronavirus, *MERS-CoV* Middle East respiratory syndrome coronavirus

coronavirus S protein vary depending on the virus, with some having the RBD at the N-terminus of S1 (MHV), while others (SARS-CoV) have the RBD at the C-terminus of S1 [29, 30]. The S-protein–receptor interaction is the primary determinant for a coronavirus to infect a host species and also governs the tissue tropism of the virus. Many coronaviruses utilize peptidases as their cellular receptor. It is unclear why peptidases are used, as entry occurs even in the absence of the enzymatic domain of these proteins. Many α-coronaviruses utilize aminopeptidase N (APN) as their receptor, SARS-CoV and HCoV-NL63 use angiotensin-converting enzyme 2 (ACE2) as their receptor, MHV enters through CEACAM1, and the recently identified MERS-CoV binds to dipeptidyl-peptidase 4 (DPP4) to gain entry into human cells (*see* Table 1 for a list of known CoV receptors).

Following receptor binding, the virus must next gain access to the host cell cytosol. This is generally accomplished by acid-dependent proteolytic cleavage of S protein by a cathepsin, TMPRRS2 or another protease, followed by fusion of the viral and cellular membranes. S protein cleavage occurs at two sites within the S2 portion of the protein, with the first cleavage important for separating the RBD and fusion domains of the S protein [31] and the second for exposing the fusion peptide (cleavage at S2′). Fusion generally occurs within acidified endosomes, but some coronaviruses, such as MHV, can fuse at the plasma membrane. Cleavage at S2′ exposes a fusion peptide that inserts into the membrane, which is followed by joining of two heptad repeats in S2 forming an antiparallel six-helix bundle [6]. The formation of this bundle allows for the mixing of viral and cellular membranes, resulting in fusion and ultimately release of the viral genome into the cytoplasm.

4.2   Replicase Protein Expression

The next step in the coronavirus lifecycle is the translation of the replicase gene from the virion genomic RNA. The replicase gene encodes two large ORFs, rep1a and rep1b, which express two co-terminal polyproteins, pp1a and pp1ab (Fig. 1). In order to express both polyproteins, the virus utilizes a slippery sequence (5′-UUUAAAC-3′) and an RNA pseudoknot that cause ribosomal frameshifting from the rep1a reading frame into the rep1b ORF. In most cases, the ribosome unwinds the pseudoknot structure, and continues translation until it encounters the rep1a stop codon. Occasionally the pseudoknot blocks the ribosome from continuing elongation, causing it to pause on the slippery sequence, changing the reading frame by moving back one nucleotide, a -1 frameshift, before the ribosome is able to melt the pseudoknot structure and extend translation into rep1b, resulting in the translation of pp1ab [32, 33]. In vitro studies predict the incidence of ribosomal frameshifting to be as high as 25 %, but this has not been determined in the context of virus infection. It is unknown exactly why these viruses utilize frameshifting to control protein expression, but it is hypothesized to either control the precise ratio of rep1b and rep1a proteins or delay the production of rep1b products until the products of rep1a have created a suitable environment for RNA replication [34].

Polyproteins pp1a and pp1ab contain the nsps 1–11 and 1–16, respectively. In pp1ab, nsp11 from pp1a becomes nsp12 following extension of pp1a into pp1b. However, γ-coronaviruses do not contain a comparable nsp1. These polyproteins are subsequently cleaved into the individual nsps [35]. Coronaviruses encode either two or three proteases that cleave the replicase polyproteins. They are the papain-like proteases (PLpro), encoded within nsp3, and a serine type protease, the main protease, or Mpro, encoded by nsp5. Most coronaviruses encode two PLpros within nsp3, except the γ-coronaviruses, SARS-CoV and MERS-CoV, which only express one PLpro [36]. The PLpros cleave the nsp1/2, nsp2/3, and nsp3/4 boundaries, while the Mpro is responsible for the remaining 11 cleavage events.

Next, many of the nsps assemble into the replicase–transcriptase complex (RTC) to create an environment suitable for RNA synthesis, and ultimately are responsible for RNA replication and transcription of the sub-genomic RNAs. The nsps also contain other enzyme domains and functions, including those important for RNA replication, for example nsp12 encodes the RNA-dependent RNA polymerase (RdRp) domain; nsp13 encodes the RNA helicase domain and RNA 5′-triphosphatase activity; nsp14 encodes the exoribonuclease (ExoN) involved in replication fidelity and N7-methyltransferase activity; and nsp16 encodes 2′-O-methyltransferase activity. In addition to the replication functions other activities, such as blocking innate immune responses (nsp1; nsp16-2′-O-methyl transferase; nsp3-deubiquitinase) have been identified for some of the nsps, while others have largely unknown functions (nsp3-ADP-ribose-1″-phosphatase; nsp15-endoribo-nuclease (NendoU)). For a list of non-structural proteins and their proposed functions, *see* Table 2. Interestingly, ribonucleases nsp15-NendoU and nsp14-ExoN activities are unique to the *Nidovirales* order and are considered genetic markers for these viruses [37].

4.3 Replication and Transcription

Viral RNA synthesis follows the translation and assembly of the viral replicase complexes. Viral RNA synthesis produces both genomic and sub-genomic RNAs. Sub-genomic RNAs serve as mRNAs for the structural and accessory genes which reside downstream of the replicase polyproteins. All positive-sense sub-genomic RNAs are 3′ co-terminal with the full-length viral genome and thus form a set of nested RNAs, a distinctive property of the order *Nidovirales*. Both genomic and sub-genomic RNAs are produced through negative-strand intermediates. These negative-strand intermediates are only about 1 % as abundant as their positive-sense counterparts and contain both poly-uridylate and anti-leader sequences [38].

Many cis-acting sequences are important for the replication of viral RNAs. Within the 5′ UTR of the genome are seven stem-loop structures that may extend into the replicase 1a gene [39–42]. The 3′ UTR contains a bulged stem-loop, a pseudoknot, and a hyper-variable region [43–46]. Interestingly, the stem-loop and the pseudoknot at the 3′ end overlap, and thus cannot form simultaneously [44, 47]. Therefore, these different structures are proposed to regulate alternate stages of RNA synthesis, although exactly which stages are regulated and their precise mechanism of action are still unknown.

Perhaps the most novel aspect of coronavirus replication is how the leader and body TRS segments fuse during production of sub-genomic RNAs. This was originally thought to occur during positive-strand synthesis, but now it is largely believed to occur during the discontinuous extension of negative-strand RNA [48]. The current model proposes that the RdRp pauses at any one of the body TRS sequences (TRS-B); following this pause the RdRp either continues elongation to the next TRS or it switches to amplifying the leader sequence at the 5′ end of the genome guided by

Table 2
Functions of coronavirus non-structural proteins (nsps)

Protein	Function	References
nsp1	Promotes cellular mRNA degradation and blocks host cell translation, results in blocking innate immune response	[125–128]
nsp2	No known function, binds to prohibitin proteins	[129, 130]
nsp3	Large, multi-domain transmembrane protein, activities include: • Ubl1 and Ac domains, interact with N protein • ADRP activity, promotes cytokine expression • PLPro/Deubiquitinase domain, cleaves viral polyprotein and blocks host innate immune response • Ubl2, NAB, G2M, SUD, Y domains, unknown functions	[131–138]
nsp4	Potential transmembrane scaffold protein, important for proper structure of DMVs	[139, 140]
nsp5	Mpro, cleaves viral polyprotein	[141]
nsp6	Potential transmembrane scaffold protein	[142]
nsp7	Forms hexadecameric complex with nsp8, may act as processivity clamp for RNA polymerase	[143]
nsp8	Forms hexadecameric complex with nsp7, may act as processivity clamp for RNA polymerase; may act as primase	[143, 144]
nsp9	RNA binding protein	[145]
nsp10	Cofactor for nsp16 and nsp14, forms heterodimer with both and stimulates ExoN and 2-O-MT activity	[146, 147]
nsp12	RdRp	[148]
nsp13	RNA helicase, 5' triphosphatase	[149, 150]
nsp14	N7 MTase and 3'-5' exoribonuclease, ExoN; N7 MTase adds 5' cap to viral RNAs, ExoN activity is important for proofreading of viral genome	[151–154]
nsp15	Viral endoribonuclease, NendoU	[155, 156]
nsp16	2'-O-MT; shields viral RNA from MDA5 recognition	[157, 158]

Ubl ubiquitin-like, *Ac* acidic, *ADRP* ADP-ribose-1'-phosphatase, *PLPro* papain-like protease, *NAB* nucleic acid binding, *SUD* SARS-unique domain, *DMVs* double-membrane vesicles, *Mpro* main protease, *RdRp* RNA-dependent RNA polymerase, *MTase* methyltransferase, *Exo N* viral exoribonuclease, *Nendo U* viral endoribonuclease, *2'-O-MT* 2'-O-methyltransferase, *MDA5* melanoma differentiation associated protein 5

complementarity of the TRS-B to the leader TRS (TRS-L). Many pieces of evidence currently support this model, including the presence of anti-leader sequence at the 3' end of the negative-strand sub-genomic RNAs [38]. However, many questions remain to fully define the model. For instance, how does the RdRp bypass all of the TRS-B sequences to produce full-length negative-strand genomic RNA? Also, how are the TRS-B sequences directed to the

TRS-L and how much complementarity is necessary [49]? Answers to these questions and others will be necessary to gain a full perspective of how RNA replication occurs in coronaviruses.

Finally, coronaviruses are also known for their ability to recombine using both homologous and nonhomologous recombination [50, 51]. The ability of these viruses to recombine is tied to the strand switching ability of the RdRp. Recombination likely plays a prominent role in viral evolution and is the basis for targeted RNA recombination, a reverse genetics tool used to engineer viral recombinants at the 3′ end of the genome.

4.4 Assembly and Release

Following replication and sub-genomic RNA synthesis, the viral structural proteins, S, E, and M are translated and inserted into the endoplasmic reticulum (ER). These proteins move along the secretory pathway into the endoplasmic reticulum–Golgi intermediate compartment (ERGIC) [52, 53]. There, viral genomes encapsidated by N protein bud into membranes of the ERGIC containing viral structural proteins, forming mature virions [54].

The M protein directs most protein–protein interactions required for assembly of coronaviruses. However, M protein is not sufficient for virion formation, as virus-like particles (VLPs) cannot be formed by M protein expression alone. When M protein is expressed along with E protein VLPs are formed, suggesting these two proteins function together to produce coronavirus envelopes [55]. N protein enhances VLP formation, suggesting that fusion of encapsidated genomes into the ERGIC enhances viral envelopment [56]. The S protein is incorporated into virions at this step, but is not required for assembly. The ability of the S protein to traffic to the ERGIC and interact with the M protein is critical for its incorporation into virions.

While the M protein is relatively abundant, the E protein is only present in small quantities in the virion. Thus, it is likely that M protein interactions provide the impetus for envelope maturation. It is unknown how E protein assists M protein in assembly of the virion, and several possibilities have been suggested. Some work has indicated a role for the E protein in inducing membrane curvature [57–59], although others have suggested that E protein prevents the aggregation of M protein [60]. The E protein may also have a separate role in promoting viral release by altering the host secretory pathway [61].

The M protein also binds to the nucleocapsid, and this interaction promotes the completion of virion assembly. These interactions have been mapped to the C-terminus of the endodomain of M with CTD of the N-protein [62]. However, it is unclear exactly how the nucleocapsid complexed with virion RNA traffics to the ERGIC to interact with M protein and become incorporated into the viral envelope. Another outstanding question is how the N protein selectively packages only positive-sense full-length genomes among the many different RNA species produced during infection. A packaging signal for MHV has been identified in the nsp15 coding sequence, but mutation of this signal does not appear to affect virus production, and a mechanism for how this packaging signal works has not been determined [22]. Furthermore, most coronaviruses do not contain similar sequences at this locus, indicating that packaging may be virus specific.

Following assembly, virions are transported to the cell surface in vesicles and released by exocytosis. It is not known if the virions use the traditional pathway for transport of large cargo from the Golgi or if the virus has diverted a separate, unique pathway for its own exit. In several coronaviruses, S protein that does not get assembled into virions transits to the cell surface where it mediates cell–cell fusion between infected cells and adjacent, uninfected cells. This leads to the formation of giant, multinucleated cells, which allows the virus to spread within an infected organism without being detected or neutralized by virus-specific antibodies.

5   Pathogenesis

5.1   Animal Coronaviruses

Coronaviruses cause a large variety of diseases in animals, and their ability to cause severe disease in livestock and companion animals such as pigs, cows, chickens, dogs, and cats led to significant research on these viruses in the last half of the twentieth century. For instance, Transmissible Gastroenteritis Virus (TGEV) and Porcine Epidemic Diarrhea Virus (PEDV) cause severe gastroenteritis in young piglets, leading to significant morbidity, mortality, and ultimately economic losses. PEDV recently emerged in North America for the first time, causing significant losses of young piglets. Porcine hemagglutinating encephalomyelitis virus (PHEV) mostly leads to enteric infection but has the ability to infect the nervous system, causing encephalitis, vomiting, and wasting in pigs. Feline enteric coronavirus (FCoV) causes a mild or asymptomatic infection in domestic cats, but during persistent infection, mutation transforms the virus into a highly virulent strain of FCoV, Feline Infectious Peritonitis Virus (FIPV), that leads to development of a lethal disease called feline infectious peritonitis (FIP). FIP has wet and dry forms, with similarities to the human disease, sarcoidosis. FIPV is macrophage tropic and it is believed that it causes aberrant cytokine and/or chemokine expression and lymphocyte depletion, resulting in lethal disease [63]. However, additional research is needed to confirm this hypothesis. Bovine CoV, Rat CoV, and Infectious Bronchitis Virus (IBV) cause mild to severe respiratory tract infections in cattle, rats, and chickens, respectively. Bovine CoV causes significant losses in the cattle industry and also has spread to infect a variety of ruminants, including elk, deer, and camels. In addition to severe respiratory disease, the virus causes diarrhea ("winter dysentery" and "shipping fever"), all leading to weight loss, dehydration, decreased milk production, and depression [63]. Some strains of IBV, a γ-coronavirus, also affect the urogenital tract of chickens causing renal disease. Infection of the reproductive tract with IBV significantly diminishes egg production, causing substantial losses in the egg-production industry each year [63]. More recently, a novel coronavirus named SW1 has been identified in a deceased Beluga whale [64]. Large numbers of virus particles were identified in the liver of the deceased whale with respiratory disease and acute liver failure. Although, electron microscopic images were not sufficient to identify the virus as a coronavirus, sequencing of the liver tissue clearly identified the virus as a coronavirus. It was subsequently determined to be a γ-coronavirus based on phylogenetic analysis but it has not yet been verified experimentally that this virus is actually a causative agent of disease in whales. In addition, there has been intense interest in identifying novel bat CoVs, since these are the likely ancestors for SARS-CoV and MERS-CoV, and hundreds of novel bat coronaviruses have been identified over the past decade [65]. Finally, another novel family of nidoviruses, *Mesoniviridae*, has been recently identified as the first nidoviruses to exclusively infect insect hosts [66, 67]. These viruses are highly divergent from other nidoviruses but are most closely related to the roniviruses. In size, they are ~20 kb, falling in between large and small nidoviruses. Interestingly, these viruses do not encode for an endoribonuclease, which is present in all other nidoviruses. These attributes suggest these viruses are the prototype of a new nidovirus family and may be a missing link in the transition from small to large nidoviruses.

The most heavily studied animal coronavirus is murine hepatitis virus (MHV), which causes a variety of outcomes in mice, including respiratory, enteric, hepatic, and neurologic infections. These infections often serve as highly useful models of disease. For instance, MHV-1 causes severe respiratory disease in susceptible A/J and C3H/HeJ mice, A59 and MHV-3 induce severe hepatitis, while JHMV causes severe encephalitis. Interestingly, MHV-3 induces cellular injury through the activation of the coagulation cascade [68]. Most notably, A59 and attenuated versions of JHMV cause a chronic demyelinating disease that bears similarities to multiple sclerosis (MS), making MHV infection one of the best models for this debilitating human disease. Early studies suggested that demyelination was dependent on viral replication in oligodendrocytes in the brain and spinal cord [69, 70]; however, more recent reports clearly demonstrate that the disease is immune-mediated. Irradiated mice or immunodeficient (lacking T and B cells) mice do not develop demyelination, but addition of virus-specific T cells restores the development of demyelination [71, 72]. Additionally, demyelination is accompanied by a large influx of macrophages and microglia that can phagocytose infected myelin [73], although it is unknown what the signals are that direct immune cells to destroy myelin. Finally, MHV can be studied under BSL2 laboratory conditions, unlike SARS-CoV or MERS-CoV, which require a BSL3 laboratory, and provides a large number of suitable animal models. These factors make MHV an ideal model for studying the basics of viral replication in tissue culture cells as well as for studying the pathogenesis and immune response to coronaviruses.

5.2 Human Coronaviruses

Prior to the SARS-CoV outbreak, coronaviruses were only thought to cause mild, self-limiting respiratory infections in humans. Two of these human coronaviruses are α-coronaviruses, HCoV-229E and HCoV-NL63, while the other two are β-coronaviruses, HCoV-OC43 and HCoV-HKU1. HCoV-229E and HCoV-OC43 were isolated nearly 50 years ago [74–76], while HCoV-NL63 and HCoV-HKU1 have only recently been identified following the SARS-CoV outbreak [77, 78]. These viruses are endemic in the human populations, causing 15–30 % of respiratory tract infections each year. They cause more severe disease in neonates, the elderly, and in individuals with underlying illnesses, with a greater incidence of lower respiratory tract infection in these populations. HCoV-NL63 is also associated with acute laryngotracheitis (croup) [79]. One interesting aspect of these viruses is their differences in tolerance to genetic variability. HCoV-229E isolates from around the world have only minimal sequence divergence [80], while HCoV-OC43 isolates from the same location but isolated in different years show significant genetic variability [81]. This likely explains the inability of HCoV-229E to cross the species barrier to infect mice while HCoV-OC43 and the closely related bovine coronavirus, BCoV, are capable of infecting mice and several ruminant species. Based on the ability of MHV to cause demyelinating disease, it has been suggested that human CoVs may be involved in the development of multiple sclerosis (MS). However, no evidence to date suggests that human CoVs play a significant role in MS.

SARS-CoV, a group 2b β-coronavirus, was identified as the causative agent of the Severe Acute Respiratory Syndrome (SARS) outbreak that occurred in 2002–2003 in the Guangdong Province of China. It is the most severe human disease caused by any coronavirus. During the 2002–2003 outbreak approximately 8,098 cases occurred with 774 deaths, resulting in a mortality rate of 9 %. This rate was much higher in elderly individuals, with mortality rates approaching 50 % in individuals over 60 years of age. Furthermore, the outbreak resulted in the loss of nearly $40 billion dollars in economic activity, as the virus nearly shut down many activities in Southeast Asia and Toronto, Canada for several months. The outbreak began in a hotel in Hong Kong and ultimately spread to more than two dozen countries. During the epidemic, closely related viruses were isolated from several exotic animals including Himalayan palm civets and raccoon dogs [82]. However, it is widely accepted that SARS-CoV originated in bats as a large number of Chinese horseshoe bats contain sequences of SARS-related CoVs and contain serologic evidence for a prior infection with a related CoV [83, 84]. In fact, two novel bat SARS-related CoVs have been recently identified that are more similar to SARS-CoV than any other virus identified to date [85]. They were also found to use the same receptor as the human virus, angiotensin converting enzyme 2 (ACE2), providing further evidence that SARS-CoV originated in bats. Although some human individuals within wet animal markets had serologic evidence of SARS-CoV infection prior to the outbreak, these individuals had no apparent symptoms [82]. Thus, it is likely that a closely related virus circulated in the wet animal markets for several years before a series of factors facilitated its spread into the larger population.

Transmission of SARS-CoV was relatively inefficient, as it only spread through direct contact with infected individuals after the onset of illness. Thus, the outbreak was largely contained within households and healthcare settings [86], except in a few cases of superspreading events where one individual was able to infect multiple contacts due to an enhanced development of high viral burdens or ability to aerosolize virus. As a result of the relatively inefficient transmission of SARS-CoV, the outbreak was controllable through the use of quarantining. Only a small number of SARS cases occurred after the outbreak was controlled in June 2003.

SARS-CoV primarily infects epithelial cells within the lung. The virus is capable of entering macrophages and dendritic cells but only leads to an abortive infection [87, 88]. Despite this, infection of these cell types may be important in inducing proinflammatory cytokines that may contribute to disease [89]. In fact, many cytokines and chemokines are produced by these cell types and are elevated in the serum of SARS-CoV infected patients [90]. The exact mechanism of lung injury and cause of severe disease in humans remains undetermined. Viral titers seem to diminish when severe disease develops in both humans and in several animal models of the disease. Furthermore, animals infected with rodent-adapted SARS-CoV strains show similar clinical features to the human disease, including an age-dependent increase in disease severity [91]. These animals also show increased levels of proinflammatory cytokines and reduced T-cell responses, suggesting a possible immunopathological mechanism of disease [92, 93].

While the SARS-CoV epidemic was controlled in 2003 and the virus has not since returned, a novel human CoV emerged in the Middle East in 2012. This virus, named Middle East Respiratory Syndrome-CoV (MERS-CoV), was found to be the causative agent in a series of highly pathogenic respiratory tract infections in Saudi Arabia and other countries in the Middle East [94]. Based on the high mortality rate of ~50 % in the early stages of the outbreak, it was feared the virus would lead to a very serious outbreak. However, the outbreak did not accelerate in 2013, although sporadic cases continued throughout the rest of the year. In April 2014, a spike of over 200 cases and almost 40 deaths occurred, prompting fears that the virus had mutated and was more capable of human-to-human transmission. More likely, the increased number of cases resulted from improved detection and reporting methods combined with a seasonal increase in birthing camels. As of August 27th, 2014 there have been a total of 855 cases of MERS-CoV, with 333 deaths and a case fatality rate of nearly 40 %, according to the European Center for Disease Prevention and Control.

MERS-CoV is a group 2c β-coronavirus highly related to two previously identified bat coronaviruses, HKU4 and HKU5 [95]. It is believed that the virus originated from bats, but likely had an intermediate host as humans rarely come in contact with bat secreta. Serological studies have identified MERS-CoV antibodies in dromedary camels in the Middle East [96], and cell lines from camels have been found to be permissive for MERS-CoV replication [97] providing evidence that dromedary camels may be the natural host. More convincing evidence for this comes from recent studies identifying nearly identical MERS-CoVs in both camels and human cases in nearby proximities in Saudi Arabia [98, 99]. In one of these studies the human case had direct contact with an infected camel and the virus isolated from this patient was identical to the virus isolated from the camel [99]. At the present time it remains to be determined how many MERS-CoV cases can be attributed to an intermediate host as opposed to human-to-human transmission. It has also been postulated that human-to-camel spread contributed to the outbreak.

MERS-CoV utilizes Dipeptidyl peptidase 4 (DPP4) as its receptor [100]. The virus is only able to use the receptor from certain species such as bats, humans, camels, rabbits, and horses to establish infection. Unfortunately for researchers, the virus is unable to infect mouse cells due to differences in the structure of DPP4, making it difficult to evaluate potential vaccines or antivirals. Recently, a small animal model for MERS-CoV has been developed using an Adenoviral vector to introduce the human DPP4 gene into mouse lungs [101]. This unique system makes it possible to test therapeutic interventions and novel vaccines for MERS-CoV in any animal sensitive to adenoviral transductions.

6 Diagnosis, Treatment, and Prevention

In most cases of self-limited infection, diagnosis of coronaviruses is unnecessary, as the disease will naturally run its course. However, it may be important in certain clinical and veterinary settings or in epidemiological studies to identify an etiological agent. Diagnosis is also important in locations where a severe CoV outbreak is occurring, such as, at present, in the Middle East, where MERS-CoV continues to circulate. The identification of cases will guide the development of public health measures to control outbreaks. It is also important to diagnose cases of severe veterinary CoV-induced disease, such as PEDV and IBV, to control these pathogens and protect food supplies. RT-PCR has become the method of choice for diagnosis of human CoV, as multiplex real-time RT-PCR assays have been developed, are able to detect all four respiratory HCoVs and could be further adapted to novel CoVs [102, 103]. Serologic assays are important in cases where RNA is difficult to isolate or is no longer present, and for epidemiological studies.

To date, there are no antiviral therapeutics that specifically target human coronaviruses, so treatments are only supportive. In vitro, interferons (IFNs) are only partially effective against coronaviruses [104]. IFNs in combination with ribavirin may have increased activity in vitro when compared to IFNs alone against some coronaviruses; however, the effectiveness of this combination in vivo requires further evaluation [105]. The SARS and MERS outbreaks have stimulated research on these viruses and this research has identified a large number of suitable antiviral targets, such as viral proteases, polymerases, and entry proteins. Significant work remains, however, to develop drugs that target these processes and are able to inhibit viral replication.

Only limited options are available to prevent coronavirus infections. Vaccines have only been approved for IBV, TGEV, and Canine CoV, but these vaccines are not always used because they are either not very effective, or in some cases have been reported to be involved in the selection of novel pathogenic CoVs via recombination of circulating strains. Vaccines for veterinary pathogens, such as PEDV, may be useful in such cases where spread of the virus to a new location could lead to severe losses of veterinary animals. In the case of SARS-CoV, several potential vaccines have been developed but none are yet approved for use. These vaccines include recombinant attenuated viruses, live virus vectors, or individual viral proteins expressed from DNA plasmids. Therapeutic SARS-CoV neutralizing antibodies have been generated and could be retrieved and used again in the event of another SARS-CoV outbreak. Such antibodies would be most useful for protecting healthcare workers. In general, it is thought that live attenuated vaccines would be the most efficacious in targeting coronaviruses. This was illustrated in the case of TGEV, where an attenuated variant, PRCV, appeared in Europe in the 1980s. This variant only caused mild disease and completely protected swine from TGEV. Thus, this attenuated virus has naturally prevented the reoccurrence of severe TGEV in Europe and the U.S. over the past 30 years [106]. Despite this success, vaccine development for coronaviruses faces many challenges [107]. First, for mucosal infections, natural infection does not prevent subsequent infection, and so vaccines must either induce better immunity than the original virus or must at least lessen the disease incurred during a secondary infection. Second, the propensity of the viruses to recombine may pose a problem by rendering the vaccine useless and potentially increasing the evolution and diversity of the virus in the wild [108]. Finally, it has been shown in FIPV that vaccination with S protein leads to enhanced disease [109]. Despite this, several strategies are being developed for vaccine development to reduce the likelihood of recombination, for instance by making large deletions in the nsp1 [110] or E proteins [111], rearranging the 3′ end of the genome [112], modifying the TRS sequences [113], or using mutant viruses with abnormally high mutation rates that significantly attenuate the virus [114].

Owing to the lack of effective therapeutics or vaccines, the best measures to control human coronaviruses remain a strong public health surveillance system coupled with rapid diagnostic testing and quarantine when necessary. For international outbreaks, cooperation of governmental entities, public health authorities, and health care providers is critical. During veterinary outbreaks that are readily transmitted, such as PEDV, more drastic measures such as destruction of entire herds of pigs may be necessary to prevent transmission of these deadly viruses.

7   Conclusion

Over the past 50 years the emergence of many different coronaviruses that cause a wide variety of human and veterinary diseases has occurred. It is likely that these viruses will continue to emerge and to evolve and cause both human and veterinary outbreaks owing to their ability to recombine, mutate, and infect multiple species and cell types.

Future research on coronaviruses will continue to investigate many aspects of viral replication and pathogenesis. First, understanding the propensity of these viruses to jump between species, to establish infection in a new host, and to identify significant reservoirs of coronaviruses will dramatically aid in our ability to predict when and where potential epidemics may occur. As bats seem to be a significant reservoir for these viruses, it will be interesting to determine how they seem to avoid clinically evident disease and become persistently infected. Second, many of the non-structural and accessory proteins encoded by these viruses remain uncharacterized with no known function, and it will be important to identify mechanisms of action for these proteins as well as defining their role in viral replication and pathogenesis. These studies should lead to a large increase in the number of suitable therapeutic targets to combat infections. Furthermore, many of the unique enzymes encoded by coronaviruses, such as ADP-ribose-1″-phosphatase, are also present in higher eukaryotes, making their study relevant to understanding general aspects of molecular biology and biochemistry. Third, gaining a complete picture of the intricacies of the RTC will provide a framework for understanding the unique RNA replication process used by these viruses. Finally, defining the mechanism of how coronaviruses cause disease and understanding the host immunopathological response will significantly improve our ability to design vaccines and reduce disease burden.

References

1. Zhao L, Jha BK, Wu A et al (2012) Antagonism of the interferon-induced OAS-RNase L pathway by murine coronavirus ns2 protein is required for virus replication and liver pathology. Cell Host Microbe 11:607–616. doi:10.1016/j.chom.2012.04.011

2. Barcena M, Oostergetel GT, Bartelink W et al (2009) Cryo-electron tomography of mouse hepatitis virus: insights into the structure of the coronavirion. Proc Natl Acad Sci U S A 106:582–587

3. Neuman BW, Adair BD, Yoshioka C et al (2006) Supramolecular architecture of severe acute respiratory syndrome coronavirus revealed by electron cryomicroscopy. J Virol 80:7918–7928

4. Beniac DR, Andonov A, Grudeski E et al (2006) Architecture of the SARS coronavirus prefusion spike. Nat Struct Mol Biol 13:751–752. doi:10.1038/nsmb1123

5. Delmas B, Laude H (1990) Assembly of coronavirus spike protein into trimers and its role in epitope expression. J Virol 64:5367–5375

6. Bosch BJ, van der Zee R, de Haan CA et al (2003) The coronavirus spike protein is a class I virus fusion protein: structural and functional characterization of the fusion core complex. J Virol 77:8801–8811

7. Collins AR, Knobler RL, Powell H et al (1982) Monoclonal antibodies to murine hepatitis virus-4 (strain JHM) define the viral glycoprotein responsible for attachment and cell–cell fusion. Virology 119:358–371

8. Abraham S, Kienzle TE, Lapps W et al (1990) Deduced sequence of the bovine coronavirus spike protein and identification of the internal proteolytic cleavage site. Virology 176:296–301

9. Luytjes W, Sturman LS, Bredenbeek PJ et al (1987) Primary structure of the glycoprotein E2 of coronavirus MHV-A59 and identification of the trypsin cleavage site. Virology 161:479–487

10. de Groot RJ, Luytjes W, Horzinek MC et al (1987) Evidence for a coiled-coil structure in the spike proteins of coronaviruses. J Mol Biol 196:963–966

11. Armstrong J, Niemann H, Smeekens S et al (1984) Sequence and topology of a model intracellular membrane protein, E1 glycoprotein, from a coronavirus. Nature 308:751–752

12. Nal B, Chan C, Kien F et al (2005) Differential maturation and subcellular localization of severe acute respiratory syndrome coronavirus surface proteins S, M and E. J Gen Virol 86:1423–1434. doi:10.1099/vir.0.80671-0

13. Neuman BW, Kiss G, Kunding AH et al (2011) A structural analysis of M protein in coronavirus assembly and morphology. J Struct Biol 174:11–22. doi:10.1016/j.jsb.2010.11.021

14. Godet M, L'Haridon R, Vautherot JF et al (1992) TGEV corona virus ORF4 encodes a membrane protein that is incorporated into virions. Virology 188:666–675

15. DeDiego ML, Alvarez E, Almazan F et al (2007) A severe acute respiratory syndrome coronavirus that lacks the E gene is attenuated in vitro and in vivo. J Virol 81:1701–1713

16. Nieto-Torres JL, Dediego ML, Verdia-Baguena C et al (2014) Severe acute respiratory syndrome coronavirus envelope protein ion channel activity promotes virus fitness and pathogenesis. PLoS Pathog 10:e1004077. doi:10.1371/journal.ppat.1004077

17. Chang CK, Sue SC, Yu TH et al (2006) Modular organization of SARS coronavirus

nucleocapsid protein. J Biomed Sci 13:59–72. doi:10.1007/s11373-005-9035-9

18. Hurst KR, Koetzner CA, Masters PS (2009) Identification of in vivo-interacting domains of the murine coronavirus nucleocapsid protein. J Virol 83:7221–7234. doi:10.1128/JVI.00440-09

19. Stohlman SA, Lai MM (1979) Phosphoproteins of murine hepatitis viruses. J Virol 32:672–675

20. Stohlman SA, Baric RS, Nelson GN et al (1988) Specific interaction between coronavirus leader RNA and nucleocapsid protein. J Virol 62:4288–4295

21. Molenkamp R, Spaan WJ (1997) Identification of a specific interaction between the coronavirus mouse hepatitis virus A59 nucleocapsid protein and packaging signal. Virology 239:78–86

22. Kuo L, Masters PS (2013) Functional analysis of the murine coronavirus genomic RNA packaging signal. J Virol 87:5182–5192. doi:10.1128/JVI.00100-13

23. Hurst KR, Koetzner CA, Masters PS (2013) Characterization of a critical interaction between the coronavirus nucleocapsid protein and nonstructural protein 3 of the viral replicase-transcriptase complex. J Virol 87:9159–9172. doi:10.1128/JVI.01275-13

24. Sturman LS, Holmes KV, Behnke J (1980) Isolation of coronavirus envelope glycoproteins and interaction with the viral nucleocapsid. J Virol 33:449–462

25. Klausegger A, Strobl B, Regl G et al (1999) Identification of a coronavirus hemagglutinin-esterase with a substrate specificity different from those of influenza C virus and bovine coronavirus. J Virol 73:3737–3743

26. Cornelissen LA, Wierda CM, van der Meer FJ et al (1997) Hemagglutinin-esterase, a novel structural protein of torovirus. J Virol 71:5277–5286

27. Kazi L, Lissenberg A, Watson R et al (2005) Expression of hemagglutinin esterase protein from recombinant mouse hepatitis virus enhances neurovirulence. J Virol 79:15064–15073

28. Lissenberg A, Vrolijk MM, van Vliet AL et al (2005) Luxury at a cost? Recombinant mouse hepatitis viruses expressing the accessory hemagglutinin esterase protein display reduced fitness in vitro. J Virol 79:15054–15063

29. Kubo H, Yamada YK, Taguchi F (1994) Localization of neutralizing epitopes and the receptor-binding site within the amino-terminal 330 amino acids of the murine coronavirus spike protein. J Virol 68:5403–5410

30. Cheng PK, Wong DA, Tong LK et al (2004) Viral shedding patterns of coronavirus in patients with probable severe acute respiratory syndrome. Lancet 363:1699–1700

31. Belouzard S, Chu VC, Whittaker GR (2009) Activation of the SARS coronavirus spike protein via sequential proteolytic cleavage at two distinct sites. Proc Natl Acad Sci U S A 106:5871–5876. doi:10.1073/pnas.0809524106

32. Baranov PV, Henderson CM, Anderson CB et al (2005) Programmed ribosomal frameshifting in decoding the SARS-CoV genome. Virology 332:498–510. doi:10.1016/j.virol.2004.11.038

33. Brierley I, Digard P, Inglis SC (1989) Characterization of an efficient coronavirus ribosomal frameshifting signal: requirement for an RNA pseudoknot. Cell 57:537–547

34. Araki K, Gangappa S, Dillehay DL et al (2010) Pathogenic virus-specific T cells cause disease during treatment with the calcineurin inhibitor FK506: implications for transplantation. J Exp Med 207:2355–2367

35. Ziebuhr J, Snijder EJ, Gorbalenya AE (2000) Virus-encoded proteinases and proteolytic processing in the Nidovirales. J Gen Virol 81:853–879

36. Mielech AM, Chen Y, Mesecar AD et al (2014) Nidovirus papain-like proteases: multifunctional enzymes with protease, deubiquitinating and deISGylating activities. Virus Res. doi:10.1016/j.virusres.2014.01.025

37. Snijder EJ, Bredenbeek PJ, Dobbe JC et al (2003) Unique and conserved features of genome and proteome of SARS-coronavirus, an early split-off from the coronavirus group 2 lineage. J Mol Biol 331:991–1004

38. Sethna PB, Hofmann MA, Brian DA (1991) Minus-strand copies of replicating coronavirus mRNAs contain antileaders. J Virol 65:320–325

39. Brown CG, Nixon KS, Senanayake SD et al (2007) An RNA stem-loop within the bovine coronavirus nsp1 coding region is a cis-acting element in defective interfering RNA replication. J Virol 81:7716–7724. doi:10.1128/JVI.00549-07

40. Guan BJ, Wu HY, Brian DA (2011) An optimal cis-replication stem-loop IV in the 5′ untranslated region of the mouse coronavirus genome extends 16 nucleotides into open reading frame 1. J Virol 85:5593–5605. doi:10.1128/JVI.00263-11

41. Liu P, Li L, Keane SC et al (2009) Mouse hepatitis virus stem-loop 2 adopts a uYNMG(U)a-like tetraloop structure that is highly functionally tolerant of base substitutions. J Virol 83:12084–12093. doi:10.1128/JVI.00915-09

42. Raman S, Bouma P, Williams GD et al (2003) Stem-loop III in the 5′ untranslated region is a cis-acting element in bovine coronavirus defective interfering RNA replication. J Virol 77:6720–6730

43. Liu Q, Johnson RF, Leibowitz JL (2001) Secondary structural elements within the 3′ untranslated region of mouse hepatitis virus strain JHM genomic RNA. J Virol 75:12105–12113. doi:10.1128/JVI.75.24.12105-12113.2001

44. Goebel SJ, Miller TB, Bennett CJ et al (2007) A hypervariable region within the 3′ cis-acting element of the murine coronavirus genome is nonessential for RNA synthesis but affects pathogenesis. J Virol 81:1274–1287. doi:10.1128/JVI.00803-06

45. Williams GD, Chang RY, Brian DA (1999) A phylogenetically conserved hairpin-type 3′ untranslated region pseudoknot functions in coronavirus RNA replication. J Virol 73:8349–8355

46. Hsue B, Masters PS (1997) A bulged stem-loop structure in the 3′ untranslated region of the genome of the coronavirus mouse hepatitis virus is essential for replication. J Virol 71:7567–7578

47. Hsue B, Hartshorne T, Masters PS (2000) Characterization of an essential RNA secondary structure in the 3′ untranslated region of the murine coronavirus genome. J Virol 74:6911–6921

48. Sawicki SG, Sawicki DL, Siddell SG (2007) A contemporary view of coronavirus transcription. J Virol 81:20–29

49. Bentley K, Keep SM, Armesto M et al (2013) Identification of a noncanonically transcribed subgenomic mRNA of infectious bronchitis virus and other gammacoronaviruses. J Virol 87:2128–2136. doi:10.1128/JVI.02967-12

50. Keck JG, Makino S, Soe LH et al (1987) RNA recombination of coronavirus. Adv Exp Med Biol 218:99–107

51. Lai MM, Baric RS, Makino S et al (1985) Recombination between nonsegmented RNA genomes of murine coronaviruses. J Virol 56:449–456

52. Krijnse-Locker J, Ericsson M, Rottier PJM et al (1994) Characterization of the budding compartment of mouse hepatitis virus: evidence that transport from the RER to the Golgi complex requires only one vesicular transport step. J Cell Biol 124:55–70

53. Tooze J, Tooze S, Warren G (1984) Replication of coronavirus MHV-A59 in sac− cells: determination of the first site of budding of progeny virions. Eur J Cell Biol 33:281–293

54. de Haan CA, Rottier PJ (2005) Molecular interactions in the assembly of coronaviruses. Adv Virus Res 64:165–230

55. Bos EC, Luytjes W, van der Meulen HV et al (1996) The production of recombinant infectious DI-particles of a murine coronavirus in the absence of helper virus. Virology 218:52–60

56. Siu YL, Teoh KT, Lo J et al (2008) The M, E, and N structural proteins of the severe acute respiratory syndrome coronavirus are required for efficient assembly, trafficking, and release of virus-like particles. J Virol 82:11318–11330. doi:10.1128/JVI.01052-08

57. Raamsman MJ, Locker JK, de Hooge A et al (2000) Characterization of the coronavirus mouse hepatitis virus strain A59 small membrane protein E. J Virol 74:2333–2342

58. Corse E, Machamer CE (2000) Infectious bronchitis virus E protein is targeted to the Golgi complex and directs release of virus-like particles. J Virol 74:4319–4326

59. Fischer F, Stegen CF, Masters PS et al (1998) Analysis of constructed E gene mutants of mouse hepatitis virus confirms a pivotal role for E protein in coronavirus assembly. J Virol 72:7885–7894

60. Boscarino JA, Logan HL, Lacny JJ et al (2008) Envelope protein palmitoylations are crucial for murine coronavirus assembly. J Virol 82:2989–2999. doi:10.1128/JVI.01906-07

61. Ye Y, Hogue BG (2007) Role of the coronavirus E viroporin protein transmembrane domain in virus assembly. J Virol 81:3597–3607. doi:10.1128/JVI. 01472-06

62. Hurst KR, Kuo L, Koetzner CA et al (2005) A major determinant for membrane protein interaction localizes to the carboxy-terminal domain of the mouse coronavirus nucleocapsid protein. J Virol 79:13285–13297

63. Perlman S, Netland J (2009) Coronaviruses post-SARS: update on replication and pathogenesis. Nat Rev Microbiol 7:439–450

64. Mihindukulasuriya KA, Wu G, St LJ et al (2008) Identification of a novel coronavirus from a beluga whale by using a panviral microarray. J Virol 82:5084–5088

65. He B, Zhang Y, Xu L et al (2014) Identification of diverse alphacoronaviruses

and genomic characterization of a novel severe acute respiratory syndrome-like coronavirus from bats in china. J Virol 88:7070–7082. doi:10.1128/JVI.00631-14

66. Nga PT, Parquet Mdel C, Lauber C et al (2011) Discovery of the first insect nidovirus, a missing evolutionary link in the emergence of the largest RNA virus genomes. PLoS Pathog 7:e1002215. doi:10.1371/journal.ppat.1002215

67. Lauber C, Ziebuhr J, Junglen S et al (2012) Mesoniviridae: a proposed new family in the order Nidovirales formed by a single species of mosquito-borne viruses. Arch Virol 157:1623–1628. doi:10.1007/s00705-012-1295-x

68. Levy GA, Liu M, Ding J et al (2000) Molecular and functional analysis of the human prothrombinase gene (HFGL2) and its role in viral hepatitis. Am J Pathol 156:1217–1225

69. Lampert PW, Sims JK, Kniazeff AJ (1973) Mechanism of demyelination in JHM virus encephalomyelitis. Acta Neuropathol 24:76–85

70. Weiner LP (1973) Pathogenesis of demyelination induced by a mouse hepatitis virus (JHM virus). Arch Neurol 28:298–303

71. Wu GF, Dandekar AA, Pewe L et al (2000) CD4 and CD8 T cells have redundant but not identical roles in virus-induced demyelination. J Immunol 165:2278–2286

72. Wang F, Stohlman SA, Fleming JO (1990) Demyelination induced by murine hepatitis virus JHM strain (MHV-4) is immunologically mediated. J Neuroimmunol 30:31–41

73. Wu GF, Perlman S (1999) Macrophage infiltration, but not apoptosis, is correlated with immune-mediated demyelination following murine infection with a neurotropic coronavirus. J Virol 73:8771–8780

74. McIntosh K, Becker WB, Chanock RM (1967) Growth in suckling-mouse brain of "IBV-like" viruses from patients with upper respiratory tract disease. Proc Natl Acad Sci U S A 58:2268–2273

75. Bradburne AF, Bynoe ML, Tyrell DAJ (1967) Effects of a "new" human respiratory virus in volunteers. Br Med J 3:767–769

76. Hamre D, Procknow JJ (1966) A new virus isolated from the human respiratory tract. Proc Soc Exp Biol Med 121:190–193

77. Woo PC, Lau SK, Chu CM et al (2005) Characterization and complete genome sequence of a novel coronavirus, coronavirus HKU1, from patients with pneumonia. J Virol 79:884–895

78. van der Hoek L, Pyrc K, Jebbink MF et al (2004) Identification of a new human coronavirus. Nat Med 10:368–373

79. van der Hoek L, Sure K, Ihorst G et al (2005) Croup is associated with the novel coronavirus NL63. PLoS Med 2:e240

80. Chibo D, Birch C (2006) Analysis of human coronavirus 229E spike and nucleoprotein genes demonstrates genetic drift between chronologically distinct strains. J Gen Virol 87:1203–1208

81. Vijgen L, Keyaerts E, Lemey P et al (2005) Circulation of genetically distinct contemporary human coronavirus OC43 strains. Virology 337:85–92

82. Guan Y, Zheng BJ, He YQ et al (2003) Isolation and characterization of viruses related to the SARS coronavirus from animals in southern China. Science 302:276–278

83. Lau SK, Woo PC, Li KS et al (2005) Severe acute respiratory syndrome coronavirus-like virus in Chinese horseshoe bats. Proc Natl Acad Sci U S A 102:14040–14045

84. Li W, Shi Z, Yu M et al (2005) Bats are natural reservoirs of SARS-like coronaviruses. Science 310:676–679

85. Ge XY, Li JL, Yang XL et al (2013) Isolation and characterization of a bat SARS-like coronavirus that uses the ACE2 receptor. Nature 503:535–538. doi:10.1038/nature12711

86. Peiris JS, Yuen KY, Osterhaus AD et al (2003) The severe acute respiratory syndrome. N Engl J Med 349:2431–2441

87. Peiris JS, Chu CM, Cheng VC et al (2003) Clinical progression and viral load in a community outbreak of coronavirus-associated SARS pneumonia: a prospective study. Lancet 361:1767–1772

88. Spiegel M, Schneider K, Weber F et al (2006) Interaction of severe acute respiratory syndrome-associated coronavirus with dendritic cells. J Gen Virol 87:1953–1960

89. Law HK, Cheung CY, Ng HY et al (2005) Chemokine upregulation in SARS coronavirus infected human monocyte derived dendritic cells. Blood 106:2366–2376

90. Lau YL, Peiris JSM (2005) Pathogenesis of severe acute respiratory syndrome. Curr Opin Immunol 17:404–410

91. Roberts A, Paddock C, Vogel L et al (2005) Aged BALB/c mice as a model for increased severity of severe acute respiratory syndrome in elderly humans. J Virol 79:5833–5838

92. Zhao J, Zhao J, Perlman S (2010) T cell responses are required for protection from clinical disease and for virus clearance in severe acute respiratory syndrome coronavirus-infected mice. J Virol 84:9318–9325

93. Zhao J, Zhao J, Legge K et al (2011) Age-related increases in PGD(2) expression impair respiratory DC migration, resulting in dimin-

ished T cell responses upon respiratory virus infection in mice. J Clin Invest 121:4921–4930. doi:10.1172/JCI59777

94. Zaki AM, van Boheemen S, Bestebroer TM et al (2012) Isolation of a novel coronavirus from a man with pneumonia in Saudi Arabia. N Engl J Med 367:1814–1820. doi:10.1056/NEJMoa1211721

95. van Boheemen S, de Graaf M, Lauber C et al (2012) Genomic characterization of a newly discovered coronavirus associated with acute respiratory distress syndrome in humans. MBio 3. doi:10.1128/mBio.00473-12

96. Meyer B, Muller MA, Corman VM et al (2014) Antibodies against MERS coronavirus in dromedary camels, United Arab Emirates, 2003 and 2013. Emerg Infect Dis 20:552–559. doi:10.3201/eid2004.131746

97. Eckerle I, Corman VM, Muller MA et al (2014) Replicative capacity of MERS coronavirus in livestock cell lines. Emerg Infect Dis 20:276–279. doi:10.3201/eid2002.131182

98. Memish ZA, Cotten M, Meyer B et al (2014) Human infection with MERS coronavirus after exposure to infected camels, Saudi Arabia, 2013. Emerg Infect Dis 20:1012–1015. doi:10.3201/eid2006.140402

99. Azhar EI, El-Kafrawy SA, Farraj SA et al (2014) Evidence for camel-to-human transmission of MERS coronavirus. N Engl J Med. doi:10.1056/NEJMoa1401505

100. Raj VS, Mou H, Smits SL et al (2013) Dipeptidyl peptidase 4 is a functional receptor for the emerging human coronavirus-EMC. Nature 495:251–254. doi:10.1038/nature12005

101. Zhao J, Li K, Wohlford-Lenane C et al (2014) Rapid generation of a mouse model for Middle East respiratory syndrome. Proc Natl Acad Sci U S A 111:4970–4975. doi:10.1073/pnas.1323279111

102. Emery SL, Erdman DD, Bowen MD et al (2004) Real-time reverse transcription-polymerase chain reaction assay for SARS-associated coronavirus. Emerg Infect Dis 10:311–316. doi:10.3201/eid1002.030759

103. Gaunt ER, Hardie A, Claas EC et al (2010) Epidemiology and clinical presentations of the four human coronaviruses 229E, HKU1, NL63, and OC43 detected over 3 years using a novel multiplex real-time PCR method. J Clin Microbiol 48:2940–2947. doi:10.1128/JCM.00636-10

104. Cinatl J, Morgenstern B, Bauer G et al (2003) Treatment of SARS with human interferons. Lancet 362:293–294

105. Stockman LJ, Bellamy R, Garner P (2006) SARS: systematic review of treatment effects. PLoS Med 3:e343

106. Laude H, Van Reeth K, Pensaert M (1993) Porcine respiratory coronavirus: molecular features and virus-host interactions. Vet Res 24:125–150

107. Saif LJ (2004) Animal coronavirus vaccines: lessons for SARS. Dev Biol (Basel) 119:129–140

108. Wang L, Junker D, Collisson EW (1993) Evidence of natural recombination within the S1 gene of infectious bronchitis virus. Virology 192:710–716

109. Vennema H, de Groot RJ, Harbour DA et al (1990) Early death after feline infectious peritonitis virus challenge due to recombinant vaccinia virus immunization. J Virol 64:1407–1409

110. Zust R, Cervantes-Barragan L, Kuri T et al (2007) Coronavirus non-structural protein 1 is a major pathogenicity factor: implications for the rational design of coronavirus vaccines. PLoS Pathog 3:e109

111. Netland J, DeDiego ML, Zhao J et al (2010) Immunization with an attenuated severe acute respiratory syndrome coronavirus deleted in E protein protects against lethal respiratory disease. Virology 399:120–128. doi:10.1016/j.virol.2010.01.004

112. de Haan CA, Volders H, Koetzner CA et al (2002) Coronaviruses maintain viability despite dramatic rearrangements of the strictly conserved genome organization. J Virol 76:12491–12502

113. Yount B, Roberts RS, Lindesmith L et al (2006) Rewiring the severe acute respiratory syndrome coronavirus (SARS-CoV) transcription circuit: engineering a recombination-resistant genome. Proc Natl Acad Sci U S A 103:12546–12551

114. Graham RL, Becker MM, Eckerle LD et al (2012) A live, impaired-fidelity coronavirus vaccine protects in an aged, immunocompromised mouse model of lethal disease. Nat Med 18:1820–1826. doi:10.1038/nm.2972

115. Yeager CL, Ashmun RA, Williams RK et al (1992) Human aminopeptidase N is a receptor for human coronavirus 229E. Nature 357:420–422. doi:10.1038/357420a0

116. Hofmann H, Pyrc K, van der Hoek L et al (2005) Human coronavirus NL63 employs the severe acute respiratory syndrome coronavirus receptor for cellular entry. Proc Natl Acad Sci U S A 102:7988–7993

117. Delmas B, Gelfi J, L'Haridon R et al (1992) Aminopeptidase N is a major receptor for the

entero-pathogenic coronavirus TGEV. Nature 357:417–420. doi:10.1038/357417a0

118. Li BX, Ge JW, Li YJ (2007) Porcine aminopeptidase N is a functional receptor for the PEDV coronavirus. Virology 365:166–172. doi:10.1016/j.virol.2007.03.031

119. Tresnan DB, Levis R, Holmes KV (1996) Feline aminopeptidase N serves as a receptor for feline, canine, porcine, and human coronaviruses in serogroup I. J Virol 70: 8669–8674

120. Benbacer L, Kut E, Besnardeau L et al (1997) Interspecies aminopeptidase-N chimeras reveal species-specific receptor recognition by canine coronavirus, feline infectious peritonitis virus, and transmissible gastroenteritis virus. J Virol 71:734–737

121. Nedellec P, Dveksler GS, Daniels E et al (1994) Bgp2, a new member of the carcinoembryonic antigen-related gene family, encodes an alternative receptor for mouse hepatitis viruses. J Virol 68:4525–4537

122. Williams RK, Jiang GS, Holmes KV (1991) Receptor for mouse hepatitis virus is a member of the carcinoembryonic antigen family of glycoproteins. Proc Natl Acad Sci U S A 88: 5533–5536

123. Schultze B, Herrler G (1992) Bovine coronavirus uses N-acetyl-9-O-acetylneuraminic acid as a receptor determinant to initiate the infection of cultured cells. J Gen Virol 73(Pt 4): 901–906

124. Li W, Moore MJ, Vasilieva N et al (2003) Angiotensin-converting enzyme 2 is a functional receptor for the SARS coronavirus. Nature 426:450–454

125. Huang C, Lokugamage KG, Rozovics JM et al (2011) Alphacoronavirus transmissible gastroenteritis virus nsp1 protein suppresses protein translation in mammalian cells and in cell-free HeLa cell extracts but not in rabbit reticulocyte lysate. J Virol 85:638–643. doi:10.1128/JVI.01806-10

126. Kamitani W, Huang C, Narayanan K et al (2009) A two-pronged strategy to suppress host protein synthesis by SARS coronavirus Nsp1 protein. Nat Struct Mol Biol 16:1134–1140. doi:10.1038/nsmb.1680

127. Kamitani W, Narayanan K, Huang C et al (2006) Severe acute respiratory syndrome coronavirus nsp1 protein suppresses host gene expression by promoting host mRNA degradation. Proc Natl Acad Sci U S A 103: 12885–12890

128. Tanaka T, Kamitani W, DeDiego ML et al (2012) Severe acute respiratory syndrome coronavirus nsp1 facilitates efficient propaga-

tion in cells through a specific translational shutoff of host mRNA. J Virol 86:11128–11137. doi:10.1128/JVI.01700-12

129. Graham RL, Sims AC, Brockway SM et al (2005) The nsp2 replicase proteins of murine hepatitis virus and severe acute respiratory syndrome coronavirus are dispensable for viral replication. J Virol 79:13399–13411. doi:10.1128/JVI.79.21.13399-13411.2005

130. Cornillez-Ty CT, Liao L, Yates JR 3rd et al (2009) Severe acute respiratory syndrome coronavirus nonstructural protein 2 interacts with a host protein complex involved in mitochondrial biogenesis and intracellular signaling. J Virol 83:10314–10318. doi:10.1128/JVI.00842-09

131. Chatterjee A, Johnson MA, Serrano P et al (2009) Nuclear magnetic resonance structure shows that the severe acute respiratory syndrome coronavirus-unique domain contains a macrodomain fold. J Virol 83:1823–1836

132. Egloff MP, Malet H, Putics A et al (2006) Structural and functional basis for ADP-ribose and poly(ADP-ribose) binding by viral macro domains. J Virol 80:8493–8502. doi:10.1128/JVI.00713-06

133. Eriksson KK, Cervantes-Barragan L, Ludewig B et al (2008) Mouse hepatitis virus liver pathology is dependent on ADP-ribose-1″-phosphatase, a viral function conserved in the alpha-like supergroup. J Virol 82:12325–12334. doi:10.1128/JVI.02082-08

134. Frieman M, Ratia K, Johnston RE et al (2009) Severe acute respiratory syndrome coronavirus papain-like protease ubiquitin-like domain and catalytic domain regulate antagonism of IRF3 and NF-kappaB signaling. J Virol 83:6689–6705

135. Neuman BW, Joseph JS, Saikatendu KS et al (2008) Proteomics analysis unravels the functional repertoire of coronavirus nonstructural protein 3. J Virol 82:5279–5294

136. Serrano P, Johnson MA, Almeida MS et al (2007) Nuclear magnetic resonance structure of the N-terminal domain of nonstructural protein 3 from the severe acute respiratory syndrome coronavirus. J Virol 81:12049–12060

137. Serrano P, Johnson MA, Chatterjee A et al (2009) Nuclear magnetic resonance structure of the nucleic acid-binding domain of severe acute respiratory syndrome coronavirus nonstructural protein 3. J Virol 83:12998–13008. doi:10.1128/JVI.01253-09

138. Ziebuhr J, Thiel V, Gorbalenya AE (2001) The autocatalytic release of a putative RNA virus transcription factor from its polyprotein precursor involves two paralogous papain-like

proteases that cleave the same peptide bond. J Biol Chem 276:33220–33232. doi:10.1074/jbc.M104097200

139. Clementz MA, Kanjanahaluethai A, O'Brien TE et al (2008) Mutation in murine coronavirus replication protein nsp4 alters assembly of double membrane vesicles. Virology 375:118–129

140. Gadlage MJ, Sparks JS, Beachboard DC et al (2010) Murine hepatitis virus nonstructural protein 4 regulates virus-induced membrane modifications and replication complex function. J Virol 84:280–290. doi:10.1128/JVI.01772-09

141. Lu Y, Lu X, Denison MR (1995) Identification and characterization of a serine-like proteinase of the murine coronavirus MHV-A59. J Virol 69:3554–3559

142. Oostra M, Hagemeijer MC, van Gent M et al (2008) Topology and membrane anchoring of the coronavirus replication complex: not all hydrophobic domains of nsp3 and nsp6 are membrane spanning. J Virol 82:12392–12405

143. Zhai Y, Sun F, Li X et al (2005) Insights into SARS-CoV transcription and replication from the structure of the nsp7-nsp8 hexadecamer. Nat Struct Mol Biol 12:980–986

144. Imbert I, Guillemot JC, Bourhis JM et al (2006) A second, non-canonical RNA-dependent RNA polymerase in SARS coronavirus. EMBO J 25:4933–4942

145. Egloff MP, Ferron F, Campanacci V et al (2004) The severe acute respiratory syndrome-coronavirus replicative protein nsp9 is a single-stranded RNA-binding subunit unique in the RNA virus world. Proc Natl Acad Sci U S A 101:3792–3796

146. Bouvet M, Debarnot C, Imbert I et al (2010) In vitro reconstitution of SARS-coronavirus mRNA cap methylation. PLoS Pathog 6:e1000863. doi:10.1371/journal.ppat.1000863

147. Decroly E, Debarnot C, Ferron F et al (2011) Crystal structure and functional analysis of the SARS-coronavirus RNA cap 2'-O-methyltransferase nsp10/nsp16 complex. PLoS Pathog 7:e1002059. doi:10.1371/journal.ppat.1002059

148. Xu X, Liu Y, Weiss S et al (2003) Molecular model of SARS coronavirus polymerase: implications for biochemical functions and drug design. Nucleic Acids Res 31:7117–7130

149. Ivanov KA, Thiel V, Dobbe JC et al (2004) Multiple enzymatic activities associated with severe acute respiratory syndrome coronavirus helicase. J Virol 78:5619–5632

150. Ivanov KA, Ziebuhr J (2004) Human coronavirus 229E nonstructural protein 13: characterization of duplex-unwinding, nucleoside triphosphatase, and RNA 5'-triphosphatase activities. J Virol 78:7833–7838. doi:10.1128/JVI.78.14.7833-7838.2004

151. Eckerle LD, Becker MM, Halpin RA et al (2010) Infidelity of SARS-CoV Nsp14-exonuclease mutant virus replication is revealed by complete genome sequencing. PLoS Pathog 6:e1000896. doi:10.1371/journal.ppat.1000896

152. Eckerle LD, Lu X, Sperry SM et al (2007) High fidelity of murine hepatitis virus replication is decreased in nsp14 exoribonuclease mutants. J Virol 81:12135–12144

153. Minskaia E, Hertzig T, Gorbalenya AE et al (2006) Discovery of an RNA virus 3'->5' exoribonuclease that is critically involved in coronavirus RNA synthesis. Proc Natl Acad Sci U S A 103:5108–5113. doi:10.1073/pnas.0508200103

154. Chen Y, Cai H, Pan J et al (2009) Functional screen reveals SARS coronavirus nonstructural protein nsp14 as a novel cap N7 methyltransferase. Proc Natl Acad Sci U S A 106:3484–3489. doi:10.1073/pnas.0808790106

155. Bhardwaj K, Sun J, Holzenburg A et al (2006) RNA recognition and cleavage by the SARS coronavirus endoribonuclease. J Mol Biol 361:243–256. doi:10.1016/j.jmb.2006.06.021

156. Ivanov KA, Hertzig T, Rozanov M et al (2004) Major genetic marker of nidoviruses encodes a replicative endoribonuclease. Proc Natl Acad Sci U S A 101:12694–12699

157. Decroly E, Imbert I, Coutard B et al (2008) Coronavirus nonstructural protein 16 is a cap-0 binding enzyme possessing (nucleoside-2'O)-methyltransferase activity. J Virol 82:8071–8084

158. Zust R, Cervantes-Barragan L, Habjan M et al (2011) Ribose 2'-O-methylation provides a molecular signature for the distinction of self and non-self mRNA dependent on the RNA sensor Mda5. Nat Immunol 12:137–143

Part I

Detection, Diagnosis and Evolution of Coronaviruses

Chapter 2

Identification of a Novel Coronavirus from Guinea Fowl Using Metagenomics

Mariette F. Ducatez and Jean-Luc Guérin

Abstract

While classical virology techniques such as virus culture, electron microscopy, or classical PCR had been unsuccessful in identifying the causative agent responsible for the fulminating disease of guinea fowl, we identified a novel avian gammacoronavirus associated with the disease using metagenomics. Next-generation sequencing is an unbiased approach that allows the sequencing of virtually all the genetic material present in a given sample.

Key words Next-generation sequencing, Illumina, Ultracentrifugation, Nuclease treatment, Random RT-PCR

1 Introduction

The field of pathogen discovery gained a completely new dimension when new genomics tools, next-generation sequencing (NGS), became available. Isolation and identification by electronic microscopy used to be the gold standard techniques to identify a new pathogen. However, some pathogens are difficult to culture, and/or to separate from co-infecting agents. PCR has improved pathogen discovery, but even when pan-species/Genus/family PCRs may be developed (which is not always possible, especially for highly variable RNA viruses), the primers chosen do condition the nucleic acids that will be amplified: it is still a "biased" technique with which only known (or closely related to known) pathogens whose presence was suspected may be detected [1, 2]. NGS allows for (1) massive sequencing of genetic material and (2) unbiased sequencing, to a much lower cost per sequenced base than Sanger sequencing technology. Here we describe an NGS technique that can be used to identify novel pathogens, which we recently used to identify the pathogen responsible for the guinea fowl fulminating disease [3].

Helena Jane Maier et al. (eds.), *Coronaviruses: Methods and Protocols*, Methods in Molecular Biology, vol. 1282,
DOI 10.1007/978-1-4939-2438-7_2, © Springer Science+Business Media New York 2015

2 Materials

2.1 Materials for Preparation of Specimens and Concentrating Viral Particles

1. PBS.
2. PBS containing 100 U/ml Penicillin and 0.1 mg/ml streptomycin.
3. 0.45 μm filter.
4. RNAse.
5. DNAse (10 U/μl).
6. 10× DNase buffer: 200 mM Tris–HCl (pH 8.4), 20 mM $MgCl_2$, 500 mM KCl.

2.2 Materials for RNA Extraction and Amplification

1. TRIzol reagent or similar.
2. Chloroform.
3. RNA extraction kit, e.g., Nucleospin RNA virus kit (Macherey-Nagel) or similar.
4. Reverse transcription kit, e.g., RevertAid kit (Thermofisher) or similar.
5. 10 mM dNTPs.
6. RNase inhibitor.
7. 20 μM tagged random hexamer: 454-A: 5′ ATG-GTC-GTC-GTA-GGC-TGC-TCN-NNN-NNN-N 3′ (*see* **Note 1**).
8. PCR kit, e.g., Phusion (NEB) or similar.
9. 20 μM PCR primer (tag only): 454-A: 5′ ATG-GTC-GTC-GTA-GGC-TGC-TC 3′.
10. Thermocycler.
11. 1 % agarose gel.
12. 1× TBE buffer: 89 mM Tris, 89 mM orthoboric acid, 2 mM EDTA.
13. Sybr® safe (Life technologies) or similar.
14. 1 kb plus DNA ladder.
15. Gel purification kit.
16. PicoGreen® quantitation assay (Life technologies) or similar.

2.3 Next-Generation Sequencing

1. Miseq reagent kit (Illumina).
2. Miseq (Illumina).

3 Methods

Figure 1 summarizes the methodology used for the identification of a novel guinea fowl gammacoronavirus.

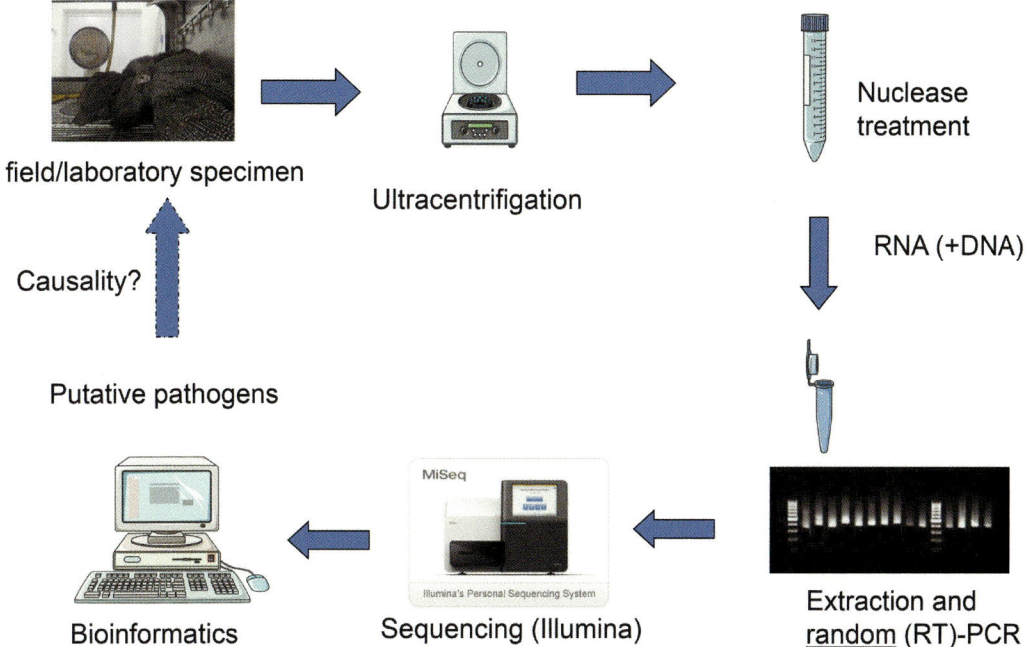

Fig. 1 High-throughput unbiased sequencing of intestinal contents of guinea fowls. The scheme represents the main steps of the novel guinea fowl gammacoronavirus detection by NGS

3.1 Sample Preparation

1. Pool intestinal contents of experimentally infected guinea fowl poults and resuspend in 500 μl PBS with penicillin and streptomycin. Vortex.

2. Filter (0.45 μm filter) the solution to eliminate eukaryotic- and bacterial-cell-sized particles.

3. Centrifuge the digestive content at $10,000 \times g$ for 30 min twice to clarify the solution and collect the supernatant in a new tube.

3.2 Concentration of Viral Particles

1. Pellet the concentrated material by ultracentrifugation at $100,000 \times g$ for 2 h.

2. Treat with RNAse and DNAse to remove non-particle-protected nucleic acids: make a mix of 500 μl of sample, 10 μl DNAse (100 U), 12 μl RNAse (20 μg/μl), 60 μl 10× DNAse buffer, 16 μl PBS. Incubate the mix for 20 min at 37 °C and then 10 min at 75 °C to stop the reaction.

3.3 RNA Extraction and Amplification (See Note 2)

1. Add 750 μl TRIzol to 250 μl sample from Subheading 3.2, **step 2** and incubate 5 min at room temperature.

2. Add 200 μl chloroform, vortex vigorously, and incubate 10 min at room temperature.

3. Centrifuge for 15 min at $11,000 \times g$ at 4 °C.

4. Collect the top aqueous phase.

5. Extract RNA from 150 μl of the collected aqueous phase on a silicate column with a RNA extraction kit.

6. Perform a reverse transcription reaction using random primers by mixing 7.5 μl RNA, 5 μl tagged random hexamer.

7. Incubate at 65 °C for 5 min then keep on ice.

8. Add 4 μl 5× reaction buffer, 0.5 μl RNase inhibitor, 2 μl dNTP, and 1 μl RevertAid reverse transcriptase.

9. Incubate at 25 °C for 10 min, 42 °C for 60 min, and 70 °C for 10 min.

10. Reactions can now be stored at –20 °C or used immediately for PCR.

11. Perform a random PCR by mixing 10 μl 5× Phusion HF reaction buffer, 1 μl dNTP, 0.5 μl tag only primer, 5 μl cDNA, 0.5 μl Phusion polymerase, and 33 μl water.

12. Perform PCR using the following cycle: 98 °C 30 s followed by 40 cycles of 98 °C 10 s, 65 °C 20 s, and 72 °C 30 s followed by a final incubate of 72 °C 10 min.

13. Analyze the PCR products on a 1 % agarose gel, migrate for 1 h at 60 V.

14. Excise the 300 bp bands and perform a gel purification with a commercial kit.

15. Quality assessment of the prepared library: quantify the DNA generated by a fluorescence-based method (PicoGreen® quantitation assay) and aim at 1 μg DNA as input; check the DNA quality and aim for a 260:280 ratio >1.8.

3.4 Cluster Generation Using the Miseq Reagent Kit (Illumina)

1. Hybridize sample to flow cell.

2. Amplify sample (bridge amplification).

3. Linearize fragments.

4. Block fragments.

5. Hybridize sequencing primer.

3.5 Sequencing on the Miseq (Illumina)

1. The library DNA fragment act as a template, from which a complementary strand is synthesized.

2. ddNTPs are added one by one (one cycle = one ddNTP added, a picture taken and defluoration of the ddNTP to be able to add a new ddNTP the next cycle) by a DNA polymerase. The addition of ddNTP is digitally recorded as sequence data cycle after cycle.

3.6 Data Analysis

1. Preprocess the data to remove adapter sequences and demultiplexing using splitbc (several samples can be multiplexed and run together on the MiSeq Illumina sequencer to reduce cost).

2. Preprocess the data to remove low-quality reads and compiling paired sequences using illuminapairedend.

3. Map the data to a reference genome or *de novo* align the sequence reads (alignment with bwa [4], consensus computed with the SAMtools [5] software package. Display the results with the IGV [6] browser).

4. Analyze the compiled sequence with the GAAS software (http://gaas.sourceforge.net/) with an expected value of 10^{-3}.

4 Notes

1. Victoria et al. [7] described a panel of tagged primers (named with alphabet letter), and while we selected "454-A" for the present study, any of the tagged primers described in [7] could be used instead (454-B, 454-C, etc).

2. To identify pathogens with DNA genomes, a DNA extraction would be performed, followed by a Klenow step before the random PCR:

    (a) DNA extraction: High Pure template preparation kit (Roche) can be used, following the manufacturer's instructions.

    (b) Klenow step on DNA: mix 2.5 µl tagged random primer, 3 µl 10× buffer, 1 µl 0.5 mM dNTP, 10 µl DNA, 1 µl 0.5 U/µl DNA pol1 and 12.5 µl water. Incubate at room temperature for 1 h.

    (c) Proceed from Subheading 3.3, **step 10**.

References

1. Padmanabhan R, Mishra AK, Raoult D et al (2013) Genomics and metagenomics in medical microbiology. J Microbiol Methods 95:415–424

2. Miller RR, Montoya V, Gardy JL et al (2013) Metagenomics for pathogen detection in public health. Genome Med 5:81

3. Liais E, Croville G, Mariette J et al (2014) Novel avian coronavirus and fulminating disease in guinea fowl, France. Emerg Infect Dis 20:105–108

4. Li H, Durbin R (2009) Fast and accurate short read alignment with Burrows-Wheeler transform. Bioinformatics 25:1754–1760

5. Li H, Handsaker B, Wysoker A et al (2009) The Sequence alignment/map format and SAMtools. Bioinformatics 25:2078–2079

6. Robinson JT, Thorvaldsdóttir H, Winckler W et al (2011) Integrative genomics viewer. Nat Biotechnol 29:24–26

7. Victoria JG, Kapoor A, Li L et al (2009) Metagenomic analyses of viruses in stool samples from children with acute flaccid paralysis. J Virol 83:4642–4651

Chapter 3

Serological Diagnosis of Feline Coronavirus Infection by Immunochromatographic Test

Tomomi Takano and Tsutomu Hohdatsu

Abstract

The immunochromatographic assay (ICA) is a simple antibody–antigen detection method, the results of which can be rapidly obtained at a low cost. We designed an ICA to detect anti-feline coronavirus (FCoV) antibodies. A colloidal gold-labeled recombinant FCoV nucleocapsid protein (rNP) is used as a conjugate. The Protein A and affinity-purified cat anti-FCoV IgG are blotted on the test line and the control line, respectively, of the nitrocellulose membrane. The specific detection of anti-FCoV antibodies was possible in all heparin-anticoagulated plasma, serum, whole blood, and ascitic fluid samples from anti-FCoV antibody positive cats, and nonspecific reaction was not noted in samples from anti-FCoV antibody negative cats.

Key words Feline coronavirus, Immunochromatographic assay, Serological diagnosis

1 Introduction

Feline coronavirus (FCoV) is composed of nucleocapsid (N) proteins, membrane (M) proteins, and spike (S) proteins. FCoV has been classified into serotypes I and II according to the amino acid sequence of its S protein [1, 2]. Both serotypes consist of two biotypes: feline infectious peritonitis virus (FIPV) and feline enteric coronavirus (FECV). FECV infection is asymptomatic in cats, whereas FIPV infection causes lethal disease: FIP [3]. FIPV (virulent FCoV) has been proposed to arise from FECV (avirulent FCoV) due to a mutation [4–6]; however, the exact mutation and inducing factors have not yet been clarified.

It is normally comprehensively diagnosed based on the clinical condition, hematological profile, and results of FCoV genomic RNA and anti-FCoV antibody measurements in cats suspected of FIP [7]. An indirect immunofluorescence assay (IFA), and enzyme-linked immunosorbent assay (ELISA) are used to measure FCoV-antibodies. IFA and ELISA are highly sensitive and specific, but are cumbersome, expensive, and time-consuming.

Helena Jane Maier et al. (eds.), *Coronaviruses: Methods and Protocols*, Methods in Molecular Biology, vol. 1282, DOI 10.1007/978-1-4939-2438-7_3, © Springer Science+Business Media New York 2015

A simple and rapid method is necessary to prevent an epidemic of FCoV infection. A low-cost method is also needed to measure anti-FCoV antibodies in cats maintained in multi-cat environments. The most appropriate diagnostic method meeting these conditions may be immunochromatographic assay (ICA). The detection of anti-FCoV antibodies using ICA requires no special device or reagent, and the results can be simply and rapidly obtained. In this chapter, we describe our protocol for the preparation of the ICA to detect anti-FCoV antibodies using recombinant FCoV N protein [8].

2   Materials

2.1   Recombinant FCoV N Protein (rNP)

1. Plasmid DNA: pGEX4T-1 (GE Healthcare) with the N gene of the type I FIPV KU-2 strain (Gene Accession No. AB086881.1) (*see* **Note 1**).

2. Competent *Escherichia coli*, e.g., strain BL-21.

3. LB broth: 1.6 % (w/v) Bacto Tryptone, 1.0 % (w/v) Bacto Yeast Extract, and 0.5 % (w/v) NaCl in ddH$_2$O, and adjusted to pH 7.0 with 5 N NaOH.

4. 100 mg/ml ampicillin sodium in water.

5. 100 mM isopropyl β-D-L-thiogalactopyranoside (IPTG) in water.

6. 10 mg/ml lysozyme in water.

7. 100 mM phenylmethylsulfonyl fluoride (PMSF) in methanol.

8. 1 mg/ml DNase I in 0.15 M NaCl.

9. 0.1 mg/ml sodium deoxycholate in water.

10. Elution buffer: 0.3 % (w/v) reduced glutathione in 0.1 M Tris–HCl, pH 8.0.

11. Sonicator.

2.2   Capture Agent

1. 0.5 mg/ml purified IgG from serum of FCoV-infected cat (*see* **Notes 2** and **3**).

2. 2.0 mg/ml monoclonal antibody (mAb) YN-2 (anti-FCoV N protein; IgG2a) (*see* **Note 4**).

3. 0.1 mg/ml Protein A.

2.3   ICA Test Strip

1. Sample pad and absorbent pad: C083 Cellulose Fiber Sample Pad Strips (Millipore).

2. Nitrocellulose membrane: Hi-Flow Plus 240 Membrane Cards (Millipore).

3. Automatic cutter, e.g., CM4000 (BioDot) or scissors.

4. Dispensing machine, e.g., XYZ3050 (BioDot) or fine-point brush.

2.4   Colloidal
Gold-Labeled rNP

1. Diluting/preserving solution: 20 mM sodium tetraborate, 1 % (w/v) bovine serum albumin (BSA), and 0.1 % (w/v) NaN_3 in water.

2. Colloidal gold solution (40 nm).

3. 10 % (w/v) BSA in water.

4. Borax containing 10 % BSA.

3   Methods

3.1   Preparation
of rNP

1. Incubate BL-21 cells containing pGEX4T-1 with the N gene of the type I FIPV KU-2 strain overnight at 37 °C in 10 ml of LB broth containing 100 µg/ml ampicillin (LB/AMP broth).

2. Dilute overnight cultures 1:100 in 100 ml of fresh LB/AMP broth and grow to OD_{600} of 0.4–0.5.

3. Induce expression of GST-tagged rNP by adding 100 µl of 0.1 mM IPTG to the culture.

4. Incubate for 24 h at 25 °C in shaking incubator.

5. Centrifuge at $12,000 \times g$ for 15 min and resuspend the cell pellet in 30 ml of PBS.

6. Add 0.8 ml of 10 mg/ml lysozyme and 0.4 ml of 100 mM PMSF to the suspension and mix.

7. Incubate on ice for 20 min.

8. Add 0.3 ml of 0.1 mg/ml sodium deoxycholate.

9. Lyse bacterial cells by three 30 s pulses of sonication on ice using a sonicator.

10. Add 0.2 ml of 1 mg/ml DNase I to the cell lysate and incubate for 30 min at 25 °C.

11. To remove bacterial debris, centrifuge cell lysate at $12,000 \times g$ for 15 min at 4 °C.

12. Wash glutathione sepharose beads by adding 40 ml of PBS per 6.7 ml of original slurry of glutathione sepharose (*see* **Note 5**).

13. Add supernatant to the washed glutathione sepharose beads, and rotate overnight at 4 °C.

14. Wash sepharose beads with bound GST-tagged rNP in PBS at for 5 h at 4 °C with rotation.

15. Spin down the beads at $700 \times g$ for 5 min at RT, and resuspend in 40 ml of PBS per 6.7 ml of original slurry.

16. Spin down the beads at $700 \times g$ for 5 min at RT.

17. Elute GST-tagged rNP from beads using 10 ml of Elution buffer at RT, and 1 ml of fractions (0.2 ml/tube) are collected into test tubes (*see* **Note 6**) and analyzed by SDS-PAGE and Western immunoblotting, using standard procedures.

18. Eluted peak fractions (tube no. 18–22) are pooled and dialyzed against PBS using dialysis tubing for overnight.

3.2 Preparation of the ICA Test Strip

The ICA test strip consists of three main components: a sample pad, nitrocellulose membrane, and absorbent pad (Fig. 1a, b).

1. Stick the sample pad and absorbent pad onto the nitrocellulose membrane using adhesive tape, overlapping by 3 mm.

2. Cut this sheet into 2 cm strips using an automatic cutter or scissors.

3. Dispense 20–30 μl of protein A onto the test line of the nitrocellulose membrane using a dispensing machine or a fine-point brush (*see* **Note 7**).

4. Dispense 20–30 μl of affinity-purified IgG from serum of FCoV-infected cat or mAb YN-2 onto the control line of the nitrocellulose membrane using a dispensing machine or a fine-point brush.

5. Dry the membrane for 30 min at room temperature and then cut into 0.5 cm strips using an automatic cutter or scissors.

3.3 Preparation of Colloidal Gold-Labeled rNP

1. Dilute rNP in PBS to 0.5 mg/ml.

2. Add 30 μl of the diluted rNP to 1 ml of colloidal gold solution.

3. Stir well and incubate for 30 min at room temperature.

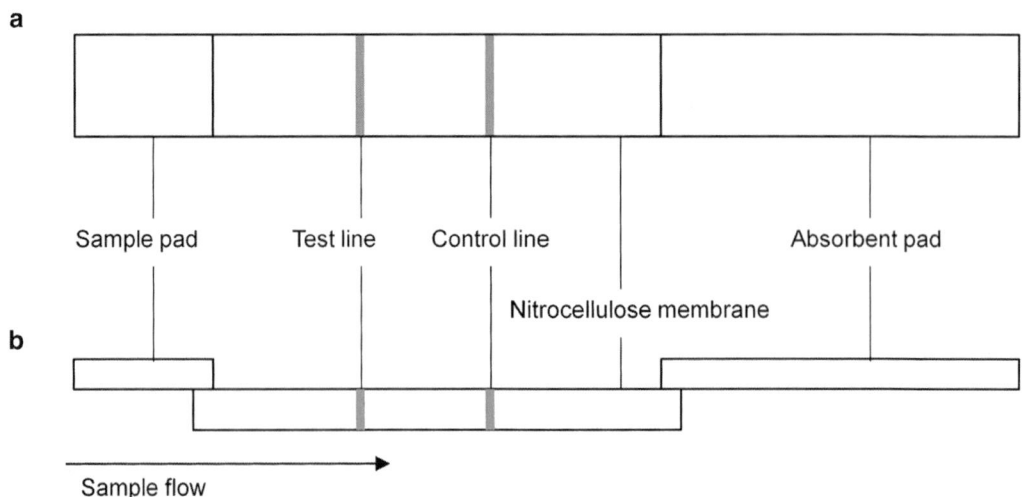

Fig. 1 Schematic diagrams of ICA test strip. (a) Top view and (b) Side view

4. Add 100 μl of 20 mM Borax containing 10 % BSA.

5. Incubate for 30 min at room temperature.

6. Centrifugation at $22,000 \times g$ for 10 min and resuspend the pellet in 0.75 ml of 20 mM Borax containing 10 % BSA.

3.4 Procedure for ICA Test

A schematic of the principle of the ICA test is provided in Fig. 2a.

1. Dilute the sample (i.e., plasma, serum, and effusive fluid) 80 times with eluent solution (*see* **Note 8**).

2. Mix 40 μl of this dilution with 20 μl of the colloidal gold-labeled rNP (from Subheading 3.3, **step 6**) in the well of a 96-well plate.

3. Insert the ICA test strip into the well of the 96-well plate and allow mixture to be absorbed (Fig. 2b).

4. The test line or/and control line will appear after 10 min at room temperature (Fig. 2c).

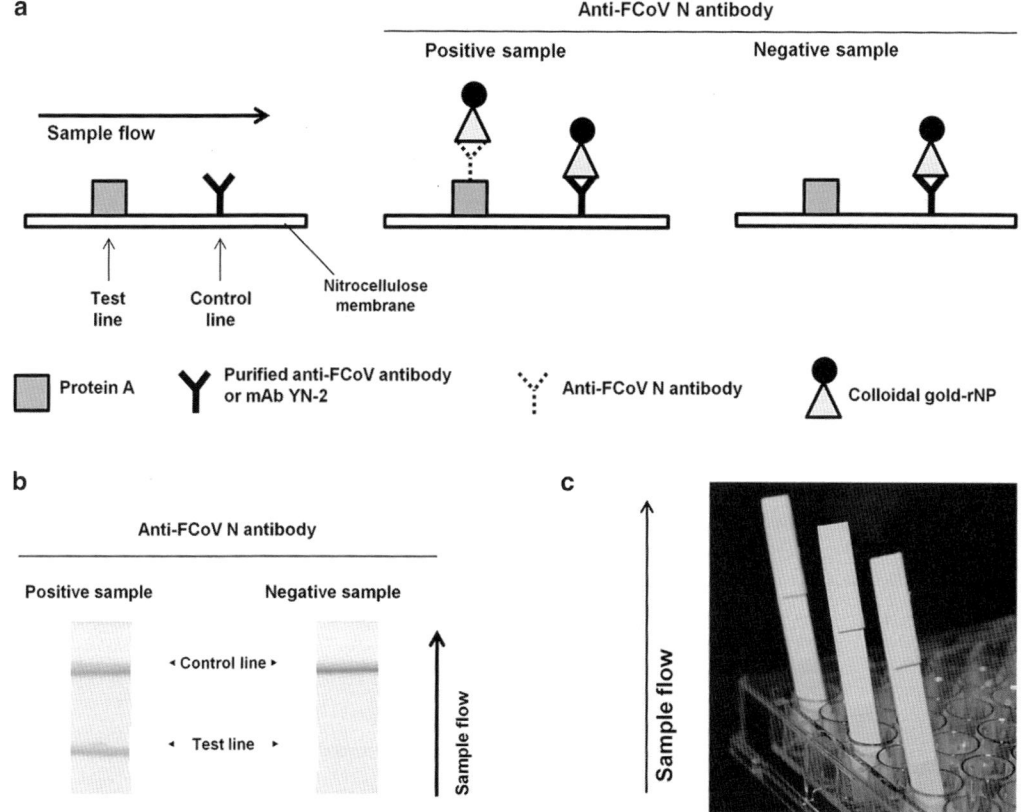

Fig. 2 Principle of the ICA test and example results. (a) Outline of the principle of anti-FCoV antibody detection. (b) The sample pad at the end of the ICA test strip is dipped in the sample mixture. (c) Typical positive and negative results. C: control line (mAb YN-2). T: test line (Protein A)

4   Notes

1. Based on our experience, N protein is more efficiently expressed in *E. coli* than other structural proteins (S protein and M protein) of FCoV.

2. Feline IgG is isolated by affinity chromatography on a Protein G column.

3. Serum and plasma from FCoV-infected cats are good source of cat anti-FCoV IgG. However, large amount of serum and plasma are needed for developing ICA test. It is practically difficult to obtain large amounts of serum and plasma from FCoV-infected cats. Therefore, we recommend using anti-FCoV mAb instead of cat anti-FCoV IgG.

4. The Hybridoma producing mAb YN-2 was prepared following the method reported by Hohdatsu et al. [1]. In our experience, the mAb YN-2 has a higher affinity for the colloidal gold-labeled rNP than other anti-FCoV N mAb (e.g., mAb E22-2).

5. Glutathione sepharose beads are just added to PBS and then they are ready to use.

6. GST-tagged rNP was eluted from the beads by drip-through at a constant flow (2.0–3.0 ml/min).

7. Generally, reagents are dispensed by the dispensing machine on a nitrocellulose membrane. However, this machine is very expensive. If you intend to develop an ICA kit on a trial basis, we recommend using a fine-point brush instead of the machine.

8. The specific detection of anti-FCoV antibodies was possible in all heparin-anticoagulated plasma, serum, whole blood, and ascitic fluid samples from anti-FCoV positive cats. On the other hand, the nonspecific test line formation was noted in EDTA- or sodium citrate-anticoagulated plasma of anti-FCoV negative cats.

Acknowledgments

We thank all the colleagues of our laboratory for assistance on ICA development. This work was in part supported by KAKENHI (Grants-in-Aid for Scientific Research (B), No. 25292183) from the Ministry of Education, Culture, Sports, Science, and Technology, and the Research Fund of Petience Medical Corporation.

References

1. Hohdatsu T, Okada S, Koyama H (1991) Characterization of monoclonal antibodies against feline infectious peritonitis virus type II and antigenic relationship between feline, porcine, and canine coronaviruses. Arch Virol 117:85–95

2. Motokawa K, Hohdatsu T, Aizawa C et al (1995) Molecular cloning and sequence determination of the peplomer protein gene of feline infectious peritonitis virus type I. Arch Virol 140:469–480

3. Pedersen NC (2009) A review of feline infectious peritonitis virus infection: 1963-2008. J Feline Med Surg 11:225–258

4. Brown MA (2011) Genetic determinants of pathogenesis by feline infectious peritonitis virus. Vet Immunol Immunopathol 143:265–268

5. Chang HW, Egberink HF, Halpin R et al (2012) Spike protein fusion peptide and feline coronavirus virulence. Emerg Infect Dis 18:1089–1095

6. Chang HW, de Groot RJ, Egberink HF et al (2010) Feline infectious peritonitis: insights into feline coronavirus pathogenesis and epidemiology based on genetic analysis of the viral 3c gene. J Gen Virol 91:415–420

7. Addie D, Belák S, Boucraut-Baralon C et al (2009) Feline infectious peritonitis. ABCD guidelines on prevention and management. J Feline Med Surg 11:594–604

8. Takano T, Ishihara Y, Matsuoka M et al (2013) Use of recombinant nucleocapsid proteins for serological diagnosis of feline coronavirus infection by three immunochromatographic tests. J Virol Methods 196:1–6

Chapter 4

Estimation of Evolutionary Dynamics and Selection Pressure in Coronaviruses

Muhammad Munir and Martí Cortey

Abstract

Evolution of coronaviruses is facilitated by the strong selection, large population size, and great genetic diversity within the susceptible hosts. This predisposition is primarily due to high error rate, and limited proofreading capability of the viral polymerase and by recombination. These characteristics make coronaviruses an interesting model system to study the mechanisms involved in viral evolution and the ways viruses adapt to switch host or to gain novel functions. Here we describe the protocol to estimate selection pressures for the spike gene and evolutionary dynamics of bovine coronaviruses.

Key words Coronaviruses, Evolution, Genetics, Emergence, Selection, S gene

1 Introduction

Coronaviruses encode the largest positive sense single-stranded RNA genomes known, ranging from 27 to 31 Kb in length. Although coronaviruses have been shown to possess proofreading ability [1], relatively high mutation rates mean that coronaviruses are one of the most diverse, genetically distinct, and recently emerging groups of viruses. The emergence of these viruses are mainly triggered by the virus evolution which could occur due to high mutational rates, selection pressure on genetic diversity, inter- and intra-host selection, frequency of recombination, and genetic drifts during transmission bottlenecks. Within subfamily *Coronaviridae*, *Alphacoronaviruses*, and *Betacoronaviruses* infect and cause diseases in mammals, whereas *Gammacoronaviruses* are mainly avian specific [2].

Bovine coronaviruses (BCoVs), together with human coronavirus OC43 (HCoV-OC43), equine coronavirus (ECoV), and porcine hemagglutinating encephalomyelitis virus (PHEV), belong to the virus species Betacoronavirus1 of the lineage A of the genus Betacoronavirus [3]. BCoV causes infections both in respiratory and enteric systems in cattle of all ages. Like other coronaviruses,

Helena Jane Maier et al. (eds.), *Coronaviruses: Methods and Protocols*, Methods in Molecular Biology, vol. 1282,
DOI 10.1007/978-1-4939-2438-7_4, © Springer Science+Business Media New York 2015

BCoV exhibit high genetic mutations (one mutation per genome per replication round) [4, 5]. The nucleotide (nt) substitutions per site per year were found to be 1.3×10^{-4}, 6.1×10^{-4}, and 3.6×10^{-4} for RNA-dependent RNA polymerase (RdRp), S, and N genes, respectively [6–8]. Due to their evolutionary potential, BCoVs have been isolated from humans (BCoV-like human enteric coronavirus HECV-4408/US/94) and a recently isolated canine respiratory coronavirus (CRCoV) has also shown a high genetic similarity to Betacoronavirus1 [9, 10].

Taken together, experimental data and mathematical models have reinforced the need for studying coronavirus dynamics and evolution, which could provide bases for effective control measures. Recent availability of quantitative deep-sequencing methodologies has provided data that can be modelled for future prediction of transmission dynamics and to estimate relevant parameters. In this protocol, we used publically available S gene data on BCoV, as prototype coronavirus, and analyzed to predict epidemiological linkage, mutation-prone sites and evolution in the S gene of BCoV. The same protocol is applicable to other genes of the coronaviruses and viruses of other families.

2 Materials

To perform in silico analysis of the S genes of BCoV, the following equipment will be required (*see* **Note 1**):

1. Mac OS X with minimum 2.4 GHz processor and 2 GB RAM.
2. TextEdit (TextWrangler) stable release 1.8 or latest.
3. BioEdit version v7.2.5.
4. MrBayes version 3.2.2 or latest.
5. A Perl script for generating suitable file formats.
6. BEAST version 1.8.0 or latest.
7. BEAUti version 1.7 or latest.
8. Tracer version 1.6 or latest.
9. FigTree v1.2.3 or latest.
10. An appropriate Internet access.

3 Methods

The following procedures are adapted for Mac OS but are equally applicable for other systems.

3.1 Phylodynamics

1. Define objectives (*see* **Note 2**).
2. Construct dataset and label it as BCoV_S genes.fas (*see* **Note 3**).

3. Open the downloaded file (BCoV_S genes.fas) in TextEdit and edit the sequence titles. The sequence titles can be arranged depending upon objective in mind and the availability of downstream analysis tools. One accepted way of labelling the sequence title will be to arrange them in host/isolate_ID/genotype/country/year (accession number). Remove all illegal characters along with empty spaces and replace them with underscore/understrike (_). However, do not remove any greater than signs (>), which will destroy the .fasta format and may require rebuilding of the data set [11]. To do so, use the "Find" and "Replace with" options in the TextEdit, which can be opened with "cmd+F" command in Mac OS X. Save the file before closing the dataset.

4. Open the file in BioEdit and click on Accessory Application -> ClustalW Multiple Alignment. Save the newly opened aligned file and label it as BCoV_S genes_align.fas (*see* **Note 4**).

5. Convert the .fas file (BCoV_S_genes_align.fas) to a .nex file (BCoV_S genes_align.nex): Use either a Perl script (available to freely download at https://github.com/drmuhammadmunir/perl/blob/master/ConvertFastatoPhylip) or using trial version of CodonCode Aligner (www.codoncode.com/aligner/) (*see* **Note 5**).

6. Move BCoV_S_genes_align.nex file into the folder of MrBayes. Detailed description of the program can be found on the webpage of the program (http://mrbayes.sourceforge.net/). Briefly, open terminal and type "mb" to start the MrBayes software (double click on MrBayes application icon in Window). The following instructions should appear:

MrBayes v3.2.1 x64

(Bayesian Analysis of Phylogeny)

Distributed under the GNU General Public License

Type "help" or "help <command>" for information on the commands that are available.

Type "about" for authorship and general information about the program.

MrBayes >

7. To execute the file into the program, type "execute <Space>filename" (e.g., execute BCoV_S_genes_align.nex) then press "Enter". The message "Reached end of file" indicates successful execution of the file and the program is ready to run. In any error, either follow the instructions mentioned in the error or rebuilt datasets. The most common error is the presence of illegal characters such as pipeline sign (|), colon (:), semicolon (;) slash/stroke/solidus (/), apostrophe (' '),

quotation marks (' ', " ", ' ', " "), and brackets ([ ], (), { }, < >), among others. Therefore remove these from the fasta file as described before.

8. To set the evolutionary model to the GTR substitution, type "lset nst=6 rates=invgamma" after the MrBayes > prompt then press "Enter". The message "Successfully set likelihood model parameters" indicates the success in model setup.

9. To set the sample collection (200) from posterior probability distribution, diagnostic calculation every 1,000 generations, and print and sample frequency to 100, type "mcmc ngen=20000 samplefreq=100 printfreq=100 diagnfreq=1000" after the MrBayes > prompt then press "Enter". Program will start calculating the split frequency depending on the speed of the operating system and the size of the dataset. Note the message "Average standard deviation of split frequencies". If it is below 0.01 after 2,000 generations, type "yes" after "Continue the analysis? (yes/no)" prompt to set more generations. Continue this until the split frequency drops below 0.01. Once reached, type "no" which leads the users to MrBayes > prompt.

10. To summarize the parameter, type "sump" then press "Enter".

11. To summarize the tree type "sumt" then press "Enter". This command will save the tree with extension "nex.con.tre" (i.e., BCoV_S genes_align.nex.con.tre) in the MrBayes folder where the original file (BCoV_S genes_align.nex) was kept. The tree can be opened and annotated in the FigTree.

12. Open the desired file (BCoV_S_genes_align.nex.con.tre) after launching FigTree.

13. Label your sequences by searching your sequence-tag, such as isolate name or country, in the search button when "Taxa" is selected. Similarly, select "Nod" or "Clade" to label the respective items (*see* **Note 6**).

14. After annotation, save your tree using File -> Export Graphics -> PDF (or other desired file format from the list) -> OK path. The resulting file can be used for further editing or for presentation [11].

3.2 Selection Pressures

3.2.1 SNAP

1. To analyze the occurrences of synonymous (dS) and non-synonymous (dN) substitutions in the S gene, use the same fasta file (BCoV_S_genes_align.fas) that was generated for phylodynamics (*see* **Note 7**).

2. Open the SNAP tool freely available at http://www.hiv.lanl.gov/content/sequence/SNAP/SNAP.html and paste the sequence or upload the dataset.

3. Both accumulated (cumulated dN-dS) and per codon (dN-dS) selection sites can be calculated by the generated table of SNAP.

4. Since the selections are calculated on every nucleotide, sites under positive or negative selection can be highlighted (*see* **Note 8**).

3.2.2 Datamonkey

1. Alternatively and to verify the robustness of the data generated by the SNAP, the same alignment can be used to calculate selection pressure using GTR (general time reversible) substitution model on a neighbor-joining phylogenetic tree by the Datamonkey Web server (Freely available at http://www.datamonkey.org/dataupload.php).

2. The program uses the computational engine of the HyPhy package [12] to estimate dN–dS with a variety of evolutionary models and can analyze selection even in the presence of recombination (*see* **Note 9**).

3.3 Evolutionary Dynamics

3.3.1 XML File Generation

1. Within the BEAST package, open BEAUti program (Bayesian Evolutionary Analysis Utility) and import Nexus (BCoV_S_genes_align.nex) or Fasta (BCoV_S_genes_align.fas) file of the data set. Remember to execute the data by "File -> Import Data -> Open".

2. Several parameters of the BEAST run (i.e., the date of the sequences, the substitution model, the rate variation among sites, the length of the MCMC chain) can then be adjusted according to specific need [13] (*see* **Note 10**).

3. Once all desired parameters are set, finally, click on the "Generate BEAST File" to generate .XML file which will be used as input for BEAST analysis.

4. Label the file as BCoV_S_genes_align.xml for consistency.

3.3.2 BEAST Analysis

This is a brief explanation in order to run BEAST program and summarize results using TRACER.

1. Move the .xml files (BCoV_S_genes_align.xml) into the BEAST folder.

2. Open the BEAST program (double-click), a white screen on JAVA environment will appear, wait for several seconds until a second screen appears.

3. Choose the file to analyze in this second screen. Before beginning the analyses enable the "Allow overwriting of log files" option. Then press "Run" and the analysis will begin.

4. After few moments, depending upon the processing capacity of the operating system and the size of the data, the chain will begin to run. There will be seven columns that extend vertically. Every column is one of the parameters that are being estimated; however, the first and the last column are crucial to observe. The first column is the generation being sampled in every moment (every chain has ten million steps) and the last

column shows how many millions of states will be run per hour (remember, ten million steps per chain). Depending on the length of the chain, the length of the sequences, and the number of sequences to be analyzed, it may take variable time to complete the run.

5. Once the chain has run, it is required to store the parameters. Close the BEAST window and open the BEAST folder. Every time a chain is run, two files are generated: .xml file and several ends (.log and .tre). Once the first run is complete, change the name of the .log and .tre files. For example, after the completion of a run for BCoV_S_gene_align.xml, BCoV_S_gene_align.log and BCoV_S_gene_align.tre files will be generated. Rename these two files to BCoV_S_gene_align1.log and BCoV_S_gene_align1.tre.

6. Run the BCoV_S_gene_align.xml at least for two more times (**steps 2–5**).

7. Finally, three different .log files and .tre files will be available labelled as BCoV_S_gene_align1.log, BCoV_S_gene_align2.log, and BCoV_S_gene_align3.log. These three files contain the estimations of the substitution rate that have to be summarized in TRACER.

8. To summarize the run, open the TRACER program and select the option "File" and "Input Trace file", and open the first …1.log file from the folder, followed by the addition of the second (…2.log) file. Finally add the third (…3.log) log file.

9. The estimations of the parameters are viewable in the graphic interface. Select the option "Combined" from the Trace Files (Upper left) and the estimations that will appear on the traces table are the main estimations for all the parameters (*see* **Note 11**). Generally, the desired parameters are:

Tree Model Root: This is the number of years that passed after the most recent common ancestor (TMRCA). Subtract this number from the most modern date to yield the TMRCA for the dataset.

Clock Rate: This is directly the rate of evolution in substitution/site/year.

4   Notes

1. This protocol is optimized for Mac OS X; however, all the software packages and tools used here are also available for Windows which can be installed using recommended methodologies. All the software used here are Open Access, which do not require any subscription for any operating systems.

These software packages are only for demonstration purposes, and there may be alternative solutions for the same purpose. The overall time of the data analysis depends upon processing power of the operating system and the number and length of sequences in the dataset.

2. The same phylogenetic tree can be used for different interpretations. Failing to create a proper objective can lead to drawing incorrect conclusions from phylogenetic studies. It is therefore essential to define the objective for the downstream analyses before initiating the study.

3. Construction of datasets depends on the objectives. One of the most common interests of bioinformaticians is to determine the epidemiological linking of the query sequence to that of sequences reported from the world and are available in the public domains. For this purpose, the Basic Local Alignment Search Tool (BLAST) is the most widely used tool, primarily owing to its speed of execution. Search the nucleotide sequences with objective-based keyword such as "Bovine Coronaviruses S gene". Manual editing and investigations of the downloaded sequences are always suggested. Notably, BLAST-Explorer is primarily aimed at helping the construction of sequence datasets for further phylogenetic study, and it can also be used as a standard BLAST server with enriched output. Use BLAST or BLAST-Explorer or other suitable database for construction of datasets.

4. There are different algorithms for DNA sequence alignment with variable degrees of utility. In this protocol, ClustalW was used for simplicity. Any other algorithm can be used depending upon the preferences and interest.

5. Nexus format is required input for MrBayes. Different tools, both online and offline, can be used to generate appropriate nexus output. We have only presented two commonly used and easily achievable methods.

6. Detailed demonstration for tree annotation is described in our earlier publication [11].

7. The file used for phylogenetic analysis may contain all available sequences in the public domain, which increases the size of the file significantly. However, depending upon the objective in mind, the datasets can be modified accordingly. For the larger datasets, the compiled data will be emailed to the email address provided once ready. This is also important to keep a record for future use.

8. The cut point is calculated to be zero. All sites showing cumulated dN-dS values above 0 are under positive pressure whereas values below 0 are under negative pressure.

9. The nature of parameter selection and interpretation is complex and is beyond the scope of this protocol. Please consult developers' published report for thorough understating of the concepts and applications [12].

10. The parameters of the BEAST run are crucial and can determine the nature of output and may heavily influence the results. However, normally the default parameters are used [13].

11. When summarizing the BEAST results, do not use the mean as it appears in the output (this is the arithmetic mean); instead use the geometric mean that appears in the summary statistic table (Right Upper). From a methodological point of view this is much more correct (Click on every parameter in the Tracer table and the summary statistic table for every parameter will change).

Acknowledgement

The work at our laboratories is supported by the grant from Biotechnology and Biological Sciences Research Council (BBSRC) through Institute Strategic Programme Grant (BB/J004448/1).

References

1. Minskaia E, Hertzig T, Gorbalenya A et al (2006) Discovery of an RNA virus 3′→5′ exoribonuclease that is critically involved in coronavirus RNA synthesis. Proc Natl Acad Sci U S A 103:5108–5113

2. Jackwood MW, Hall D, Handel A (2012) Molecular evolution and emergence of avian gammacoronaviruses. Infect Genet Evol 12:1305–1311

3. de Groot RJ, Baker SC, Baric R et al (2012) Coronaviridae. In: King AMQ et al (eds) Virus taxonomy: classification and nomenclature of viruses: ninth report of the international committee on taxonomy of viruses. Elsevier Academic Press, Oxford, pp 806–828

4. Drake JW, Holland JJ (1999) Mutation rates among RNA viruses. Proc Natl Acad Sci U S A 96:13910–13913

5. Moya A, Holmes EC, Gonzalez-Candelas F (2004) The population genetics and evolutionary epidemiology of RNA viruses. Nat Rev Microbiol 2:279–288

6. Vijgen L, Keyaerts E, Lemey P et al (2006) Evolutionary history of the closely related group 2 coronaviruses: porcine hemagglutinating encephalomyelitis virus, bovine coronavirus, and human coronavirus OC43. J Virol 80:7270–7274

7. Woo PC, Lau SK, Lam CS et al (2012) Discovery of seven novel Mammalian and avian coronaviruses in the genus deltacoronavirus supports bat coronaviruses as the gene source of alphacoronavirus and betacoronavirus and avian coronaviruses as the gene source of gammacoronavirus and deltacoronavirus. J Virol 86:3995–4008

8. Erles K, Shiu KB, Brownlie J (2007) Isolation and sequence analysis of canine respiratory coronavirus. Virus Res 124:78–87

9. Bidokhti MR, Tråvén M, Krishna NK et al (2013) Evolutionary dynamics of bovine coronaviruses: natural selection pattern of the spike gene implies adaptive evolution of the strains. J Gen Virol 94:2036–2049

10. Zhang XM, Herbst W, Kousoulas KG et al (1994) Biological and genetic characterization of a hemagglutinating coronavirus isolated from a diarrhoeic child. J Med Virol 44:152–161

11. Munir M (2013) Bioinformatics analysis of large-scale viral sequences: From construction of data sets to annotation of a phylogenetic tree. Virulence 4:97–106

12. Pond SLK, Frost SDW, Muse SV (2005) HyPhy: hypothesis testing using phylogenies. Bioinformatics 21:676–679

13. Drummond AJ, Suchard MA, Xie D et al (2012) Bayesian phylogenetics with BEAUti and the BEAST 1.7. Mol Biol Evol 29:1969–1973

Part II

Propogation, Titration, and Purification of Coronaviruses

Chapter 5

The Preparation of Chicken Tracheal Organ Cultures for Virus Isolation, Propagation, and Titration

Ruth M. Hennion

Abstract

Chicken tracheal organ cultures (TOCs), comprising transverse sections of chick embryo trachea with beating cilia, have proved useful in the isolation of several respiratory viruses and as a viral assay system, using ciliostasis as the criterion for infection. A simple technique for the preparation of chicken tracheal organ cultures in glass test tubes, in which virus growth and ciliostasis can be readily observed, is described.

Key words Tracheal organ culture, Ciliostasis, Respiratory virus, Viral assay

1 Introduction

Tracheal organ cultures (TOCs) have been used for the study of a number of respiratory tract pathogens [1]. The first human coronavirus (HCoV) was isolated using human ciliated embryonal trachea [2], and studies on persistent infection with Newcastle disease virus [3], isolation of the Hong Kong variant of influenza A2 virus [4], and studies on the pathogenicity of mycoplasmas [5] using TOCs have all been reported. More recently, TOCs have been used in studies on the pathogenicity and induction of protective immunity by a recombinant strain of infectious bronchitis virus (IBV) [6].

Tracheal organ cultures derived from 20 day old chicken embryos are reported to be as sensitive as 9 day old embryonated eggs for the isolation and titration of IBV [7], and are more sensitive than TOCs from chickens up to 31 days of age with complete ciliostasis, the criterion for infection, being observed 3 days after infection.

With the ease of production and the proven usefulness of TOCs in virus isolation and in studies on pathogenicity and immunization strategies, their more widespread use for research into respiratory tract viruses should be considered. Whilst TOCs have been successfully prepared for assays using multiwell plates [8], the

Helena Jane Maier et al. (eds.), *Coronaviruses: Methods and Protocols*, Methods in Molecular Biology, vol. 1282,
DOI 10.1007/978-1-4939-2438-7_5, © Springer Science+Business Media New York 2015

method described below is based on that previously reported [5] and utilizes chicken embryo TOCs on a rolling culture tube assembly, where TOCs are reported to be capable of maintaining ciliary activity for longer periods than in static cultures. Debris accumulating within the TOCs rings is reduced, making observation of ciliary activity easier.

2   Materials

2.1   Preparation of Tracheal Section

1. 19- to 20-day-old embryonated eggs from specific pathogen free (SPF) chicken flock.
2. Tissue chopper: the following method assumes the use of a McIlwain mechanical tissue chopper (Mickle Laboratory Engineering Co. Ltd.).
3. Sterile curved scissors (small).
4. Sterile scissors (large).
5. Sterile forceps.
6. Sterile Whatman filter paper discs 55 mm diameter (*see* **Note 1**).
7. 70 % industrial methylated spirits (IMS).
8. Double-edged razor blades.
9. Eagle's Minimum Essential Medium (MEM) with Earle's salts, 2 mM L-glutamine, and 2.2 g/L sodium bicarbonate.
10. Penicillin + streptomycin (100,000 U of each per ml).
11. 1 M HEPES buffer prepared from HEPES (free acid) and tissue culture grade water, sterilized in an autoclave at 115 °C for 20 min.
12. Culture medium: MEM, 40 mM HEPES buffer, 250 U/ml penicillin, and 250 U/ml streptomycin.
13. Sterile Bijou bottles or similar.
14. Sterile 100- and 150-mm-diameter petri dishes.

2.2   Culture of Tracheal Sections

1. Tissue culture roller drum capable of rolling at approximately 8 revolutions/hour at 37 °C.
2. Associated rack suitable for holding 16 mm tubes on roller drum.
3. Sterile, extra-strong rimless soda glass tubes 150 mm long × 16 mm outside diameter, suitable for bacteriological work (*see* **Note 2**).
4. Sterile silicone rubber bungs 16 mm diameter at wide end, 13 mm diameter at narrow end, and 24 mm in length (*see* **Note 3**).
5. Inverted microscope (60–100× magnification).

3   Method

To calculate the number of embryonated eggs required for an assay, assume that each trachea will yield 17–20 rings. Expect a loss of up to 20 % of the cultures during the preliminary incubation step, owing to damage to the rings during preparation or spontaneous cessation of ciliary activity.

3.1   Preparation of Tracheal Sections

1. On a clean workbench spray the top of the eggs with 70 % IMS (*see* **Note 4**).

2. Using curved scissors remove the top of the shell, lift the embryo out by the wing and cut off the yolk sac. Place the embryo in a 150 mm petri dish and discard the egg and yolk sac.

3. With a sharp pair of scissors decapitate the bird, severing the spinal cord just below the back of the head and angling the cut to just below the beak (*see* **Note 5**).

4. Position the embryo on its back and, using small forceps and scissors, cut the skin along the length of the body from the neck to the abdomen. Care must be taken not to damage the underlying structures.

5. Locate the trachea and using small scissors and forceps, dissect it away from the surrounding tissues (*see* **Note 6**).

6. Cut the trachea at the levels of the carina and larynx (the larynx may have been removed on decapitation) and remove it from the embryo, placing the tissue in a Bijou bottle containing culture medium (*see* **Note 7**).

7. Repeat **steps 2–6** for all available embryos.

8. Place one trachea at a time on a disc of filter paper in a petri dish and, using two pairs of fine forceps, gently remove as much fat as possible (*see* **Note 8**).

9. Place the cleaned tracheas in a 100 mm petri dish containing culture medium.

10. Swab the tissue chopper with 70 % IMS.

11. Place two filter paper discs on top of the plastic cutting table disc and slide the assembled discs under the cutting table clips on the tissue chopper.

12. Raise the chopping arm of the of the tissue chopper and attach the razor blade.

13. Position the arm over the center of the cutting table (*see* **Note 9**).

14. Place the tracheas on to the filter paper under, and perpendicular to, the raised blade and moisten with a small amount of culture medium (*see* **Note 10**).

15. Adjust the machine to cut sections 0.5–1.0 mm thick and activate the chopping arm.

16. Once the arm has stopped moving, discard the first few rings from each end of the cut tracheas; then with a scalpel, scrape the remaining rings into a 150-mm petri dish containing culture medium.

17. With a large bore Pasteur pipette or similar gently aspirate the medium to disperse the cut tissue into individual rings.

18. Repeat **steps 11–17** until all the tracheas have been sectioned (*see* **Note 11**).

3.2 Culture of Tracheal Sections

1. With a large bore Pasteur pipette or similar dispense one TOC ring together with approximately 0.5 ml of culture medium into a glass tube (*see* **Note 12**).

2. Seal the tube with silicone bung and check that each tube contains one ring (*see* **Note 13**).

3. Put the tubes in the roller tube rack, place on the roller apparatus and set to roll at approximately 8 revolutions/hour, at approximately 37 °C. Leave the tubes rolling for 1–2 days (*see* **Note 14**).

4. Check each tube culture for complete rings and the presence of ciliary activity, using a low power inverted microscope

5. Discard any tubes in which less than 60 % of the luminal surface has clearly visible ciliary activity.

6. The remaining tubes may be used for viral assays (*see* **Note 15**).

4 Notes

1. Batches of sterile Whatman filter papers can be prepared by interleaving individual discs with slips of grease-proof paper and placing them in a glass petri dish. Wrap the dish in aluminum foil and sterilize in a hot air oven (160 °C for 1 h).

2. Batches of sterile tubes can be prepared by placing them, open end down, in suitable sized lidded tins lined with aluminum foil. Sterilize in a hot air oven as above.

3. Batches of sterile silicone rubber bungs can be prepared by placing them, narrow end down, in shallow, lidded tins. Sterilize by autoclaving at 120 °C for 20 min.

4. Preparation of TOCs can be performed on the open laboratory bench after cleaning the surfaces with 70 % IMS or any other suitable disinfectant.

5. Care must be taken at this stage not to damage the trachea.

6. The trachea can be identified by the presence of transverse ridges seen down its length owing to the underlying rings of cartilage.

7. The carina and larynx can be identified by the increased diameter at either end of the trachea.

8. To avoid damage to the trachea hold it as close to one end as possible with the first pair of forceps and use the second pair to strip away the fatty tissue.

9. At this stage gently lower the arm on to the cutting area disc, loosen the screw holding the blade slightly, check that the blade is aligned correctly (the full length of the blade must be in contact with the cutting area), tighten the screw again, and raise the arm.

10. A maximum of five tracheas can be laid side by side on the cutting bed at any one time. Gently stretch each trachea as it is placed on the cutting area, and when all five are in the correct position, wet them with a few drops of culture medium.

11. It is important to use a fresh blade and paper discs for each set of five tracheas to be sectioned and ensure used blades are disposed of in an appropriate sharps bin.

12. Check for damaged glass tubes at this stage, particularly around the rims. Discard any with cracks as these can fail when bungs are inserted, leading to injured fingers.

13. Make sure the tracheal rings are fully submerged in culture medium and not stuck on the wall of the tube. Discard any that are ragged or incomplete.

14. Make sure that the drum is aligned correctly on the apparatus and that the roller is actually moving before leaving the cultures to incubate; the speed of the roller apparatus is slow.

15. A simple quantal assay for infectivity of IBV has been described by Cook et al. [7] and is used extensively in our Institute. Five tubes of TOCs per tenfold serial dilution of virus give sufficiently accurate results for most purposes. A simplification of the method of Cook et al. [7], used for many years by Cavanagh and colleagues, is to add 0.5 ml of diluted virus per TOC tube without prior removal of the medium already in the tube. TOCs are scored as positive for virus when ciliary activity is completely abrogated. If a virus is poorly ciliostatic, its presence can be demonstrated using indirect immunofluorescence, with TOCs conveniently not fixed [9].

References

1. McGee ZA, Woods ML (1987) Use of organ cultures in microbiological research. Ann Rev Microbiol 41:291–300
2. Tyrell DAJ, Bynoe ML (1965) Cultivation of novel type of common-cold virus in organ cultures. Br Med J 5448:1467–1470
3. Cummiskey JF, Hallum JV, Skinner MS et al (1973) Persistent Newcastle disease virus infection in embryonic chicken tracheal organ cultures. Infect Immun 8:657–664
4. Higgins PG, Ellis EM (1972) The isolation of influenza viruses. J Clin Pathol 25:521–524

5. Cherry JD, Taylor-Robinson D (1970) Large quantity production of chicken embryo organ culture and use in virus and mycoplasma studies. Appl Microbiol 19:658–682

6. Hodgson T, Casais R, Dove B et al (2004) Recombinant infectious bronchitis coronavirus Baudette with the spike protein gene of the pathogenic M41 strain remains attenuated but induces protective immunity. J Virol 78:13804–13811

7. Cook JKA, Darbyshire JH, Peters RW (1976) The use of chicken tracheal organ cultures for the isolation and assay of infectious bronchitis virus. Arch Virol 50:109–118

8. Yacida S, Aoyam S, Takahashi N et al (1978) Plastic multiwell plates to assay avian infectious bronchitis virus in organ cultures of chicken embryo trachea. J Clin Microbiol 8:380–387

9. Battacharjee PS, Naylor CJ, Jones RC (1994) A simple method for immunofluorescence staining of tracheal organ cultures for the rapid identification of infectious bronchitis virus. Avian Pathol 23:471–480

Chapter 6

The Preparation of Chicken Kidney Cell Cultures for Virus Propagation

Ruth M. Hennion and Gillian Hill

Abstract

Chicken kidney (CK) cell cultures have historically proved useful for the assay of a number of viruses including coronaviruses. A technique for the preparation of such cell cultures, using a combination of manual and trypsin disaggregation of kidneys dissected from 2- to 3-week-old birds is described. This technique routinely gives high cell yield together with high viability and the resultant adherent primary cultures can be used for virus growth and plaque formation.

Key word Chicken kidney cell culture, Virus growth, Viral assay

1 Introduction

Techniques for the preparation of monolayer cultures from adult kidney cells suitable for the growth and quantitation of viruses have been available for many years; Dulbecco and Vogt [1] described the preparation of Monkey Kidney cultures in 1953 and Youngner [2] published a modification of the process in 1954. Maassab in 1959 [3] describes the preparation of Chicken Kidney monolayer cultures from 4- to 5-day-old chicks, the cultures being used for studies with some human viruses and Churchill [4] reports the use of chicken kidney tissue cultures derived from 3 to 8 week old chickens in the study of avian viruses including Infectious Bronchitis Virus (IBV). The technique for the production of kidney cell monolayer cultures from young birds, as described here, is adapted from those published by Dulbecco and Vogt [1], and Youngner [2] for monkey kidney cells.

Whilst titration of IBV in CK cells gives lower titers than those obtained in embryonated eggs [5] or tracheal organ cultures [6], the ability of CK cells to support the growth of many strains of IBV is well proven. Following adaptation in embryonated eggs, the Beaudette strain of IBV produced characteristic cytopathic effects

Helena Jane Maier et al. (eds.), *Coronaviruses: Methods and Protocols*, Methods in Molecular Biology, vol. 1282, DOI 10.1007/978-1-4939-2438-7_6, © Springer Science+Business Media New York 2015

(CPE) on first passage in CK cells, whilst the Massachusetts strain produced CPE in the second CK passage [4]. CPE consists of syncytia formation which occurs at 6 h post inoculation with the Beaudette strain [7]. The syncytia may contain as many as 20–40 or more nuclei and they quickly round up and detach from the culture surface. Growth curves of IBV in CK cells show a lag phase of 2–4 h and maximum virus yield in 18–20 h [5].

The ability of CK cells to support the growth of IBV has been utilized in wide ranging studies including the assessment of pH stability of a series of IBV strains [8], the identification of the presence of a leader sequence on IBV mRNA A [9], the demonstration that the spike protein of IBV is a determinant of cell tropism [10], the induction of protective immunity with recombinant IBV Beaudette [11] through to the identification of novel zippered ER and associated spherules induced by IBV [12].

2   Materials

2.1   Preparation of Kidney Cells

1. 2–3 Week old chicken(s) from specific pathogen free (SPF) flock killed by cervical dislocation (see **Note 1**).
2. 70 % Industrial methylated spirits (IMS).
3. Sterile instruments to include large scissors, small scissors, small forceps, and scalpels.
4. Sterile 150 mm diameter petri dish.
5. Sterile glassware including conical flasks and beakers.
6. Funnel.
7. 150 ml bottles with leakproof lids.
8. Sterile wire mesh (50 mesh × 0.200 mm diameter wire) folded into a filter shape to fit a funnel (see **Note 2**).
9. 50 ml centrifuge tubes.
10. Dulbecco's phosphate buffered saline without calcium and magnesium (PBSa).
11. Trypsin: 0.25 % porcine trypsin, glucose 0.1 %, PBSa, sterile filtered.
12. Ethylenediaminetetraacetic acid (EDTA): 0.2 mg/ml EDTA, PBSa, autoclaved at 115 °C for 20 min.
13. New born bovine serum (NBBS), heat-inactivated at 56 °C for 30 min (see **Note 3**).

2.2   Culture of Kidney Cells

1. Incubator set at 37 °C and 5 % CO_2.
2. Tissue culture grade flasks or plates.
3. Sterile 50 ml syringes.
4. Haemocytometer.

5. Trypan blue.

6. Inverted Microscope suitable for observing cell cultures.

7. Swinnex 25, reusable, syringe driven polypropylene filter unit fitted with metal gauze, 50 mesh × 0.200 mm dia wire (Swinnex Filter 1) (*see* **Notes 2** and **4**).

8. Swinnex 25, reusable, syringe driven polypropylene filter unit fitted with metal gauze 100 mesh × 0.100 mm dia wire (Swinnex Filter 2) (*see* **Notes 2** and **4**).

9. Eagles Minimum Essential Medium (EMEM) with Earle's salts, 2 mM L-glutamine and 2.2 g/L sodium bicarbonate.

10. Tryptose phosphate broth (TPB): 29.5 g/L dry broth in tissue culture grade water, autoclaved at 115 °C for 20 min.

11. HEPES buffer: 1 M HEPES (free acid) and tissue culture grade water, autoclaved at 115 °C for 20 min (HEPES).

12. Penicillin and Streptomycin at 100,000 U of each per ml (P & S).

13. Growth medium: EMEM, 10 % NBBS, 2.95 g/L TPB, 10 mM HEPES, 100 U/ml penicillin and 100 U/ml streptomycin.

3   Method

The number of kidney cells obtained from each bird will vary with the age and the strain of the birds used. We have found that the average cell yield from a 2 week old Rhode Island Red bird is approximately 2.0×10^8 cells.

3.1 Preparation of Kidney Cells

1. Aseptically prepare disaggregation mix by addition of 7.5 ml of trypsin to 80 ml of EDTA and warm to 37 °C (Trypsin/EDTA).

2. Add 50–100 ml NBBS to suitable sterile flask (*see* **Note 5**).

3. Spray work area with IMS, protect with clean paper towels and collect sterile instruments and glassware so that they are close to hand.

4. Spray the back of the birds and under the wings with IMS to clean and dampen the feathers and lay the bird, dorsal side uppermost, on the paper towels.

5. Insert the blade of a large, robust pair of scissors just below where a wing attaches to the body and sever across the body, through the spinal cord to where the second wing attaches, taking care to avoid piercing the gut.

6. From the ends of this first cut, once again taking care to avoid piercing the gut and your own hand, cut along each side of the body towards the legs and through the top of each leg.

7. Carefully fold back the cut section of the bird to reveal the internal organs. Move the intestines, which should have remained attached to this retracted section, to reveal the kidneys.

8. Remove the kidneys using a small pair of scissors and forceps and place in a 250 ml beaker of PBSa.

9. Repeat **steps 4–8** for every bird.

10. When all the kidneys required have been removed from the birds, agitate them in the beaker and discard the PBSa. Repeat this process until the wash PBSa looks clear (*see* **Note 6**).

11. Tip the drained kidneys into a large glass petri dish and using two scalpels shred and mince the kidneys into very small pieces removing as much clotted blood, connective tissue, and kidney core as possible.

12. Transfer the minced tissue into a tightly capped bottle (for example a 150 ml medical flat) and wash with approximately 80 ml of PBSa until the supernatant runs clear, allowing the tissue fragments to settle for 1 min in between the washes and discarding the PBSa washes (*see* **Note 6**).

13. Add 50–80 ml Trypsin/EDTA to drained tissue and shake moderately hard for 2 min. Allow the tissue to settle and discard the supernatant (*see* **Note 6**).

14. Add another volume of Trypsin/EDTA and shake for 4 min. Allow the tissue to settle and this time pour the supernatant into the conical flask containing NBBS. Gently swirl the flask to distribute the isolated cells in the NBBS.

15. Repeat **step 14** until no more tissue remains (*see* **Note 7**).

16. Filter the cell suspension/NBBS mix collected in **Step 14** through the metal gauze filter supported in a funnel placed in a fresh conical flask (*see* **Note 8**). Decant filtered cells into centrifuge tube(s) and centrifuge at approximately $300 \times g$ for 10 min to pellet the cells.

3.2 Culture of Kidney Cells

1. Warm growth medium to 37 °C.

2. Working in a Microbiological Safety Cabinet (Class 2), carefully discard the supernatant from the centrifuge tubes and resuspend the pelleted cells in growth medium, triturating at least five times.

3. Using a 50 ml syringe pass the cell suspension through Swinnex filter 1 then through Swinnex filter 2 collecting the filtrates from filter 1 and 2 in a fresh flask each time (*see* **Note 8**).

4. Measure and record the volume. Take 0.1 ml of cell suspension add to 0.9 ml of trypan blue and count the viable cells (*see* **Note 9**).

5. Dilute cell suspension in growth medium to the cell concentration required, seed culture flasks and place in incubator until intact monolayer forms (*see* **Note 10**).

4 Notes

1. This should be done immediately prior to the removal of the kidneys to minimize the buildup of blood clots.

2. Wire mesh is obtainable from Locker Wire Weavers, www.wiremesh.co.uk.

3. As one batch of serum may not support the growth of CK cells as well as another, we recommend batch testing of serum prior to purchase and use.

4. Swinnex filter holders, support screens and silicone gasket are obtainable from Millipore. To assemble the Swinnex filters, cut a 25 mm diameter disc from 50 mesh × 0.200 mm dia wire for Filter 1 and 100 mesh × 0.100 mm dia wire metal gauze for Filter 2. Place mesh disc and O-ring in the Swinnex 25 holder, pack in autoclavable bag, label Filter I or Filter 2 as appropriate and sterilize at 121 °C for 20 min. Filter holders and mesh discs can be reused after disassembling and washing.

5. The volume of NBBS, which is used to inactivate the Trypsin/EDTA, will depend on the number of birds. Use at least 50 ml for up to 6 birds and increase the volume by 10 ml for each extra bird. The flask volume should be at least 500 ml.

6. Whilst some kidney cells may be lost in this process, it is an effective way of removing many of the red blood cells that are still present at this stage of the preparation.

7. This may require 6–8 repeats.

8. This process helps to remove some of the larger aggregates of cells that remain in the preparation at this stage. Removing the aggregates makes the cell counting process more accurate and the resultant monolayers more evenly dispersed.

9. There will still be a number of red blood cells at this stage. These should be excluded from the cell count. The red blood cells can be distinguished from other kidney cells by their size and shape.

10. For plaque assays and viral growth seed flasks at approximately $0.3 \times 10^6/\text{cm}^2$ and incubate for 72 h prior to virus introduction.

References

1. Dulbecco R, Vogt M (1954) Plaque formation and isolation of pure lines with poliomyelitis viruses. J Exp Med 99:167–182

2. Youngner JS (1954) Monolayer tissue cultures. 1. Preparation and standardization of suspensions of trypsin-dispersed monkey kidney cells. Proc Soc Exp Biol Med 85:202–205

3. Maassab HF (1959) The propagation of multiple viruses in chick kidney cultures. Proc Natl Acad Sci U S A 45:1035–1039

4. Churchill AE (1965) The use of chicken kidney tissue culture in the study of the avian viruses of Newcastle Disease, Infectious Laryngo Tracheitis and Infectious Bronchitis. Res Vet Sci 6:162–169

5. Darbyshire JH, Cook JKA, Peters RW (1975) Comparative growth kinetic studies on Avian Infectious Bronchitis Virus in different systems. J Comp Pathol 85:623–630

6. Cook KKA, Darbyshire JH, Peters RW (1976) The use of tracheal organ cultures for the isolation and assay of avian infectious bronchitis virus. Arch Virol 50:109–118

7. Alexander DJ, Collins MS (1975) Effect of pH on the growth and cytopathogenicity of avian infectious bronchitis virus in chicken kidney cells. Arch Virol 49:339–348

8. Cowen BS, Hitchner SB (1975) pH stability studies with Avian Infectious Bronchitis Virus (Coronavirus) strains. J Virol 15:430–432

9. Brown TDK, Boursnell MEG, Binns MM (1984) A leader sequence is Present on mRNA A of Avian Infectious Bronchitis Virus. J Gen Virol 65:1437–1442

10. Casais R, Dove B, Cavanagh D et al (2003) Recombinant avian infectious bronchitis virus expressing a heterologous spike gene demonstrates that the spike protein is a determinant of cell tropism. J Virol 77:9084–9089

11. Hodgson T, Casais R, Dove B et al (2004) Recombinant Infectious Bronchitis Coronavirus Beaudette with the spike protein gene of the M41 strain remains attenuated but induces protective immunity. J Virol 78:13804–13811

12. Maier HJ, Hawes PC, Cottam EM et al (2013) Infectious bronchitis virus generates spherules from zippered endoplasmic reticulum membranes. mBio 4(5):e00801-13. doi:10.1128/mBio.00801-13

Chapter 7

Isolation and Propagation of Coronaviruses in Embryonated Eggs

James S. Guy

Abstract

The embryonated egg is a complex structure comprised of an embryo and its supporting membranes (chorioallantoic, amniotic, yolk). The developing embryo and its membranes provide the diversity of cell types that are needed for successful replication of a wide variety of different viruses. Within the family *Coronaviridae* the embryonated egg has been used as a host system primarily for two avian coronaviruses within the genus *Gammacoronavirus*, infectious bronchitis virus (IBV) and turkey coronavirus (TCoV). The embryonated egg also has been shown to be suitable for isolation and propagation of pheasant coronavirus, a proposed member of the *Gammacoronavirus* genus. IBV and pheasant coronavirus replicate well in the embryonated chicken egg, regardless of inoculation route; however, the allantoic route is favored as these viruses replicate well in epithelium lining the chorioallantoic membrane, with high virus titers found in these membranes and associated allantoic fluids. TCoV replicates only in epithelium lining the embryo intestines and bursa of Fabricius, thus amniotic inoculation is required for isolation and propagation of this virus. Embryonated eggs also provide a potential host system for detection and characterization of other, novel coronaviruses.

Key words Embryonated egg, Allantoic, Amniotic, Chicken, Turkey

1 Introduction

Embryonated eggs are utilized as a laboratory host system for primary isolation and propagation of a variety of different viruses, including the avian coronaviruses, infectious bronchitis virus (IBV), turkey coronavirus (TCoV), and pheasant coronavirus [1–4]. They have been extensively utilized for propagation of these viruses for research purposes and, in the case of IBV, for commercial production of vaccines. In addition, embryonated eggs provide a potential host system for studies aimed at identifying other, novel coronavirus species.

The embryonated egg is comprised of the developing embryo and several supporting membranes which enclose cavities or "sacs" within the egg [5]. The shell membrane lies immediately beneath the shell; this is a tough fibrinous membrane that forms the air sac in the region of the blunt end of the egg (Fig. 1). In contrast to the

Helena Jane Maier et al. (eds.), *Coronaviruses: Methods and Protocols*, Methods in Molecular Biology, vol. 1282, DOI 10.1007/978-1-4939-2438-7_7, © Springer Science+Business Media New York 2015

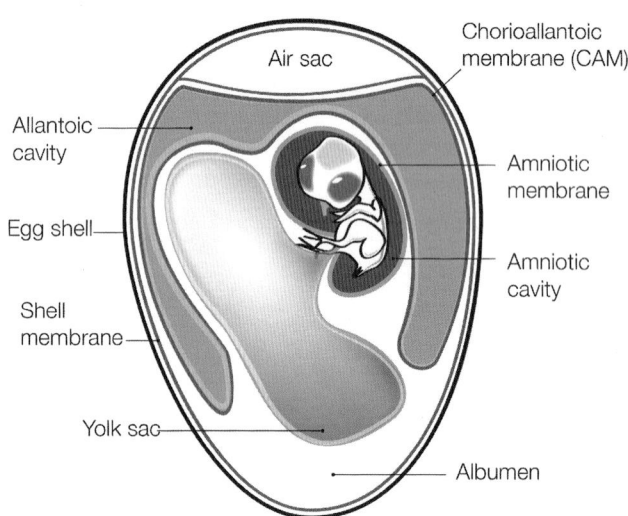

Fig. 1 Anatomical features of an embryonated chicken egg at approximately 11 days of incubation

shell membrane, chorioallantoic, amniotic, and yolk membranes are comprised largely of epithelium, and represent potential sites of coronaviral replication. The chorioallantoic membrane (CAM) lies directly beneath the shell membrane; this is a highly vascular membrane that serves as the respiratory organ of the embryo. The CAM is the largest of the embryo membranes, and it encloses the largest cavity within the egg, the allantoic cavity; in the embryonated chicken egg, this cavity contains approximately 5–10 ml of fluid, depending upon the stage of embryonation. The amniotic membrane encloses the embryo and forms the amniotic cavity; in the embryonated chicken egg, this cavity contains approximately 1 ml fluid. The yolk sac is attached to the embryo and contains the nutrients the embryo utilizes during embryonic development and the immediate post-hatch period.

The developing embryo and its membranes (CAM, amniotic, yolk) provide the diversity of cell types that are needed for successful replication of a wide variety of different viruses. Embryonated eggs may be inoculated by depositing virus directly onto the CAM, or by depositing virus within allantoic, amniotic, and yolk sacs [6]. For avian coronaviruses, inoculation of eggs by allantoic or amniotic routes has been shown to provide these viruses with access to specific cell types that support their replication [2–4]. IBV is an epitheliotropic virus that replicates in a variety of epithelial tissues in the post-hatch chicken including respiratory tract, gastrointestinal tract, kidney, bursa of Fabricius, and oviduct [7]. In the embryonated chicken egg, IBV replicates well regardless of inoculation route; however, the allantoic route is favored as the virus replicates extensively in epithelium of the CAM and high titers are shed into allantoic fluid [8]. A pheasant coronavirus has been isolated and propagated in embryonated chicken eggs using procedures similar to those utilized for IBV (allantoic route inoculation) [3]. TCoV also is epitheliotropic in post-hatch chickens and turkeys, but replicates only in epithelium lining the intestinal tract and bursa of Fabricius [1, 4, 9]. These cellular tropisms of TCoV also are observed in the embryonated egg; the virus replicates only in embryonic intestines and bursa of Fabricius, sites that are reached only via amniotic inoculation.

2 Materials

2.1 Preparation and Collection of Samples for Egg Inoculation

1. Dulbecco's modified Eagle's medium (DMEM) supplemented with 1 % fetal bovine serum (FBS) and antibiotics: penicillin 1,000 U/ml, gentamicin 0.05 mg/ml, amphotericin B 5 µg/ml. Adjust pH to 7.0–7.4 using either 1 N NaOH or 1 N HCl. Tryptose phosphate broth and other cell culture basal media (minimal essential medium, RPMI 1640, etc.) may be substituted for DMEM.

2. Sterile cotton-tipped swabs are used for collection of antemortem samples (e.g., respiratory secretions, feces, etc.). Type 4 Calgiswab (Puritan Medical Products) is useful for collection of respiratory secretions from small birds.

3. Sterile Whirl-Pak® bags (Fisher Scientific) are used for collection of tissues.

4. Tissue homogenizer. Mortar and pestle, Ten Broeck homogenizer, or Stomacher® (Fisher).

2.2 Egg Inoculation and Incubation

1. Fertile eggs are obtained, preferably, from specific-pathogen-free (SPF) flocks (e.g., Charles River/SPAFAS). Alternatively, fertile eggs may be used that are from healthy flocks free of antibody to the virus of interest (see Note 1).

2. Disinfectant: 70 % ethanol, 3.5 % iodine, 1.5 % sodium iodide.

3. A vibrating engraver (Fisher Scientific) or drill (Dremel) is used to prepare holes in egg shells. Prior to use, disinfect the tip of the engraving tool/drill to prevent contamination of the egg.

4. Plastic cement, glue, tape, or nail varnish are used to seal holes in egg shells after inoculation.

5. Egg flats.

6. Egg candlers are available from a variety of commercial sources.

7. A suitable egg incubator is needed; these are available from a variety of commercial sources. Commercially available egg incubators generally are equipped with heat source, humidifier, and a timer-based mechanical turning system.

**2.3 Collection
of Specimens
from Inoculated Eggs**

1. Sterile scissors and forceps.
2. Sterile pipettes or 5 ml syringes with 1 in., 18 gauge needles.
3. Sterile plastic tubes, e.g., 12 × 75 mm snap-cap tubes or micro-centrifuge tubes.

3 Methods

Embryonated chicken and turkey eggs are extensively utilized for isolation and propagation of IBV and TCoV, respectively [2, 4]. These same eggs and techniques may be useful for amplification of other coronaviruses, and this has been demonstrated with isolation and propagation of pheasant coronavirus in embryonated chicken eggs [3]. However, many viruses exhibit host specificity and this should be considered when attempting to isolate and propagate novel coronaviruses.

Embryonated eggs from avian species other than chickens and turkeys may be utilized; these are inoculated essentially as described for chicken and turkey eggs, primarily by making adjustments in the length of time embryos are incubated before inoculation. Embryonated chicken eggs are inoculated by the allantoic route at approximately the middle of the 21-day embryonation period, at 8–10 days of embryonation; they are inoculated by the amniotic route late in the incubation period, at 14–16 days of embryonation. Turkey and duck eggs have a 28-day embryonation period and generally are inoculated by the allantoic route at 11–14 days of embryonation, and by the amniotic route at 18–22 days of embryonation.

Embryonated chicken and turkey eggs are incubated at a temperature of 38–39 °C with a relative humidity of 83–87 %. They should be turned several times per day to ensure proper embryo development and to prevent development of adhesions between the embryo and its membranes. Fertile eggs may be stored for brief periods with minimal loss of viability [10]. Ideally, fertile eggs are stored at a temperature of 19 °C with a relative humidity of approximately 70 %. Alternatively, eggs may be stored at room temperature; these should be tilted at 45°, and daily alternated from side to side to minimize loss of embryo viability.

Indirect evidence of coronavirus replication in inoculated embryonated eggs may consist of embryo mortality or lesions in the embryos such as hemorrhage, edema or stunting; however, virus replication may occur in the absence of readily discernible effects on the embryo. Methods for specific detection of coronaviruses in inoculated embryonated eggs include electron microscopy, immunohistochemistry, and reverse transcriptase-polymerase chain reaction (RT-PCR) procedures [2, 4, 11, 12]. Electron microscopy is a particularly useful tool as this method depends solely on morphologic identification of the virus and does not require specific reagents [13]. The characteristic electron microscopic morphology of coronaviruses allows their presumptive identification in embryonic fluids (e.g., allantoic fluid) or embryo intestinal contents. A variety of immunohistochemical and RT-PCR procedures have been developed for detection of coronaviruses, and these same procedures may be useful for detection of novel coronaviruses due to antigenic and genomic similarities among coronaviruses, particularly those within the same genus [2, 4, 9, 11, 12, 14–16].

3.1 Collection of Samples for Egg Inoculation

1. Swabs used to collect clinical samples such as respiratory secretions and feces are placed in 2–3 ml of DMEM supplemented with FBS and antibiotics.

2. Tissues are collected using aseptic technique and placed in clean, tightly sealed bags (Whirl-Pak bags).

3. Clinical samples should be chilled immediately after collection and transported to the laboratory with minimal delay. Samples may be shipped on ice, dry ice or with commercially available cold packs (see **Note 2**).

3.2 Preparation of Samples for Egg Inoculation

1. Use a vortex mixer to expel material from swabs, then remove and discard swab. Clarify by centrifugation (1,000–2,000 × g for 10 min) in a refrigerated centrifuge. Filter, if needed, through a 0.45 μm filter, and store at –70 °C (see **Note 3**).

2. Tissues and feces are prepared as 10–20 % suspensions in DMEM supplemented with FBS and antibiotics. Tissues are homogenized using a mortar and pestle, Ten Broeck homogenizer, or Stomacher[R] (Fisher). Tissue and fecal suspensions are clarified by centrifugation (1,000–2,000 × g for 10 min) in a refrigerated centrifuge; this removes cellular debris and most bacteria. Filter, if needed, through a 0.45 μm filter, and store at –70 °C (see **Note 3**).

3.3 Allantoic Sac Inoculation

1. Chicken eggs (21 day embryonation period) are generally inoculated at 8–10 days of embryonation; eggs from other avian species may be used by making adjustments in the ages at which embryos are inoculated. Turkey and duck eggs (28 day embryonation period) generally are inoculated by this route at 11–14 days of embryonation.

2. Place eggs in an egg flat with the air-cell up. Candle eggs to ensure viability and mark the edge of the air-cell.

3. Disinfect the area marked on the shell and drill a small hole just above the mark so that the hole penetrates the air-cell, but not the portion of the egg below the air-cell.

4. A 1-ml syringe with a 25-gauge, 0.5 in. (12 mm) needle is used to inoculate eggs. The needle is inserted to the hub while holding the syringe vertically and 0.1–0.3 ml of inoculum is injected into the allantoic cavity.

5. Seal holes and return eggs to incubator.

6. Incubate eggs for 3–7 days. Evaluate embryos and allantoic fluid for presence of virus as described below.

3.4 Amniotic Sac Inoculation (Method A)

1. Fertile embryonated eggs are inoculated late in the incubation period. Chicken eggs are inoculated at 14–16 days of embryonation; turkey and duck eggs are inoculated at 18–22 days of embryonation.

2. Candle eggs to ensure embryo viability. Place eggs in an egg flat with the air-cell up.

3. Disinfect the shell at the top of the egg, over the center of the air-cell. Drill a small hole through the shell at center of air-cell using a vibrating engraver.

4. A 1-ml syringe with a 22-gauge, 1.5 in. (38 mm) needle is used to inoculate chicken, duck, and turkey embryos. The needle is inserted to the hub while holding the syringe vertically and 0.1–0.2 ml of inoculum is injected into the amniotic cavity (*see* **Note 4**).

5. Seal holes and return eggs to incubator.

6. Inoculated embryos are generally examined for presence of virus after incubation for 2–5 days. Evaluate inoculated embryos for presence of virus as described below.

3.5 Amniotic Sac Inoculation (Method B)

1. Fertile embryonated chicken eggs are inoculated, as above, at 14–16 days of embryonation; turkey and duck eggs at 18–22 days of embryonation. Candle eggs and mark the general location of the embryo (*see* **Note 5**).

2. Place eggs in an egg flat with the air-cell up. Disinfect the shell at the top of the egg, over the center of the air-cell. Drill a small hole through the shell at center of air-cell.

3. A 1-ml syringe with a 22-gauge, 1.5 in. (38 mm) needle is used to inoculate chicken, duck, and turkey embryos. Eggs are inoculated in a darkened room, as the embryo must be visualized for this method of amniotic inoculation. Hold the egg against an egg candler and insert the needle into the egg and toward the shadow of the embryo. As the tip of the needle approaches the embryo, a quick stab is used to penetrate the amniotic sac. Penetration of the amniotic sac may be verified by moving the needle sideways; the embryo should move as the needle moves (*see* **Note 4**).

4. Seal holes and return eggs to incubator. Inoculated embryos are generally examined for presence of virus after incubation for 2–5 days.

3.6 Collection of Allantoic Fluid from Eggs Inoculated by Allantoic Route

1. Candle eggs once daily after inoculation. Discard all eggs with embryos that die within the first 24 h after inoculation (*see* **Note 6**).

2. Collect allantoic fluid from all eggs with embryos that die >24 h after inoculation and from eggs with embryos that survive through the specified incubation period. Eggs with live embryos following the specified incubation period are refrigerated at 4 °C for at least 4 h, or overnight, prior to collection of allantoic fluid (*see* **Note 7**).

3. Place eggs in an egg flat with the air-cell up. Disinfect the portion of the egg shell that covers the air cell, and use sterile forceps to crack and remove egg shell over air cell.

4. Use forceps to gently dissect through the shell membrane and CAM to expose the allantoic fluid. Use forceps to depress membranes within the allantoic cavity so that allantoic fluid pools around the tip of the forceps. Use a pipette or syringe with needle to aspirate fluid. Place fluid in sterile, 12×75 mm snap-cap tubes, or other vials. Store at −70 °C (*see* **Note 8**).

5. Examine allantoic fluid for presence of coronavirus using electron microscopy, immunohistochemistry or RT-PCR (*see* **Note 9**).

3.7 Collection of Embryo Tissues from Eggs Inoculated by Amniotic Route

1. Candle eggs once daily after inoculation. Discard all eggs with embryos that die within the first 24 h after inoculation (*see* **Note 6**).

2. Examine all eggs with embryos that die >24 h after inoculation and eggs with embryos that survive through the specified incubation period (*see* **Note 10**).

3. Euthanize live embryos by placing eggs in a plastic bag or plastic bucket filled with carbon dioxide gas, or refrigerate (4 °C) overnight. Alternatively, embryos may be euthanized by cervical dislocation upon removal from eggs using the handles of a pair of scissors (*see* **Note 11**).

4. Place eggs in an egg flat with the air-cell up. Disinfect the portion of the egg shell that covers the air cell, and use sterile forceps to crack and remove the egg shell over air cell.

5. Use forceps to dissect through the shell membrane and CAM.

6. Grasp the embryo with sterile forceps and gently remove from the egg.

7. Remove selected tissues and/or intestinal contents from embryo for coronavirus detection using electron microscopy, immunohistochemistry, or RT-PCR (*see* **Note 12**).

4 Notes

1. Fertile eggs from non-SPF flocks may be used; however, presence of antibodies may interfere with isolation and propagation, and presence of egg-transmitted infectious agents may result in contamination of any viruses obtained with these eggs.

2. If dry ice is used, samples must be placed in tightly sealed containers to prevent inactivation of viruses from released carbon dioxide.

3. The supernatant fluid should be filtered if the specimen is feces or other sample that likely is contaminated with high concentrations of bacteria. Filtration of samples will reduce virus titer, and should be used only when necessary.

4. The accuracy of delivering an inoculum into the amniotic sac using this method may be determined by sham-inoculation of embryos with a dye such as crystal violet (0.2 % crystal violet in 95 % ethanol), then opening eggs and determining site of dye deposition.

5. A distinct advantage of Method A is that visualization of the embryo is not required. Method B requires visualization of the embryo, and this is may not be possible for embryonated eggs having a dark shell color (e.g., turkey eggs, brown chicken eggs). Method A also requires less skill for delivery of inoculum into the amniotic cavity, but is more prone to error than Method B with the possibility of inoculum being deposited at sites other than the amniotic cavity. If the embryo can be visualized, a potential advantage of Method B is more precise delivery of inoculum into the amniotic cavity as compared with Method A.

6. Embryo deaths that occur <24 h after inoculation generally are due to bacterial contamination, toxicity of the inoculum, or injury.

7. Refrigeration kills the embryo and causes the blood to clot. This prevents contamination of allantoic fluid with blood.

8. Multiple passages in embryonated eggs may be necessary for initial isolation of coronaviruses; allantoic fluid is used as inoculum for additional passages in embryonated eggs. Embryos at each passage should be evaluated for gross lesions. For IBV, embryo-lethal strains generally result in embryos with cutaneous hemorrhage; non-embryo-lethal strains result in stunting, curling, clubbing of down, or urate deposits in the mesonephros of the kidney. In some cases, virus replication in embryonated eggs may not be associated with readily detectable embryo lesions.

9. Allantoic fluids commonly are examined for presence of coronavirus using electron microscopy or RT-PCR procedures. Alternatively, immunohistochemical detection may be accomplished by staining sections of allantoic membrane or

the allantoic epithelial cells that are present in allantoic fluid (these should be collected by centrifugation prior to freezer storage of allantoic fluid).

10. TCoV rarely results in embryo mortality. Typically, only those eggs with live embryos are examined following the specified incubation period; however, the possibility of embryo-lethal viruses should not be overlooked.

11. The method of euthanasia employed will depend upon the method used to detect virus in inoculated embryos. Fresh tissues are required if immunohistochemistry is to be employed; for this, embryos should be euthanized by cervical dislocation or exposed briefly to carbon dioxide gas.

12. Intestinal contents commonly are examined for presence of coronaviruses using electron microscopy or RT-PCR procedures. Alternatively, immunohistochemical detection may be accomplished by staining sections of intestines or bursa of Fabricius.

References

1. Adams NR, Hofstad MS (1970) Isolation of transmissible enteritis agent of turkeys in avian embryos. Avian Dis 15:426–433

2. Cavanagh D, Naqi SA (2003) Infectious bronchitis. In: Saif YM, Barnes HJ, Fadly A, Glisson JR, McDougald LR, Swayne DE (eds) Diseases of Poultry, 11th edn. Iowa State University Press, Ames, IA, pp 101–120

3. Gough RE, Cox WJ, Winkler CE et al (1996) Isolation and identification of infectious bronchitis virus from pheasants. Vet Rec 138:208–209

4. Guy JS (2013) Turkey coronavirus enteritis. In: Swayne DE, Glisson JR, McDougald LR (eds) Diseases of Poultry, 13th edn. Wiley-Blackwell, Ames, IA, pp 376–381

5. Hawkes RA (1979) General principals underlying laboratory diagnosis of viral infections. In: Lennette EH, Schmidt NJ (eds) Diagnostic procedures for viral, rickettsial and chlamydial infections, 5th edn. American Public Health Association, Washington, DC, pp 1–48

6. Senne DA (2008) Virus propagation in embryonating eggs. In: Dufour-Zavala L, Swayne DE, Glisson JR, Pearson JE, Reed WM, Jackwood MW, Woolcock PR (eds) A laboratory manual for isolation and identification of avian pathogens, 5th edn. American Association of Avian Pathologists, Jacksonville, FL, pp 204–208

7. Cavanagh D (2003) Severe acute respiratory syndrome vaccine development: experiences of vaccination against avian infectious bronchitis virus. Avian Pathol 32:567–582

8. Jordan FTW, Nassar TJ (1973) The combined influence of age of embryo, temperature and duration of incubation on the replication and yield of avian infectious bronchitis virus in the developing chick embryo. Avian Pathol 2:279–294

9. Guy JS (2000) Turkey coronavirus is more closely related to avian infectious bronchitis virus that to mammalian coronaviruses: a review. Avian Pathol 29:207–212

10. Brake J, Walsh TJ, Benton CE et al (1997) Egg handling and storage. Poultry Sci 76:144–151

11. Jonassen CM, Kofstad T, Larsen IL et al (2005) Molecular identification and characterization of novel coronaviruses infecting graylag geese (*Anser anser*), feral pigeons (*Columba livia*) and mallards (*Anas platyrhynchos*). J Gen Virol 86:1597–1607

12. Stephensen CB, Casebolt DB, Gangopadhyay NN (1999) Phylogenetic analysis of a highly conserved region of the polymerase gene from 11 coronaviruses and development of a consensus polymerase chain reaction assay. Virus Res 60:181–189

13. McNulty MS, Curran WL, Todd D et al (1979) Detection of viruses in avian faeces by direct electron microscopy. Avian Pathol 8:239–247

14. Cavanagh D, Mawditt K, Welchman DB et al (2002) Coronaviruses from pheasants (*Phasianus colchicus*) are genetically closely related to coronaviruses of domestic fowl (infectious bronchitis virus) and turkeys. Avian Pathol 31:81–93

15. Guy JS, Barnes HJ, Smith LG et al (1997) Antigenic characterization of a turkey coronavirus identified in poult enteritis and mortality syndrome-affected turkeys. Avian Dis 41:583–590

16. Cavanagh D (2005) Coronaviruses in poultry and other birds. Avian Pathol 34:439–448

Chapter 8

Characterization of Human Coronaviruses on Well-Differentiated Human Airway Epithelial Cell Cultures

Hulda R. Jonsdottir and Ronald Dijkman

Abstract

The human airway serves as the entry point of human respiratory viruses, including human coronaviruses. In this chapter we outline the methods by which we establish fully differentiated airway epithelium and its use for human coronavirus propagation. Additionally, we outline methods for immunofluorescence staining of these cultures for virus detection, characterization of cell tropism, and how to perform antiviral assays and quantify viral replication.

Key words Human coronavirus, Antivirals, Cell tropism, Human airway epithelial cells, Virus detection

1 Introduction

The human airway serves as the entry point of human respiratory viruses, including human coronaviruses (HCoVs). In order to properly recapitulate the complex anatomy of the human lung specialized cell culture models have been developed to resemble both the upper and lower airways [1–3]. Primary human bronchial epithelial cells cultured in an air–liquid interface (ALI) system serve as a universal platform to study human respiratory viruses [4–6]. These human airway epithelial (HAE) cultures morphologically and functionally resemble the upper conducting airways in vivo. In these cultures, the epithelial layer is pseudostratified and after differentiation they contain many different cell types such as basal, ciliated, and goblet cells and furthermore, generate protective mucus equivalent to that of in vivo epithelium [7].

Establishment of HAE cultures requires time and patience but the differentiated cultures allow for a number of advantageous analyses in respiratory virus research. We have adapted and optimized our methods based on previously published work [8–10]. Moreover, we have standardized methods for the propagation of human coronaviruses and evaluation of the effects of antiviral

Helena Jane Maier et al. (eds.), *Coronaviruses: Methods and Protocols*, Methods in Molecular Biology, vol. 1282,
DOI 10.1007/978-1-4939-2438-7_8, © Springer Science+Business Media New York 2015

compounds on both viral replication and cell viability. We are able to propagate all known human coronaviruses in this system and can easily evaluate their tropism by immunohistochemistry [5, 11]. In this chapter we outline the methods by which we establish fully differentiated airway epithelium and use it for human coronavirus propagation. Additionally, we outline methods for immunofluorescence staining of these cultures for virus detection, characterization of cell tropism and how to perform antiviral assays and quantify viral replication.

2   Materials

2.1   Human Airway Epithelial Cell Cultures

1. Primary human tracheobronchial epithelial cells can be obtained in accordance with local ethical guidelines from patients willing to give informed consent, who are undergoing bronchoscopy and/or surgical lung resections. Alternatively isolated primary human airway epithelial cells can be obtained commercially from a number of distributors.

2. 10× digestion solution: Minimum Essential Medium (MEM), 1 % m/v Protease from Streptomyces griseus Type XIV, 0.01 % m/v Deoxyribonuclease I from bovine pancreas.

3. Isolation/washing solution: MEM, 100 U/ml penicillin, 100 μg/ml streptomycin, 0.25 μg/ml Amphotericin B Solution, 50 μg/ml gentamicin, 100 U/ml nystatin.

4. Bronchial epithelial cell serum-free growth medium (BEGM): LHC basal medium, supplemented with the required additives (Table 1).

5. Air–liquid interface (ALI) medium: LHC basal medium and Dulbecco's Modified MEM (DMEM) mixed in a 1:1 ratio, supplemented with the required additives (Table 1).

6. 12-Well inserts, pore size 0.4 μM and 12-well cluster plates or 12-well deep well cluster plates.

7. 24-Well inserts pore size 0.4 μM and 24-well cluster plates.

8. Human collagen Type I + III, Vitrocol 100.

9. Collagen Type IV from human placenta reconstituted in 5 ml filter-sterilized water with 0.25 % acetic acid. Dissolve for a few hours at 37 °C, occasionally swirling. Once dissolved, increase volume to 20 ml and maintain acetic acid concentration at 0.25 %, mix gently by pipetting. Filter-sterilize the solution through a 0.22 μm filter, and store at −20 °C in aliquots of 800 μl per eppendorf tube. The stock solution is stable for at least 1 year at −20 °C.

Table 1
Preparation of stock additives for BEGM and ALI medium

Component	Stock concentration	Comment
Bovine Serum Albumin (BSA)	300× 150 mg/ml	*See* **Note 1**
Bovine pituitary extract (BPE)	1,000× ±14 mg/ml	
Insulin	2,000×, 10 mg/ml	Store at +4 °C
Transferrin (TF)	1,000×, 10 mg/ml	
Hydrocortisone (H)	1,000×, 0.072 mg/ml	
Triiodothyronine (T3)	1,000×, 0,067 mg/ml	
Epinephrine (EP)	1,000×, 0.6 mg/ml	
Epidermal Growth Factor (EGF)	1,000× or 50,000×, 25 µg/ml	1,000× for BEGM, 50,000× for ALI medium.
Retinoic acid (RA)	1,000×, 5×10^{-5} M	Light sensitive. *See* **Note 2**
Phosphorylethanolamine (PE)	1,000×, 70 mg/ml	
Ethanolamine (EA)	1,000×, 30 µl/ml	
Stock 11 (S11)	1,000×, 0,863 mg/ml	
Stock 4 (S4)	1,000×	*See* **Note 3**
Trace Elements (TR)	1,000×	*See* **Note 4**
Penicillin/Streptomycin (P/S)	100× 10,000 U/ml of penicillin and 10,000 µg/ml of streptomycin	Store at +4 °C
Gentamicin	1,000×, 50 mg/ml	Store at +4 °C. *See* **Note 5**
Amphotericin B	1,000×, 50 mg/ml	*See* **Note 5**

All additives should be aliquoted and stored at –20 °C unless stated otherwise

2.2 Human Coronavirus Propagation

1. Apical wash solution: Hank's Balanced Salt Solution (HBSS), without calcium and magnesium.

2. Virus transport medium (VTM): MEM, 25 mM HEPES-buffered, 0.5 % gelatin, 100 U/ml penicillin, 100 µg/ml streptomycin.

3. Aerosol barrier pipette tips and 1.5 ml Eppendorf Safe-Lock Tubes™.

2.3 Immuno-fluorescence Analysis

1. Fixation solution: 4 % formalin solution, neutral buffered (Formafix).

2. Confocal staining buffer (CB): 50 mM ammonium chloride (NH$_4$Cl), 0.1 % saponin, and 2 % IgG and protease-free BSA dissolved in 500 ml of phosphate buffered saline (PBS,

Table 2
Primary antibodies

Antibody	Target	Dilution	Host	Comment
Anti-β-Tubulin IV	Cilia	1:400	Mouse, IgG1	Clone ONS.1A6
anti-ZO1	Tight junctions	1:200–400	Goat	Directed against C-terminal domain
anti-dsRNA	dsRNA	1:500–1,000	Mouse, IgG2a	Clone J2
Anti-CD13	CD13/APN	1:200	Sheep	Receptor 229E
Anti-CD26	CD26/DPPIV	1:200	Goat	Receptor MERS
Anti-ACE2	ACE2	1:200	Goat	Receptor SARS and NL63
intravenous immunoglobulin (IVIG)	Viral proteins	1:1,000	Human	
Anti-β-Tubulin	Cilia	1:400	Mouse, IgG1	Clone Tub2.1, Cy3 conjugate

pH 7.4). Filter-sterilize (0.2 μm filter) solution and prepare aliquots of 40 ml and store at –20 °C.

3. Primary antibodies: *see* Table 2.
4. Fluorescent DNA dyes: DAPI or Hoechst 33528.
5. Wash solution: Phosphate buffered saline, pH 7.4, without calcium and magnesium.
6. Scalpel (No.10).
7. Rat-tooth forceps.
8. Fluorescence Mounting Medium.
9. Gyro-rocker.

2.4 Antiviral Assays

1. Inhibitors: e.g. K22 [12], recombinant Interferon Alpha and Lambda proteins [13].
2. CellTiter-Glo® Luminescent Cell Viability Assay (Promega).

2.5 Virus Detection

2.5.1 Renilla Luciferase Assay

1. Renilla Luciferase Assay System (Promega).
2. White, non-transparent 96-well plates.
3. Gyro-rocker.
4. Luminometer.

2.5.2 Plaque Assay

1. Huh-7 cells.
2. Medium: DMEM, high glucose, 100 U/ml penicillin, 100 μg/ml streptomycin, 1 mM Sodium Pyruvate, 5 % heat-inactivated FBS.

3. Overlay medium: 2.4 g of Avicel RC-581 (FMC biopolymer) dissolved in 100 ml of distilled water and autoclaved for 20 min at 121 °C. 2.7 g of DMEM powder (high glucose) dissolved in 90 ml of distilled water and the pH adjusted to 7.4 with 1 M NaOH. Fill volume up to 100 ml and filter-sterilize (0.2 μm filter). Freshly prepare a 1:1 mixture of Avicel (2.4 %) and 2× DMEM solution, supplemented with 10 % FBS and 100 U/ml penicillin, 100 μg/ml streptomycin.

4. Crystal-violet solution: 25 g of Crystal Violet, 40 g NaCl dissolved in 2,500 ml of 99 % Ethanol. Add 2,250 ml of distilled water and 250 ml of 37 % formaldehyde. Mix solution overnight at room temperature (*see* **Note 6**).

2.5.3 Quantitative Reverse Transcriptase PCR

1. Nucleospin RNA isolation kit (Machery Nagel).

2. Moloney Murine Leukemia Virus Reverse Transcriptase (M-MLV RT).

3. Random primers.

4. RNAse-free water.

5. FastStart Universal SYBR Green Master reaction mixture (Roche).

6. Positive control; in vitro transcribed RNA of target gene or plasmid DNA containing target gene.

3 Methods

Carry out all procedures in a biosafety cabinet according to local biosafety regulations.

3.1 Human Airway Epithelial Cell Cultures

3.1.1 Collagen Type I and III Coating of Cell Culture Flasks

Cell culture flasks are coated for 2 h with a mixture of Type I and III collagen that is necessary to efficiently expand the number of primary airway epithelial cells.

1. Use filter-sterilized dH_2O (0.22 μm) to prepare a 1:75 dilution of Vitrocol 100.

2. Use 4 ml per 75 cm^2, make sure the entire surface is covered with the collagen solution.

3. Incubate for 2 h at 37 °C.

4. Aspirate remaining liquid and wash twice with 10 ml of PBS to remove traces of acetic acid.

5. Culture flasks can be directly used. *Optional*: Store coated flasks at +4 °C for a maximum of 6 weeks.

3.1.2 Collagen Type IV Coating of Inserts

The inserts need to be coated overnight with collagen type IV, necessary for development and long-term maintenance of differentiated primary airway epithelial cell cultures.

1. Mix 7.2 ml of filter-sterile dH_2O with 800 μl of Collagen Type IV solution (0.5 mg/ml).

2. Apply 150 μl per 12-well inserts, or 50 μl per 24-well inserts. After completing one plate, make sure that the entire surface of each well is covered with the 1:10 collagen solution.

3. Air-dry the inserts overnight in a laminar flowhood, and afterwards expose them to UV-light (type C) for 30 min.

4. To remove traces of acetic acid wash inserts twice with at least 500 μl of PBS.

5. After these steps, coated inserts can be used directly. *Optional*: Store at +4 °C (wrapped in foil) for a maximum of 6 weeks. Repeat UV-exposure and washing steps before use.

3.1.3 Isolation of Primary Human Tracheal and/or Bronchial Cells

Primary epithelial cells can be isolated from whole lung tissue resections of tracheal and/or bronchial origin according to the following protocol. Smaller lung tissue resections can be processed with the same protocol. All procedures are performed at room temperature unless stated otherwise.

1. Trim the bronchial tissue free of connective tissue and fat using forceps and scissors or a scalpel. If needed, cut the bronchial tissue into 2 cm segments.

2. Wash the cleaned tissue three times in washing solution.

3. Fill the desired number of 50 ml tubes with 30 ml of wash solution and transfer as many tissue segments as possible into a single tube, until the volume reaches 36 ml. Then add 4 ml of 10× digestion solution to each tube, to end volume 40 ml (40 mg Protease/0.4 mg DNase).

4. Place tubes on a rocking platform/tube roller at +4 °C and incubate for 48 h.

5. Place the 50 ml tube containing the digested tissue on ice and add 4 ml of heat-inactivated FBS to each tube (to a final concentration of 10 % ($^v/_v$)), to neutralize protease activity. Invert tubes three times.

6. Pour solution along with the tissue onto a large petri dish, and gently scrape off the epithelium from the collagen-cartilage surface, using a scalpel in the reverted angle. Pool solutions containing dissociated cells into a 50 ml conical tube and wash the petri dish once with PBS.

7. Centrifuge for 5 min at $500 \times g$. Wash cells once with HBSS and resuspend cells in BEGM to a concentration of, approximately, 5×10^6 cells/ml.

8. Count cells using a hemocytometer and seed into collagen coated flasks with 20 ml of pre-warmed BEGM. An appropriate amount of cells for T75 flasks ranges between 0.5 and 1.0×10^6 cells.

9. Change medium the next day to remove red blood cells and any unattached epithelial cells.

10. To prevent acidification of the medium change it every 2–3 days, until 80–90 % confluence.

3.1.4 Establishment of Fully Differentiated HAE Cultures

When the primary cells have reached 80–90 % confluence in the expansion phase one can dissociate and seed the dedifferentiated primary cells on collagen type IV coated inserts, according to the following protocol. All procedures are performed at room temperature unless stated otherwise.

1. Remove BEGM and transfer it into a 50 ml tube and wash the cell monolayer twice with 12 ml of HBSS.

2. Dissociate the bronchial cells for 3 min at 37 °C in a humidified 5 % CO_2 incubator with the appropriate amount of trypsin ($25\ cm^2$: 1 ml, $75\ cm^2$: 3 ml). If needed tap the flask to dissociate the cells (*see* **Note 7**).

3. Collect the cells in the previously collected BEGM and centrifuge for 5 min at $500 \times g$.

4. Carefully discard the supernatant and resuspend cells in HBSS and centrifuge the suspension for 5 min at $500 \times g$.

5. Discard the supernatant and resuspend cells in pre-warmed ALI medium and count using a hemocytometer.

6. For generation of differentiated HAE cultures the number of cells seeded should be $1.0–2.0 \times 10^5$ cells per 12-well insert in 500 µl, or $0.3–0.6 \times 10^5$ cells per 24-well insert in 200 µl of ALI medium. A single $75\ cm^2$ flask should provide enough cells for preparing 48 individual 12-well inserts or 96 individual 24-well inserts.

7. Fill the basolateral compartment of the plates with 1 ml of ALI medium (500 µl for 24-well inserts), and transfer 500 µl (200 µl for 24-well inserts) of diluted cell suspension to the upper chamber of the collagen coated inserts and incubate overnight at 37 °C in a humidified 5 % CO_2 incubator. Cells are now in liquid–liquid interface.

8. The next day, medium in the apical compartment must be changed to remove any unattached cells. Discard the old medium and wash the apical surface with 500 µl HBSS and apply 500 µl of pre-warmed ALI medium to the apical side. Adjust volume to 200 µl for 24-well inserts.

9. To prevent acidification of the medium it should be changed every 2–3 days until cells have reached complete confluence (*see* **Note 8**).

10. During media change in liquid–liquid interface change apical medium first (as described in **step 8**) followed by exchange of medium in the basolateral compartment.

11. To establish air–liquid interface, aspirate apical side medium, once cells have reached complete confluence, and wash twice with HBSS (500 μl for 12-well inserts and 200 μl for 24-well inserts).

12. Incubate cultures for a few hours at 37 °C in a humidified 5 % CO_2 incubator and monitor if seeping of basolateral medium into the apical compartment occurs. If no seepage occurs cultures can be maintained at air–liquid interface. Otherwise cultures have to be cultured at liquid–liquid interface for another day.

13. Incubate cultures for 4–6 weeks to allow differentiation. Appearance of active ciliated cells can be used as an indicator of differentiation. During the extended culture time medium must be changed regularly (every 2–3 days). If desired, inserts can be transferred to deep well plates that only require medium renewal every 7 days.

14. After differentiation HAE cultures are suitable for human coronavirus propagation.

3.2 Human Coronavirus Propagation

1. Wash the apical surface of the HAE culture twice with 500 μl of HBSS solution prior to inoculation with human coronavirus specimen to remove excess of mucus.

2. Dilute the clinical material or virus supernatant in HBSS and inoculate 200 μl dropwise to the apical surface and incubate for 2 h at either 33 °C or 37 °C (*see* **Note 9**), in a humidified 5 % CO_2-incubator. *Optional*: Centrifuge inoculum solution for 4 min at $1,500 \times g$ at room temperature to remove cell debris prior to inoculation.

3. Collect the inoculum and transfer it to a container and store at −80 °C for later analysis, and wash the apical surface three times with 500 μl HBSS. *Optional*: Transfer the collected inoculum into an equal volume of VTM.

4. Incubate the infected cultures for the desired amount of time at the appropriate temperature in a humidified 5 % CO_2-incubator, e.g. 48 h at 33 °C for HCoV-229E.

5. Apply 200 μl of HBSS dropwise to the apical surface 10 min prior to the desired collection time and incubate in the humidified 5 % CO_2-incubator. Then collect progeny virus and transfer it to a container and store at −80 °C for later analysis. *Optional*: Transfer the collected progeny virus into an equal volume of VTM.

**3.3 Immuno-
fluorescence Analysis**

All incubation steps are performed at room temperature on a gyro-rocker (20–30 rpm), unless stated otherwise

1. After the apical washing has been collected the apical surface is washed twice with 500 μl of PBS before cells are fixed with formalin-solution for later immunofluorescence analysis.

2. Apply 500 μl of 4 % formalin-solution to the apical compartment and 1 ml to the basolateral. Incubate for 15–30 min.

3. Remove the formalin-solution and wash both compartments three times with equal volumes of PBS.

4. Transfer the fixed HAE cultures to a new conventional 12-well plate.

5. Discard washing solution and apply 500 μl and 1 ml of confocal buffer (CB) solution to apical and basolateral compartments, respectively.

6. Incubate fixed cultures for 30–60 min to block non-specific binding of antibodies (*see* **Note 10**).

7. Remove the CB solution from the apical and basolateral compartments.

8. From this stage one should only apply CB solution to the apical compartment.

9. Wash the apical surface once with 500 μl of CB solution for 5 min.

10. Apply primary antibodies (*see* Table 2) diluted in 250 μl CB solution dropwise to the apical surface and incubate for 120 min.

11. Wash the apical surface three times with 500 μl of CB solution for 5 min (*see* **Note 11**).

12. Apply the appropriately diluted conjugated secondary antibodies in 250 μl CB solution dropwise to the apical surface and incubate for 60 min.

13. Wash the apical surface twice with 500 μl of CB solution for 5 min.

14. Incubate cells with nucleic acid counter stain solution diluted in 250 μl of CB solution for 5 min.

15. Wash the apical surface once with 500 μl of CB solution for 5 min.

16. Lastly, wash the apical surface twice with 500 μl of PBS for 5 min to remove residual saponin and restore cell membrane integrity.

17. Before removing the washing solution, apply mounting medium on a glass slide (use 1–2 drops). Remove any air bubbles.

18. Excise the membrane from the plastic holder and carefully place the basolateral side of the membrane on top of the mounting medium, without generating air bubbles.

19. Then slowly add one drop of mounting medium on top of each membrane.

20. Slowly place the coverslip, in a tilted fashion, on top of the membrane without generating air bubbles.

21. Allow the mounting medium to polymerize for 30 min, after which the slide can directly be analyzed.

3.4 Antiviral Assays

1. Pre-warm ALI medium to 37 °C.

3.4.1 Treatment

2. Mix antiviral compounds (e.g. K22, recombinant interferons) in various concentrations or by serial dilution in ALI medium. Include non-treated controls. Also, to exclude viral inhibition by solvents (e.g. DMSO) include solvent controls.

3. For evaluation of either prophylactic or therapeutic effects of antivirals, the HAE cultures can be incubated with the compounds diluted in the basolateral medium prior to, during or after infection.

4. Infect cultures apically with human coronaviruses as described in Subheading 3.2.

5. Collect apical washings in HBSS as described in Subheading 3.2 for viral quantification by plaque assay and cells for viral quantification by Renilla Luciferase Assay or qRT-PCR.

3.4.2 Cytotoxicity Assay

1. Thaw CellTiter-Glo buffer and equilibrate both buffer and CellTiter-Glo substrate to room temperature.

2. Transfer the buffer to the amber bottle containing the substrate to reconstitute the enzyme. Mix by gently swirling the bottle.

3. Wash the apical side of the HAE cultures three times with 500 μl HBSS to remove excess mucus.

4. Apply 50 μl of HBSS to the apical side and mix with equal volume of reconstituted CellTiter-Glo enzyme solution (optimized for 24-well inserts, for other insert sizes adjust buffer amount accordingly) and incubate for 5 min at room temperature on a gyro-rocker to induce cell lysis.

5. Next incubate the plate for 10 min at room temperature to allow for stabilization of the luminescence signal.

6. Transfer 20 μl of cell lysate to a white, non-transparent 96-well plate for analysis.

7. Record luminescence (*see* **Note 12**). To account for background signal include empty wells in your analysis.

3.5   Virus Detection

*3.5.1   Renilla Luciferase Assay (See **Note 13**)*

1. Thaw Renilla Luciferase Assay buffer and dilute 1:5 in water.

2. Wash HAE inserts with HBSS three times prior to cell lysis.

3. Incubate inserts with 80 μl of Renilla lysis buffer on a gyro-rocker for 30 min at room temperature (optimized for 24-well inserts, adjust lysis buffer amount accordingly for other insert sizes).

4. During incubation, thaw Renilla Assay buffer.

5. Transfer the cell lysate to a 96-well plate.

6. Transfer 20 μl of the lysate to a white, non-transparent 96-well plate for analysis.

7. Add Renilla substrate at 1:200 dilution to the required amount of Renilla Assay buffer (100 μl per sample). Protect from light (*see* **Note 14**).

8. Program your luminometer settings with 10 s measure time followed by a 2 s delay. 100 μl of assay buffer should be dispensed into each well. If the luminometer is not equipped with an injector the assay buffer can be added manually using a multichannel pipette.

9. To adjust samples for background include empty wells in your analysis.

10. Plot your values as Log_{10} RLU (Relative Light Units).

3.5.2   Plaque Assay

The current protocol is optimized for HCoV-229E, but can easily be adapted to any other cell line and coronavirus strain.

1. Seed 150,000 target cells in a 12-well cluster plates with 1 ml of complete medium per well and incubate overnight at 37 °C in a humidified 5 % CO_2-incubator.

2. Make 6 tenfold serial dilutions of the harvested virus supernatants in 1 ml and inoculate the cells.

3. Incubate inoculum for 2 h at 33 °C in a humidified 5 % CO_2-incubator before removing the serial diluted virus inoculums from the cells and replace with 1 ml of overlay medium.

4. Incubate titration plates for 3–4 days at 33 °C in a humidified 5 % CO_2-incubator.

5. Remove overlay and wash wells twice with water to remove residual Avicel.

6. Subsequently add approximately 0.5–1 ml of crystal violet solution to each well and incubate for 10 min.

7. Remove crystal violet solution and wash the cells once with water and allow the plates to air-dry before counting the number of plaques.

3.5.3 Quantitative Reverse Transcriptase PCR

1. Isolate viral RNA with NucleoSpin RNA kit according to the manufacturer's protocol and elute in the appropriate amount of RNase-free water.

2. For reverse transcription use M-MLV reverse transcriptase (100 U), M-MLV buffer, and random primers and 10 μl of extracted total RNA in a total volume of 20 μl, at 37 °C for 60 min. *Optional*: include serial dilutions of in vitro transcribed RNA of the target gene for virus yield quantification.

3. To quantify viral HCoV RNA yields from contemporary strains use the FastStart Universal SYBR Green Master reaction mixture. Amplify 2 μl of cDNA according to the manufacturer's protocol, using the previously published sense and antisense strain-specific primers (*see* **Note 15**). Measurements and analysis can, for instance, be done on a LightCycler 480 II instrument, using the LightCycler 480 software package (Roche). Use the following cycle profile of 10 min at 95 °C followed by 45 cycles of 10 s at 95 °C, 20 s at 60 °C, and 30 s at 72 °C followed by a melting curve step to confirm product specificity.

4   Notes

1. Dissolve 5 g of BSA, globulin free, powder in 20 ml PBS in a 50 ml tube (do not vortex). Place the tube on a shaker/roller-bank for 2–4 (max 24) hours at +4 °C, until the BSA is completely dissolved. Add the volume up to 34 ml, mix gently by inverting the tube three times. Filter-sterilize the solution through a 0.22 μm filter, and store at –20 °C in aliquots of 3.5 ml in 15 ml tubes. Invert the tube three times before usage.

2. Dissolve 12 mg of Retinoic Acid (RA) in 40 ml absolute EtOH in a 50 ml tube wrapped in aluminum foil, the RA–EtOH stock (1×10^{-3} M) should be stored at –20 °C. To prepare the 1,000× stock, first confirm the RA concentration of the ethanol stock by diluting it 1:100 in absolute EtOH. Measure the absorbance at 350 nm using a spectrophotometer and a 1 cm light path quartz cuvette (or NanoDrop with 0.1 cm light path), blanked on 100 % EtOH. The absorbance of the diluted stock should equal 0.45 (0.045 on a NanoDrop). RA absorbance readings below 0.18 should be discarded. If the absorbance equals 0.45, add 3 ml of RA–EtOH stock solution to 53 ml PBS and add 4 ml of BSA 150 mg/ml stock. For absorbance values less than 0.45, calculate the needed volume of RA–EtOH stock as 1.35/absorbance and adjust the PBS volume appropriately. The 1,000× stock solution should be stored at –20 °C in aliquots of 1 ml per eppendorf tube.

3. Dissolve 42 mg ferrous sulfate, 12.2 g magnesium chloride, and 1.62 g calcium chloride-dihydrate in 80 ml H_2O, add

500 μl concentrated hydrochloric acid (HCl). Filter-sterilize the solution through a 0.22 μm filter, and store at −20 °C in aliquots of 1,100 μl per eppendorf tube.

4. Prepare seven separate 25 or 50 ml stock solutions (*see* Table 3a and b) in H_2O. Filter-sterilize (0.22 μm) each component after preparation. Afterwards, transfer an aliquot of 50 μl from each separate component into 49,600 μl filter-sterilized water (0.22 μm) and add a volume of 50 μl concentrated HCl solution. Mix the solution well through gentle vortexing and filter-sterilize the solution through a 0.22 μm filter, and store at −20 °C in aliquots of 1,100 μl per eppendorf tube.

5. Gentamicin and Amphotericin B should be omitted from ALI medium. These antibiotics are only required in BEGM medium right after cell isolation to prevent contamination.

6. For preparation of crystal violet solution safety glasses and protective clothing should be worn. Any spillage must be cleaned with 96 % ethanol.

7. Cells might take longer to dissociate from the bottom of the flask due to the collagen coating. If the cells are not dissociated after 3 min additional rounds of 1-min incubations can be performed until all cells have detached.

8. The seeded primary cells should reach confluence on the inserts within 1 week. If this takes longer the success rate of establishing well differentiated HAE cultures declines exponentially.

Table 3
Stock solutions for trace elements

(a)

Component	Formula	Amount/25 ml	Comment
Selenium	$NaSeO_3$	130.0 mg	Solution stable for 30 days at +4 °C
Silicone	$Na_2SiO_3 \cdot 9H_2O$	3.55 g	
Molybdenum	$(NH_4)_6Mo_7O_{24} \cdot 4H_2O$	31.0 mg	
Vanadium	NH_4VO_3	14.75 mg	Heat >100 °C to dissolve

(b)

Component	Formula	Amount/50 ml	Comment
Nickel	$NiSO_4 \cdot 6H_2O$	13.0 mg	
Tin	$SnCl_2 \cdot 2H_2O$	5.5 mg	
Manganese	$MnCl_2 \cdot 4H_2O$	10.0 mg	

9. Human coronavirus NL63, 229E, HKU1, and OC43 are predominantly found in the upper respiratory tract and are therefore incubated at 33 °C. Both MERS-CoV and SARS-CoV are predominantly found in the lower respiratory tract and are therefore incubated at 37 °C.

10. The fixed HAE cultures can be kept for 1–3 months at 4 °C if the CB is filter-sterilized (0.2 µM) and all the procedure were performed under sterile conditions. After cold storage it is preferential to acclimatize the fixed cultures for 15 min to room temperature on a gyro-rocker (20–30 rpm) prior to continuation of the staining protocol.

11. To prevent bleaching of the fluorophores one should cover the inserts from daylight exposure during each incubation step.

12. Luminometer settings depend on the manufacturer. However, a measurement time of 1–2 s per well has proved effective.

13. For this assay cultures must be infected with coronaviruses expressing a Renilla Luciferase reporter gene.

14. If your luminometer is equipped with an injector you must remember to account for priming by increasing the volume of required Renilla Assay buffer by 2–3 ml.

15. Primers targeting HCoV-NL63, HCoV-HKU1, HCoV-229E, and HCoV-OC43 have been characterized and described [4, 14, 15].

Acknowledgement

This work was supported by the 3R Research Foundation Switzerland (project 128-11).

References

1. de Jong PM, van Sterkenburg MA, Hesseling SC et al (1994) Ciliogenesis in human bronchial epithelial cells cultured at the air-liquid interface. Am J Respir Cell Mol Biol 10(3):271–277

2. Lin H, Li H, Cho HJ et al (2007) Air-liquid interface (ALI) culture of human bronchial epithelial cell monolayers as an in vitro model for airway drug transport studies. J Pharm Sci 96(2):341–350

3. Fuchs S, Hollins AJ, Laue M et al (2003) Differentiation of human alveolar epithelial cells in primary culture: morphological characterization and synthesis of caveolin-1 and surfactant protein-C. Cell Tissue Res 311(1):31–45

4. Pyrc K, Sims AC, Dijkman R et al (2010) Culturing the unculturable: human coronavirus HKU1 infects, replicates, and produces progeny virions in human ciliated airway epithelial cell cultures. J Virol 84(21):11255–11263

5. Kindler E, Jonsdottir HR, Muth D et al (2013) Efficient replication of the novel human betacoronavirus EMC on primary human epithelium highlights its zoonotic potential. MBio 4(1):e00611–e00612

6. Thompson CI, Barclay WS, Zambon MC et al (2006) Infection of human airway epithelium by human and avian strains of influenza a virus. J Virol 80(16):8060–8068

7. Bernacki SH, Nelson AL, Abdullah L et al (1999) Mucin gene expression during differentiation of human airway epithelia in vitro. Muc4 and muc5b are strongly induced. Am J Respir Cell Mol Biol 20(4):595–604

8. Gray TE, Guzman K, Davis CW et al (1996) Mucociliary differentiation of serially passaged normal human tracheobronchial epithelial cells. Am J Respir Cell Mol Biol 14(1):104–112

9. Lechner J, LaVeck M (1985) A serum-free method for culturing normal human bronchial epithelial cells at clonal density. J Tissue Cult Methods 9(2):43–48

10. Fulcher ML, Gabriel S, Burns KA et al (2005) Well-differentiated human airway epithelial cell cultures. Methods Mol Med 107:183–206

11. Dijkman R, Jebbink MF, Koekkoek SM et al (2013) Isolation and characterization of current human coronavirus strains in primary human epithelial cell cultures reveal differences in target cell tropism. J Virol 87(11):6081–6090

12. Lundin A, Dijkman R, Bergstrom T et al (2014) Targeting membrane-bound viral RNA synthesis reveals potent inhibition of diverse coronaviruses including the middle East respiratory syndrome virus. PLoS Pathog 10(5):e1004166

13. Hamming OJ, Terczynska-Dyla E, Vieyres G et al (2013) Interferon lambda 4 signals via the IFNlambda receptor to regulate antiviral activity against HCV and coronaviruses. EMBO J 32(23):3055–3065

14. Schildgen O, Jebbink MF, de Vries M et al (2006) Identification of cell lines permissive for human coronavirus NL63. J Virol Methods 138(1–2):207–210

15. Vijgen L, Keyaerts E, Moes E et al (2005) Development of one-step, real-time, quantitative reverse transcriptase PCR assays for absolute quantitation of human coronaviruses OC43 and 229E. J Clin Microbiol 43(11):5452–5456

Chapter 9

Quantification of Infectious Bronchitis Coronavirus by Titration In Vitro and In Ovo

Joeri Kint, Helena Jane Maier, and Erik Jagt

Abstract

Quantification of the number of infectious viruses in a sample is a basic virological technique. In this chapter we provide a detailed description of three techniques to estimate the number of viable infectious avian coronaviruses in a sample. All three techniques are serial dilution assays, better known as titrations.

Key words Titration, EID_{50}, $TCID_{50}$, Plaque-forming units

1 Introduction

Technological advances in particle analysis have made it possible to quantify the number of virus particles in a sample with increasing accuracy. Techniques such as specialized flow cytometry [1], dynamic light scattering [2], quantitative capillary electrophoresis [3], and fluorescence correlation spectroscopy [4] can determine the number of particles in a sample within hours. The choice of technique depends on the sort of virus and the matrix in which it is suspended. All aforementioned techniques differentiate particles on the basis of physical properties such as size or antibody affinity. As a consequence particle analysis cannot differentiate infectious from noninfectious virus particles.

There is only one technique available that can reliably quantify the number of infectious particles in a sample. This technique, developed many decades ago [5], exploits the fact that virus can propagate in biological systems such as embryonated eggs or cell cultures. Propagation of a virus is generally accompanied by changes in cell morphology (referred to as cytopathic effect or CPE), which can be visualized using a microscope, or even by eye. Some viruses do not induce CPE, in which case an antibody based assay (immunofluorescence or ELISA) is needed to determine presence or absence of virus. During a titration assay, tissue cultures

Helena Jane Maier et al. (eds.), *Coronaviruses: Methods and Protocols*, Methods in Molecular Biology, vol. 1282,
DOI 10.1007/978-1-4939-2438-7_9, © Springer Science+Business Media New York 2015

or embryonated eggs are incubated with tenfold serial dilutions of a virus containing sample and several days later the cytopathic effect is scored. From these scores, the virus titer is calculated using the methods described by Spearman and Kaerber [6, 7] or Reed and Muench [8]. The virus titer is defined as the reciprocal of the dilution at which 50 % of the inoculated embryos or tissue cultures show CPE. In this chapter we use the method of Spearman and Kaerber to calculate the titer, as this calculation can cope with unequal group sizes. Unequal group sizes frequently arise when eggs are lost to aspecific death of the embryo or bacterial infection.

Coronaviruses in general have a narrow host range and many clinical isolates only replicate in primary cells. Replication of most field isolates of infectious bronchitis coronavirus is restricted to embryonated eggs or tracheal organ culture. Most isolates, however, can be adapted to propagation in primary chicken kidney (CK) cells. Adaptation typically requires several passages and selects for viral subpopulations and can induce mutations [9]. Passaging of IBV in either embryonated eggs or primary cell cultures leads to attenuation of the virus in vivo [10–12]. The most striking example is the IBV Beaudette strain, which has been passaged hundreds of times in eggs and primary chicken kidney cells [13, 14]. Although IBV Beaudette propagates very well on eggs, CK cells and even in Vero cells, the virus is highly attenuated in vivo and vaccination using Beaudette provides little protection against infection with pathogenic strains of IBV [15].

For quantification of IBV field isolates, embryonated chicken eggs are the most suitable substrate. A protocol on titration of IBV on embryonated chicken eggs is provided in the first part of this protocol. IBV strains which have been adapted to grow in cultures of primary chicken cells can be titrated on these cells using either the $TCID_{50}$ method or plaque titration. Protocols for both methods are provided in this chapter. Plaque-forming unit (PFU) titration yields more accurate and reproducible results then the $TCID_{50}$ method, it is however more labor intensive. Both methods are presented in this chapter.

2   Materials

2.1   Titration of Avian Infectious Bronchitis Virus in Fertilized Eggs by EID_{50}

1. Fertilized specific pathogen free (SPF) eggs, 9–11-day-old (*see* **Note 1**).

2. Diluent: 2.5 % w/v tryptose, 1,000 U/ml penicillin, 1,000 µg/ml streptomycin.

3. Disinfectant: 70 % alcohol in water.

4. Egg shell drill or punch.

5. Sterile 1 ml syringes.

6. Needles, preferably 25 G; 16 mm.

7. Hobby glue or melted wax to seal the inoculation site.

8. Egg candling light.

9. Egg incubator with rocker.

2.2 Tissue Culture Infective Dose (TCID$_{50}$) Titration

1. 96-well plates containing 80–100 % confluent CK cells.

2. Titration medium: 1:1 mix of medium M199: Ham's F-10 nutrient mixture supplemented with 0.5 % fetal calf serum (FCS), 0.1 % w/v tryptose phosphate broth, 0.1 % w/v sodium bicarbonate, 0.1 % w/v HEPES, 100 U/ml penicillin, 100 µg/ml streptomycin.

3. Multistepper pipette (Socorex or equivalent).

4. Inverted microscope.

5. Positive control sample with known titer.

2.3 Plaque-Forming Unit Titration

1. Six-well plates containing 60–90 % confluent CK cells.

2. 1×BES cell culture medium: EMEM, 10 % w/v tryptose phosphate broth, 0.2 % w/v bovine serum albumin (BSA), 20 mM N,N bis(2-hydroxethyl)-2-aminoethanesulfonic acid (BES), 0.4 % w/v sodium bicarbonate, 2 mM L-glutamine, 250 U/ml nystatin, 100 U/ml penicillin, 100 µg/ml streptomycin.

3. 2×BES cell culture medium: 2×EMEM, 20 % w/v tryptose phosphate broth, 0.4 % w/v BSA, 40 mM BES, 0.8 % w/v sodium bicarbonate, 4 mM L-glutamine, 500 U/ml nystatin, 200 U/ml penicillin, 200 µg/ml streptomycin.

4. Sterile phosphate buffered saline (PBS).

5. 2 % w/v agarose in water (autoclaved).

6. 10 % w/v formaldehyde in PBS.

7. 0.1 % w/v crystal violet in water.

8. Microwave.

9. Water bath.

10. Small spatula.

3 Methods

3.1 Titration of Avian Infectious Bronchitis Virus in Fertilized Eggs by EID$_{50}$

1. Candle the eggs using the candling light and draw a line on the shell marking the edge of the air sac. Draw an X approximately 5 mm above this line, which marks the inoculation site.

2. Assign ten eggs per dilution and select those dilutions (at least three) that include the 50 % end point of the sample.

3. Prepare tenfold serial dilutions in diluent.

4. Disinfect the eggs by spraying them with disinfectant.

5. After the eggs have dried drill or pierce a hole in the egg shell at the marked inoculation site.

6. Inoculate ten eggs per dilution each with 0.2 ml volume via the allantoic cavity by holding the syringe and needle vertically and by inserting the needle approximately 16 mm into the egg (Fig. 1a, b).

7. After inoculation, the hole in each egg is sealed with hobby glue or melted wax.

8. Incubate the eggs in an egg incubator with rocker at 37.8 °C and 60–65 % humidity.

9. Candle the eggs after 24 h of incubation. Embryo mortality occurring up till 24 h post inoculation is considered nonspecific and therefore these eggs are discarded.

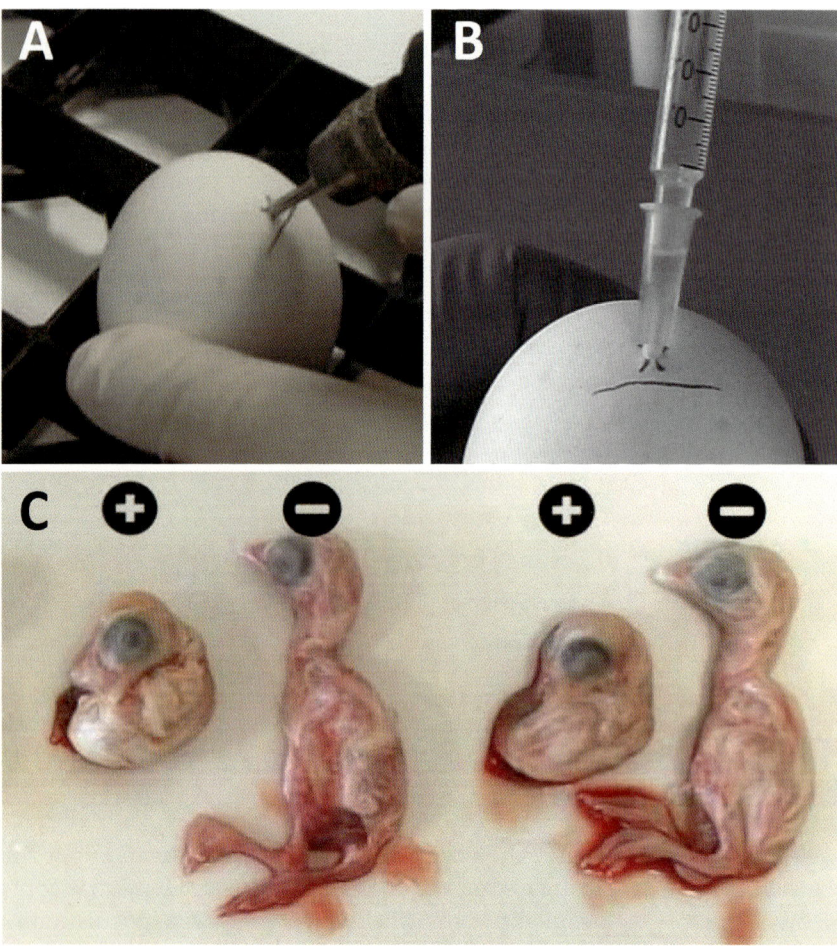

Fig. 1 Egg inoculation into the allantoic cavity. (**a**) Drill or pierce a hole in the egg 5 mm above the edge of the air sac. (**b**) Inoculate each egg with a 0.2 ml volume by inserting the needle approximately 16 mm into the egg. (**c**) Typical IBV induced malformations 2 days post infection with strain 4/91. *Plus* indicates embryos infected with IBV, *minus* indicates non-infected embryos

10. Incubate the eggs for 6 more days in an egg incubator with rocker at 37.8 °C and 60–65 % humidity. Candle the eggs at the end of the incubation period to identify embryos that have died. Subsequently, macroscopically evaluate all surviving embryos for the presence of lesions characteristic for IBV infection (stunting and curling; Fig. 1c). Embryos that have died and embryos that exhibit lesions characteristic for IBV infection are considered positive (*see* **Note 2**).

11. Any embryo with IBV specific CPE is regarded positive. Virus titers in the original sample, expressed as ^{10}log EID_{50}/ml are calculated using the method described by Spearman and Kaerber [6, 7], using the following formula:

$$\text{Titre} = \left(x_0 - \frac{d}{2} + d \sum \frac{r}{n} \right)$$

In which:

x_0: logarithm of the inverse value of the lowest dilution at which all embryos are positive.

d: logarithm of the dilution factor (d=1 when using tenfold serial dilutions).

n: number of eggs used per dilution.

r: number of positive eggs at that dilution.

12. Example of calculation of virus titer
Using the result of the titration depicted in Fig. 2, the virus titer is calculated as follows:

$$\text{Titre}\left(^{10}\text{logEID}_{50} / 0.2\,\text{ml} \right) = \left(4 - \frac{1}{2} + 1 \times \left(\frac{5}{10} + \frac{1}{9} \right) \right) =$$

$$5.1 ^{10}\log \text{EID}_{50} / 0.2\text{ml} = 10^{5.1}/0.2 = 5.8 \,^{10}\log \text{EID}_{50} / \text{ml}$$

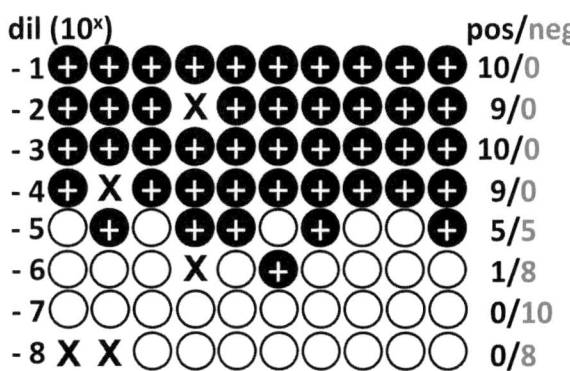

Fig. 2 Schematic result of an egg titration. Each *circle* represents one egg and *crosses* indicate aspecific death. *Plus signs* indicate embryos with IBV specific malformations

3.2 Tissue Culture Infective Dose (TCID$_{50}$) Titration

1. CK cells are seeded in 96-well plates at 7.5×10^4 cells/well, 1 or 2 days before the titration. One plate is needed per sample. At the time of titration, the monolayer should be nearly confluent.

2. Prepare tenfold serial dilutions of the samples in titration medium (*see* **Note 3**).

3. Empty the medium from the 96-well plate containing CK cells into a waste container and gently tap the plate dry on a stack of tissues.

4. Fill the wells of column 1 and 12 of the 96-well plate with 100 μl/well titration medium. These are the negative control wells (Fig. 3a).

5. Starting with the highest dilution, dispense 100 μl/well in row H using the multistepper pipette. Proceed with filling the descending dilutions in rows G till A.

6. Incubate the 96-well plates for 3–4 days at 37 °C and 5 % CO_2.

7. After 3–4 days incubation, score all wells for IBV specific CPE using a microscope. Although the CPE may vary per IBV

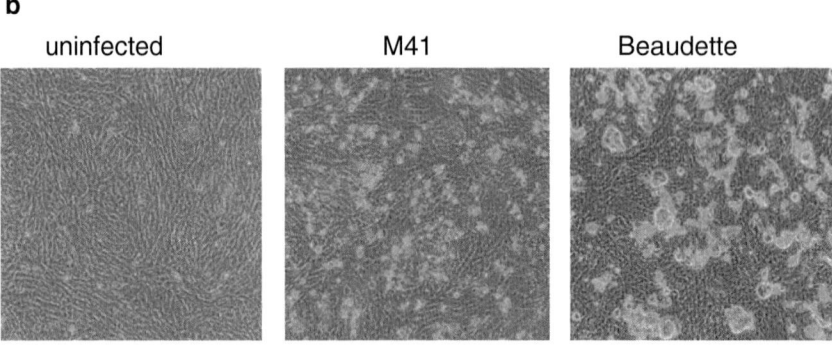

Fig. 3 Layout of sample dilutions in the 96-well titration plate. *NC* negative control (**a**). Typical CPE of IBV 2 days post infection of CK cells with M41 or Beaudette (**b**)

strain, it is generally characterized by clusters of rounded cells on top of the monolayer. At low dilutions the monolayer may be partly destroyed, exemplified by the IBV Beaudette strain (Fig. 3b).

8. Titers are calculated using the method described by Spearman and Kaerber (Subheading 3.1, **step 11**) and are expressed in $^{10}\log (TCID_{50})/ml$.

3.3 Plaque-Forming Unit Titration

3.3.1 Infection of Cells

1. CK cells are seeded into six-well plates 3 days prior to titration. When performing the titration, the monolayer should 60–90 % confluent.

2. Prepare tenfold serial dilutions of virus in $1 \times BES$.

3. Remove media from cells and wash once with sterile PBS.

4. Remove PBS from the cells and add 500 µl of diluted virus to each well. Duplicate wells should be inoculated for each dilution.

5. Incubate cells at 37 °C for 1 h to allow virus attachment.

6. Melt 2 % agar in a microwave and then transfer to a 42 °C water bath. Allow the agar to equilibrate in temperature.

7. Mix the partially cooled agar with $2 \times BES$ pre-warmed to 37 °C to generate $1 \times BES + 1$ % agar. Keep at 42 °C until needed to prevent premature setting (*see* **Note 4**).

8. Remove virus inoculum and overlay cells with 2.5 ml of the $1 \times BES$/agar mix.

9. Leave cells at room temperature for approximately 5 min until agar has solidified.

10. Incubate at 37 °C and 5 % CO_2 for 3 days for plaques to develop.

3.3.2 Staining Cells and Determining Titer

1. Overlay agar with 1 ml per well 10 % formaldehyde in PBS.

2. Incubate at room temperature for 1 h.

3. Remove formaldehyde and ensure disposal according to local regulations.

4. Using a small spatula, flick off the agar from the cells (*see* **Note 5**).

5. Wash cells by shaking the plate upside down in a sink containing water.

6. Add 0.5 ml 0.1 % crystal violet to each well.

7. Incubate at room temperature for 10 min.

8. Remove crystal violet and dispose of according to local regulations.

9. Wash plate by shaking upside down in a sink of water.

10. Pat plate dry and leave upside down at room temperature to fully dry.

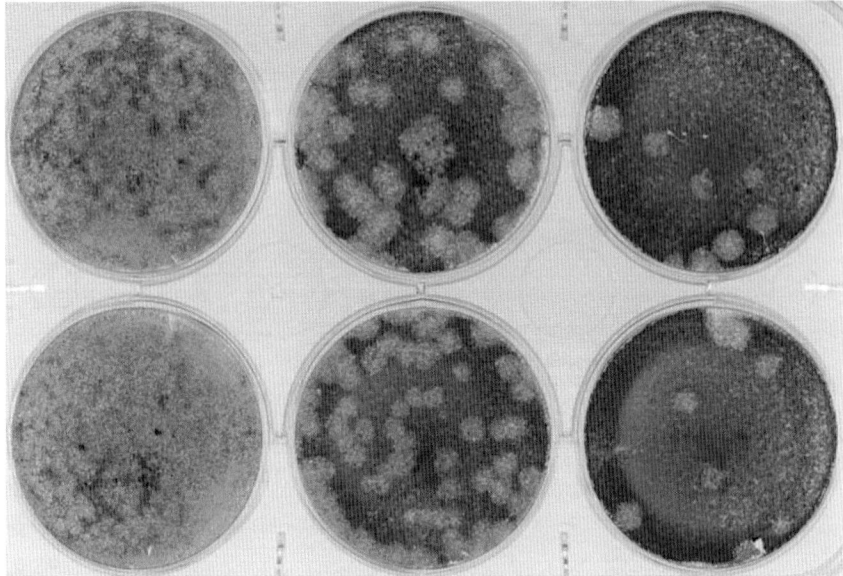

Fig. 4 Example plaque assay plate. CK cells were infected with tenfold serial dilutions of IBV in duplicate and incubated at 37 °C for 3 days. Cells were then fixed and stained with 0.1 % crystal violet

11. Plaques should be clearly visible as holes in the monolayer (Fig. 4). Count the number of plaques per well at the dilution with clearly defined, individual (not overlapping) plaques (typically 10–50 plaques/well). Ensure duplicate wells are counted and an average taken.

12. Determine titer using the following equation:

$$\text{Titre}\left(\text{PFU}\,/\,\text{ml}\right) = \frac{\text{average number of plaques}}{\text{dilution factor} \times \text{inoculum volume}\left(0.5\text{ml}\right)}$$

13. For most accurate results, the plaque assay should be repeated three times and the average titer determined.

4 Notes

1. SPF eggs should be used for titration of IBV, as non-SPF eggs may contain IBV specific antibodies that can interfere with the replication of IBV.

2. Non-egg adapted IBV isolates may induce very little IBV specific aberrations of the embryos. When titrating such viruses, the eggs are incubated for an additional 2–3 days post inoculation. Candle the eggs at the end of the incubation period to identify embryos that have died. Subsequently, collect allantoic fluid of each of the surviving embryos and test them in a monoclonal based antigen capture ELISA as

described in refs. [16, 17]. Dead embryos and those of which the allantoic fluid contains IBV as established by ELISA are considered positive.

3. IBV titers usually do not exceed $10^{8.8}$ $TCID_{50}$/ml. Therefore 10^{-8} should be adequate as highest dilution. When the virus titer of the sample is known, select a number of tenfold dilutions (at least three) that include the 50 % end-point dilution. If the titer is unknown, select a broader range of tenfold dilutions that most likely include the 50 % end-point dilution. For a $TCID_{50}$ titration typically all the wells of the 96-well plate are used

4. Alternative methods also exist for mixing media and agar. If there is concern regarding the overlay setting too quickly or risk of contamination from the water bath, hot agar can be mixed directly with cold media (4 °C). Once the mixture feels warm to the touch, rather than hot, it can be added to cells.

5. The simplest method for removing agar from the cells is to hold the plate upside down with the lid removed. The small spatula is inserted between the agar and the wall of the well. Once the base of the well is reached, a small amount of pressure is applied to remove the agar, being careful not to scrape off the cells. The whole agar plug should then fall out easily.

References

1. Ferris MM, Stepp PC, Ranno KA et al (2011) Evaluation of the Virus Counter® for rapid baculovirus quantitation. J Virol Methods 171:111–116

2. Driskell JD, Jones CA, Tompkins SM et al (2011) One-step assay for detecting influenza virus using dynamic light scattering and gold nanoparticles. Analyst 136:3083–3090

3. Mironov GG, Chechik AV, Ozer R et al (2011) Viral quantitative capillary electrophoresis for counting intact viruses. Anal Chem 83:5431–5435

4. Schwille P, Bieschke J, Oehlenschläger F (1997) Kinetic investigations by fluorescence correlation spectroscopy: the analytical and diagnostic potential of diffusion studies (evolutionary biotechnology—from theory to experiment). Biophys Chem 66:211–228

5. Dulbecco R (1952) Production of plaques in monolayer tissue cultures by single particles of an animal virus. Proc Natl Acad Sci U S A 38:747–752

6. Spearman C (1908) The method of 'right and wrong cases' ('constant stimuli') without Gauss's formulae. Br J Psychol 2:227–242

7. Kaerber G (1931) Beitrag zur kollektiven behandlung pharmakologischer reihenversuche. Arch Exp Pathol Pharmacol 162:480–487

8. Reed L, Muench H (1938) A simple method of estimating fifty percent endpoints. Am J Hyg 27:493–497

9. Gillette KG (1973) Plaque formation by infectious bronchitis virus in chicken embryo kidney cell cultures. Avian Dis 17:369–378

10. Gelb J Jr, Cloud SS (1983) Effect of serial embryo passage of an Arkansas-type avian infectious bronchitis virus isolate on clinical response, virus recovery, and immunity. Avian Dis 27:679–687

11. Huang YP, Wang CH (2006) Development of attenuated vaccines from Taiwanese infectious bronchitis virus strains. Vaccine 24:785–791

12. Jackwood MW, Hilt DA, Brown TP (2003) Attenuation, safety, and efficacy of an infectious bronchitis virus GA98 serotype vaccine. Avian Dis 47:627–632

13. Yachida S, Aoyama S, Takahashi N et al (1979) Growth kinetics of embryo- and organ-culture adapted Beaudette strain of infectious bronchitis virus in embryonated chicken eggs. Avian Dis 23:127–131

14. Cunningham CH, Spring MP, Nazerian K (1972) Replication of avian infectious bronchitis virus in African green monkey kidney cell line VERO. J Gen Virol 16:423–427

15. Hodgson T, Casais R, Dove B et al (2004) Recombinant infectious bronchitis coronavirus Beaudette with the spike protein gene of the pathogenic M41 strain remains attenuated but induces protective immunity. J Virol 78:13804–13811

16. Ignjatovic EJ, Ashton F (1996) Detection and differentiation of avian infectious bronchitis viruses using a monoclonal antibody-based. Avian Pathol 25:721–736

17. Naqi SA, Karaca K, Bauman B (1993) A monoclonal antibody-based antigen capture enzyme-linked immunosorbent assay for identification of infectious bronchitis virus serotypes. Avian Pathol 22:555–564

Chapter 10

Purification of Coronavirus Virions for Cryo-EM and Proteomic Analysis

Stuart Dent and Benjamin W. Neuman

Abstract

Purification of intact enveloped virus particles can be useful as a first step in understanding the structure and function of both viral and host proteins that are incorporated into the virion. Purified preparations of virions can be used to address these questions using techniques such as mass spectrometry proteomics. Recent studies on the proteome of coronavirus virions have shown that in addition to the structural proteins, accessory and non-structural virus proteins and a wide variety of host cell proteins associate with virus particles. To further study the presence of virion proteins, high-quality sample preparation is crucial to ensure reproducible analysis by the wide variety of methods available for proteomic analysis.

Key words Coronavirus, Cryo-EM, Proteomic, Virus purification, Density gradient centrifugation

1 Introduction

The most important factor in Cryo-EM and proteomic studies of coronavirus virions is high-quality sample preparation. A useful guideline in planning to make a purified coronavirus sample is that the final preparation should be concentrated to at least 10^{10} virions per milliliter if possible. For a simple proteomic analysis, less is sufficient, but as further digestions and purifications are often performed, higher amounts of infectious virus are recommended to recover enough sample for study, and to make it possible to recover a fairly complete virion proteome [1]. Likewise, the accuracy of cryo-EM results depends on both the concentration and intactness of the purified virions.

This requires a concentration step, since coronavirus growth seldom surpasses 10^8 PFU/ml, and can be much lower in some cases. It is also worth noting that cleaved viral spike proteins are sensitive to S1 shedding, though this susceptibility can vary considerably among viruses and even strains of the same viral species. The method of concentration and purification described here seeks to minimize virion disruption and spike loss, and has been used

Helena Jane Maier et al. (eds.), *Coronaviruses: Methods and Protocols*, Methods in Molecular Biology, vol. 1282, DOI 10.1007/978-1-4939-2438-7_10, © Springer Science+Business Media New York 2015

successfully with several coronaviruses and other enveloped viruses derived from cells and in ovo culture. The method described here is a modification of our previously published method of concentrating samples for electron microscopy, and would still be useful for that purpose [2].

Serum used as part of cell culture medium can act as a carrier, potentially increasing virus yield, but can also lead to clumping and can add to the background of the sample. If purification in the presence of serum fails despite adequate starting virus titer, a serum-free preparation might be advantageous. For this reason an alternative serum-free purification protocol has been provided below. After concentration, but before further proteomics analysis, electron microscopy to look for virion density and spike coverage can be a useful quality control tool. The reliability of results from virion purification will depend on the percentage of cell-derived exosomal vesicles that are co-purified with the virions, although the contributions of exosome-derived proteins can be estimated by performing the same purification in parallel on uninfected cell culture supernatant. It is hoped that these techniques will facilitate further examination of coronavirus, torovirus, arterivirus, mesonivirus, and ronivirus virion proteomics.

2   Materials

2.1 Virus Purification and Concentration

1. Vero cells.

2. Virus of interest.

3. DMEM: Dulbecco's Modified Eagle's Medium, 10 % fetal bovine serum, penicillin and streptomycin (100 U/ml each), 10 mM HEPES, adjusted to pH 6.7 using NaOH.

4. Polyethylene glycol-8000 or 10000, white flake type (PEG-8000 Ultra for Molecular Biology) (*see* **Note 1**).

5. NaCl, crystalline, high quality.

6. HEPES-saline: 0.9 % NaCl (w/v), 1 mM HEPES, pH adjusted to 6.7 using HCl, vacuum-sterilized through a 0.22 μm pore size filter.

7. 3× HEPES-saline: 2.7 % NaCl, 30 mM HEPES, pH adjusted to 6.7 using HCl.

8. 50 % (w/w) sucrose: 50 g sucrose, 50 ml HEPES-saline, vacuum-sterilized through a 0.22 μm pore size membrane. Dilute with additional HEPES-saline to prepare 10, 20, and 30 % sucrose solutions.

9. If samples are to be inactivated by chemical fixation, prepare 25 % neutral buffered formalin: 10 ml of formalin (37–40 % formaldehyde), 5 ml of 3× HEPES-saline.

10. Centrifuges and rotors: A low-speed centrifuge rotor with a capacity ≥1 l (Sorvall GSA or GS-3, for example) and a high-speed centrifuge rotor with a total capacity ≥100 ml (SW32.1 Ti, Beckman-Coulter, for example).

11. Tracking dye: 10× SYBR-gold dye (Life Technologies, sold as 10,000×), which will bind ssRNA in virions readily, and can be used to locate your virus pellet, and to distinguish the viral component of complex pellets that contain impurities with different sedimentation rates.

2.2 Serum-Free Virus Purification and Concentration

1. Virus production serum-free medium (VP-SFM; Life Technologies) supplemented with penicillin and streptomycin (100 U/ml each).

2. All other reagents and equipment as in Subheading 2.1.

2.3 Quality Control EM

1. Formvar/carbon coated 200-mesh or 300-mesh EM grids.

2. Negative stain: 2 % uranyl acetate in water, pH adjusted to 6.5 with 1 M NaOH, filtered freshly through a 0.22 μm hole size membrane immediately before use, stored in a brown glass bottle away from light.

3. Parafilm M.

4. Fine forceps for EM grid manipulation.

5. Filter paper.

3 Methods

Since the quality of the virus preparation is the most important component of proteomics studies, two purification protocols are listed below. Either can yield high-quality coronavirus, but Subheading 3.1 is generally preferable because the serum proteins function as a "carrier" during the PEG precipitation step. Serum-free purification (Subheading 3.2) can be used with cell and virus combinations that tend to produce overly viscous purified virions, or as a means to reduce the complexity of "background" proteins before proteomics analysis. For best results, the purification process should be completed in 1 day, and the virus should be used immediately.

3.1 Virus Purification and Concentration

This method is suitable for most coronaviruses that grow well in cultured cells, and has been used successfully with severe acute respiratory syndrome-coronavirus (SARS-CoV), feline coronavirus (FCoV) and mouse hepatitis virus (MHV), coronavirus virus-like particles, and torovirus in addition to several types of influenzavirus, arenavirus, and retrovirus-like particles [3, 4]. In the case of infectious bronchitis virus (IBV), this method has been used successfully for the purification of virus from embryonated chicken eggs. For the purpose of this protocol, it is assumed that IBV is being prepared on Vero cells.

1. Culture Vero cells in DMEM to approximately 70–90 % confluency (*see* **Note 2**).

2. Inoculate with IBV at a multiplicity of infection of three or more.

3. Remove the inoculum after 1 h and replace with fresh medium.

4. Remove and discard the culture medium 24 h after inoculation. Replace with fresh DMEM (*see* **Note 3**).

5. Collect the cell culture supernatant 48 h after inoculation. Store a small sample for plaque assay titration (*see* **Note 4**).

6. Prepare three-step 10–20–30 % sucrose gradients in tubes appropriate to the high-speed centrifuge rotor. If using the Beckman SW-32 Ti rotor, use ~8 ml for each step, leaving ~10 ml for sample loading and balancing. Pipette the 10 % sucrose into the bottom of the tube. Refill the pipette with 1 ml more sucrose solution than you will need, tilt the tube as much as possible without spilling the sample and place the pipette tip just above the bottom of the tube. Dispense the 20 % sucrose very slowly using the gravity-only setting—the last 1 ml will be retained in the pipette. Carefully and slowly withdraw the pipette, and put the 30 % sucrose layer underneath the 20 and 10 % layers in the same way. There will be visible lines at the border between steps if this is done correctly (*see* **Note 5**).

7. Transfer the supernatant to the largest available screw-cap centrifuge bottles that will fit your rotor, noting the total volume. Pellet cellular debris at $10,000 \times g$, 4 °C for 20 min. It is best to use a high-capacity rotor at this stage (Sorvall GSA, for example) to minimize preparation time.

8. During the centrifugation, prepare fresh screw-cap centrifuge bottles containing 10 g of dry PEG-8000 and 2.2 g of NaCl per 100 ml of culture medium to be added. Alternatively, prepare a large conical flask with sufficient PEG-8000, NaCl, and a heavyweight magnetic stir bar to bring the entire volume of virus-containing medium to a final concentration of 10 % PEG-8000, 2.2 % NaCl.

9. Chill HEPES-saline and neutral-buffered formalin on ice for later use (*see* **Note 6**).

10. After centrifugation (**step 7**), an off-white or yellow pellet of cell debris will be visible. Quickly decant the supernatant into the centrifuge bottles or conical flask prepared earlier with PEG and NaCl (*see* **Note 7**).

11. Swirl the PEG-8000/NaCl/supernatant mixture gently until the PEG crystals are fully dissolved. This can be done by hand for individual bottles, or using a magnetic stirring plate which has been incubated in a 4 °C incubator or cold room.

12. Add a clean stir bar if one is not already present and incubate at 4 °C for a further 30 min with gentle stirring.

13. Transfer the solution to centrifuge bottles, if necessary. Pellet the PEG-precipitated protein, which will also contain the virus, by centrifugation for 30 min at approximately $10,000 \times g$ at 4 °C.

14. Decant and discard supernatants *immediately* to minimize the amount of virus that is lost by resuspending. A large opaque white pellet should be present in each of the flasks following centrifugation, and may run from the bottom to top of one side if you used a fixed-angle rotor.

15. Swirl each pellet by hand in 1–3 ml of cold HEPES-saline until dissolved. Avoid passing the sample through a pipette at this step, if possible. It is critical that the PEG pellets be completely resuspended before proceeding to the next step. The resuspended pellet will be viscous if this step is done correctly (*see* **Note 8**).

16. Optionally, add one-tenth the volume of each pellet of Tracking dye. This can be left for 10 min with the resuspended PEG pellets, and will penetrate virions to fluorescently label the RNA inside without any additional permeabilization. This will make it possible to locate the virus-containing fraction or section of the pellet in any subsequent step, simply by resting the tube in a bottomless tube rack or clear beaker and illuminating the sample briefly with a UV transilluminator. This step is useful for troubleshooting the purification procedure.

17. Tilt the tube containing your gradient as much as possible without spilling, to allow the resuspended pellet to run as slowly as possible down the side of the tube and onto the top of the 10 % gradient layer. It is important not to disturb the gradient layers at this stage. In this manner, overlay the resuspended PEG pellet carefully onto the sucrose gradients. Balance with remaining sample or additional HEPES-saline.

18. Pellet the virions through the sucrose cushions by centrifugation at $100,000 \times g$ for at least 90 min at 4 °C. Since the pellet will be compact, it is not important to brake the centrifuge slowly.

19. After centrifugation, decant and discard supernatants immediately. Invert the empty tubes on an absorbent surface for 5 min, and tap gently to wick away any remaining sucrose solution that may have gathered near the rim of the centrifuge tube.

20. Resuspend the virion pellet in as small a volume of HEPES-saline as possible (typically 100 μl for a small pellet and 200 μl for a large pellet). Do not use a pipette to resuspend the virus, as this may shear spikes and damage fragile viral envelopes (*see* **Note 9**).

21. When the pellet has been resuspended, the HEPES-saline will turn somewhat opaque and milky in color. Use a P-1000 pipette tip from which the pointed end has been cut off to gently transfer the virus suspension to a cryovial with minimal shearing. Discard any insoluble material that remains, as most of the authentic virus will resuspend quickly and easily in comparison. Set aside a small sample (typically 5–10 % of your sample for a diagnostic plaque assay.

22. At this stage, the virus should be monodisperse, and can be formalin inactivated if desired. Treatment with a 1 % final concentration of ice-cold neutral buffered formalin, 2-propiolactone treatment, or gamma-irradiation should all yield intact, inactivated EM-quality particles. Samples for cryo-EM may be stored at 4 °C for up to 24 h, but should not be frozen for storage. Purified virus for mass spectrometry should not be formalin fixed, but can instead be inactivated using the solvent that will be used for mass spectrometry, provided this has been validated for your virus (*see* **Note 10**).

3.2 Serum-Free Virus Purification and Concentration

This alternative method is suitable for purification of viruses that grow to lower titers. Serum-free culture and preparation can also be used to remedy solutions that fail for cryo-EM or proteolysis due to high viscosity, non-viral protein contamination, or large amounts of insoluble material. Percentage recovery will typically be lower than that with Subheading 3.1.

1. Perform **steps 2–15** of Subheading 3.1 as above, substituting VP-SFM for DMEM starting at the time of inoculation (*see* **Note 11**).

2. The PEG-protein pellets should be white, and may be quite small and susceptible to resuspending quickly upon standing for even a few minutes. Decant and discard the supernatants immediately. Resuspend by swirling gently in 10 ml HEPES-saline (*see* **Note 12**).

3. Perform **steps 16–22** of Subheading 3.1 as above. The final translucent pellet may be small and quite difficult to see, but the presence of the virus can be confirmed using the Tracking dye and a transilluminator after removing the supernatant, if the dye was added at **step 16** of Subheading 3.1.

3.3 Quality Control for EM Analysis

The number of infectious virus particles in the final preparation should be directly assessed by plaque assay or similar means as a retrospective measure of quality. The quickest way to assess sample quality is by EM of a negatively stained sample. Ideal samples for proteomics or cryo-electron microscopy should contain a high density of virions with intact spikes, and relatively little non-viral material. This protocol describes how to prepare a negatively stained coronavirus specimen for transmission-EM.

1. Lay down a piece of paper towel on a work surface that is designated for work with slightly radioactive materials. Label the paper towel with the names of your samples.

2. Lay a piece of Parafilm M large enough to cover your labels, backing side facing up, on top of the paper towel and separate the backing from the Parafilm M at one corner. Trace lines through the backing paper on a piece of Parafilm M using the back of a pair of EM forceps to create channels for droplets of stain or wash buffer. The places where lines intersect will naturally hold droplets, and the labels will help keep the samples organized.

3. Peel off the rest of the backing paper, leaving the Parafilm stuck to the paper towel. On the Parafilm, place one ~20 μl droplet of 2 % uranyl acetate and up to three droplets of HEPES-saline wash buffer for each sample. Highly viscous samples will give a cleaner appearance after one or more wash steps, but washing is not necessary if the sample is quite pure (*see* **Note 13**).

4. Place 2–5 μl of the viral sample onto each grid, and leave each sample for 5 min to allow the sample to adsorb onto the grid.

5. If you are washing your sample, balance each grid on top of each droplet of HEPES-saline for 30 s. Do not reuse buffer or stain droplets, to avoid sample contamination. In between droplets, dry the grids. Hold the grid perpendicular to a clean piece of filter paper and touch the edge of the grid to the paper. Most of the wash buffer should be carried off.

6. After the washing steps, float each grid on a uranyl acetate stain droplet for 1 min. Longer staining will lead to more particles that appear to fill up with stain, if this is desired.

7. Touch the edge of the grid to the filter paper several times to remove excess stain (*see* **Note 14**).

8. Samples should be visualized immediately, and can be best stored for longer periods in a grid box that is kept under a vacuum.

9. Ideal preparations for both proteomics and cryo-electron microscopy will contain monodisperse virions as opposed to clumps of virions, few smooth-walled exosomal vesicles, little or no stringy released ribonucleoprotein, and will have virions covering roughly one-tenth of the grid surface area.

3.4 Quality Control for Proteomic Analysis

1. If possible, western blot analysis should be performed against the structural proteins of the virus to confirm the presence of viral proteins in the final preparation.

2. If infectivity of the virus is important, the number of infectious virus particles in the final preparation should be directly assessed by plaque assay or similar means as a retrospective measure of quality.

3. Coomassie and silver stained SDS-PAGE can also be performed to show the complexity of the sample. As recent proteomic studies have shown, the presence of host cell proteins associated with coronavirus virions is to be expected, but a prominent band between 40 and 50 kDa is indicative of N protein.

3.5 Proteomic Analysis

1. Protease digestion of purified virions can be used to remove proteins outside the coronavirus virion. This should be optimized for each virus and each protease individually. Proteinase K is common, and following digestion, repurification by pelleting through a 30 % sucrose cushion can remove the unwanted proteinase K. It should be noted that after digestion, removal of extra-virion proteins would lighten the density of the virion and therefore a longer centrifugation is required to repurify virus particles.

2. A wide variety of methods are available to study the coronavirus proteome. Gel based methods are still common and virions purified by these methods are compatible with gel based analysis. Direct LC-MS analysis of coronavirus virions is also possible. Success has been achieved by lysing the virions using Rapigest™ (Waters) followed by alkylation, reduction, and tryptic digestion before MS.

4   Notes

1. Larger molecular weight polyethylene glycol preparations tend to be too heterogeneous to give the best results.

2. Assuming an average IBV cell culture titer of ~10^6–10^7 PFU/ml, and a desired total of ~10^{10} infectious particles, preparations should be made to produce between 1 and 10 l of cell culture medium. One-well plates (available from Nalgene) can a cost- and space-effective alternative to standard tissue culture flasks for large-scale virus preparation. Due to the simplicity of the proteome of the allantoic fluid of embryonated chicken eggs and the high titer of IBV that can be recovered, lower volumes of starting material can be expected.

3. Early time points containing little virus can be discarded. Both sample purity and virus recovery are dependent on concentration.

4. Virus can be collected at 2 h intervals beginning 48 h after inoculation to increase the final yield. However, for best results, each virus sample should be processed immediately. In general, IBV inoculated into embryonated chickens eggs can be harvested from 16 h after inoculation, but can only be harvested at a single time point.

5. Virus can also be banded at the interface of a 30–50 % two-step sucrose gradient. The use of lower-concentration sucrose cushions can be used to study wild-type virions and virus particles containing defective mini-genomes that are of lower buoyant density [5].

6. For viruses with cleaved spike proteins, such as most strains of IBV, the amount of NaCl added should be reduced to 1 g/100 ml to minimize damage to the spikes.

7. Decanting quickly reduces the likelihood that some components of the pellet will resuspend. Ideally pellets should be separated from supernatants within 1 min of the end of centrifugation.

8. In general, rapid resuspension, cold temperature, and the minimization of mechanical stress will all improve the quality of the preparation. Soluble proteins contained in the pellet may alter the color to a yellowish hue.

9. The translucent virion pellet may be difficult to see and will probably not be visible before the supernatant is removed. The presence of the pellet can be confirmed by fluorescence under UV light after the supernatant is decanted, if the tracking dye was used. If a pellet is present, the viscosity of the added HEPES-saline will increase noticeably upon resuspension.

10. Inactivation techniques should be validated beforehand. Both formaldehyde and 2-propiolactone can lose effectiveness over time. Amines in some buffers such as Tris–HCl will react with formaldehyde, which is why HEPES buffer is recommended for use throughout the purification process.

11. The growth of SARS-CoV, FCoV and MHV is not affected by short-term treatment with VP-SFM, as outlined here. However, cells do grow more slowly in VP-SFM as compared to DMEM, and thus VP-SFM is not recommended for the initial cell culture step.

12. If left in contact with the supernatant, serum-free PEG pellets will resuspended much more quickly than serum-containing pellets, and the sample may be lost. It is therefore very important to decant the supernatants quickly with serum-free pellets.

13. Alternatively, the sample can be applied as a droplet on which the grid is floated for 5–10 min. Grids that continually sink in the saline or stain droplets likely have suffered extensive damage to the support surface and should not be used.

14. The effects of the electron beam on large amounts of residual stain can cause a "blowout" of the carbon-formvar support surface. It is therefore important to remove all excess uranyl acetate before visualization. If there is excess stain present, you will notice because the grid will stick to the forceps. If this happens, give the grid one more blot by gently touching the face of the grid containing the sample directly onto the filter paper, dry the forceps with another piece of filter paper, and try again.

Acknowledgments

Funding for this work was provided by BBSRC and The Pirbright Institute.

References

1. Neuman BW, Joseph JS, Saikatendu KS et al (2008) Proteomics analysis unravels the functional repertoire of coronavirus nonstructural protein 3. J Virol 82:5279–5294

2. Neuman BW, Adair BD, Yeager M et al (2008) Purification and electron cryomicroscopy of coronavirus particles. Methods Mol Biol 454:129–136

3. Neuman BW, Kiss G, Kunding AH et al (2011) A structural analysis of M protein in coronavirus assembly and morphology. J Struct Biol 174: 11–22

4. Neuman BW, Kiss G, Al-Mulla HM et al (2013) Direct observation of membrane insertion by enveloped virus matrix proteins by phosphate displacement. PLoS One 8:e57916

5. Morales L, Mateo-Gomez PA, Capiscol C et al (2013) Transmissible gastroenteritis coronavirus genome packaging signal is located at the 5' end of the genome and promotes viral RNA incorporation into virions in a replication-independent process. J Virol 87: 11579–11590

Chapter 11

Partial Purification of IBV and Subsequent Isolation of Viral RNA for Next-Generation Sequencing

Sarah M. Keep, Erica Bickerton, and Paul Britton

Abstract

RNA viruses are known for a high mutation rate and rapid genomic evolution. As such an RNA virus population does not consist of a single genotype but is rather a collection of individual viruses with closely related genotypes—a quasispecies, which can be analyzed by next-generation sequencing (NGS). This diversity of genotypes provides a mechanism in which a virus population can evolve and adapt to a changing environment. Sample preparation is vital for successful sequencing. The following protocol describes the process of generating a high-quality RNA preparation from IBV grown in embryonated eggs and then partially purified and concentrated through a 30 % sucrose cushion for NGS.

Key words Quasispecies, Next-generation sequencing, RNA, Infectious bronchitis virus (IBV)

1 Introduction

RNA viruses are known for a high mutation rate and rapid genomic evolution. As such an RNA virus population does not consist of a single genotype but is rather a collection of individual viruses with closely related genotypes—a quasispecies [1]. This diversity of genotypes provides a mechanism in which a virus population can evolve and adapt to a changing environment.

It is becoming increasingly important to understand virus population dynamics and the evolution of quasispecies. Standard RT-PCR and sequencing assembly methods in which genomic sequences are generated from the consensus of all aligned reads are not sufficient. To understand the depth of diversity in the population, next-generation sequencing (NGS), commonly referred to as deep sequencing is used. This process involves the parallel sequencing of genomic fragments, generating, depending on the protocol and sample preparation, thousands to millions of short sequencing reads in a single run which ultimately allows for greater coverage at each individual nucleotide.

Helena Jane Maier et al. (eds.), *Coronaviruses: Methods and Protocols*, Methods in Molecular Biology, vol. 1282, DOI 10.1007/978-1-4939-2438-7_11, © Springer Science+Business Media New York 2015

Recent research has utilized NGS to model the mutational dynamics of bovine coronavirus during adaptation to new host environments, revealing the presence of two distinct circulating genotypes that altered in frequency depending on the host cell type [2]. Cotten et al. [3] studied an outbreak of MERS coronavirus (MERS-CoV) using a combination of NGS from clinical samples and phylogenetic analysis to map transmission of the virus within a hospital, providing information about the emergence and evolution of MERS-CoV. A further study incorporating greater numbers of clinical samples determined the evolutionary rate of MERS-CoV and identified regions of the genome that are under positive selection pressure [4].

Sample preparation is key to successful deep sequencing. It is important to enrich the RNA of interest and then to generate a high-quality preparation which is DNA free. The following protocol describes the process of generating a high-quality RNA preparation from IBV grown in embryonated eggs, and then partially purified and concentrated through a 30 % sucrose cushion, which was suitable for use in 454 sequencing requiring a minimum quantity of 200 ng in 19 μl. The growth of IBV in embryonated eggs has been described previously and is not discussed in this protocol (*see* Chapter 7 for further information). It is important to start this protocol with allantoic fluid containing IBV of high titer with 10^6–10^7 pfu/ml preferable. The fluid must also be free of membrane or other solid masses and ideally be free from blood. If the IBV in question causes hemorrhaging of the blood vessels, it is important to harvest the allantoic fluid before this happens.

2   Materials

2.1   Partial Purification of IBV

1. 50 ml Falcon tubes.
2. 30 % sucrose (w/v) in PBS adjusted to pH 7.2 with HCl, filtered through 0.22 μm.
3. Refrigerated benchtop centrifuge.
4. SureSpin 630 rotor and Sorvall OTD65B ultracentrifuge or equivalent.
5. Beckman ultra-clear (25 × 89 mm) ultracentrifuge tubes or equivalent.

2.2   RNA Extraction

1. TRIzol reagent.
2. 75 % ethanol.
3. Isopropanol.
4. Chloroform.

3   Methods

3.1   Partial Purification of IBV

1. Place 25 ml of IBV-infected allantoic fluid into a 50 ml Falcon tube and centrifuge for 10 min, $1,150 \times g$, 4 °C in a benchtop centrifuge (*see* **Note 1**).

2. Take the supernatant and layer on top of 10 ml 30 % sucrose in an ultracentrifuge tube (*see* **Note 2**). Balance the tubes carefully.

3. Centrifuge for 4 h, $102,400 \times g$, 4 °C in an ultracentrifuge.

4. Remove the supernatant in layers, careful not to disturb the pellet (*see* **Note 3**).

5. Wipe the sides of the tube with tissue and proceed directly to the next stage, RNA extraction.

3.2   RNA Extraction

1. Add 1 ml TRIzol reagent directly to the virus pellet from **step 5**, Subheading 3.1. Carefully pipette up and down to mix (*see* **Note 4**).

2. Incubate for 5 min at room temperature (*see* **Note 5**).

3. Add 200 µl chloroform and mix by shaking for 15 s.

4. Incubate for 3 min at room temperature.

5. Centrifuge in a benchtop centrifuge for 15 min, 4 °C, $12,075 \times g$.

6. Carefully take the aqueous top layer, which should be clear, and place in a clean 1.5 ml tube (*see* **Note 6**).

7. Add 0.5 ml isopropanol and incubate for 10 min at room temperature (*see* **Note 7**).

8. Centrifuge for 10 min at $2,100 \times g$, 4 °C in a benchtop centrifuge.

9. Carefully remove the supernatant without disturbing the pellet (*see* **Note 8**).

10. Add 0.75 m 75 % ethanol to the pellet and mix by pipetting.

11. Centrifuge for 10 min, $12,075 \times g$, 4 °C in a benchtop centrifuge.

12. Remove the supernatant, very carefully, and wipe the sides of the tube with tissue (*see* **Note 9**).

13. Air-dry the pellet for 5–10 min (*see* **Note 10**).

14. Resuspend the pellet in 25 µl RNAase-free sterile water (*see* **Note 11**) and store at either –20 or –80 °C.

4   Notes

1. This first spin is to remove any solid matter from the IBV infected allantoic fluid.

2. It is important that the layers do not mix.

3. The pellet will be difficult to see and will have a spectacled appearance. The pellet should not be colored.

4. Make sure the pellet is completely resuspended in the TRIzol reagent.

5. This incubation step allows the TRIzol reagent to break down the virus particles.

6. It is important to take the top aqueous layer only. The interface or organic layer, which will have a cloudy white appearance, contains DNA and protein. It is therefore best to be cautious when taking the top layer and is preferable to leave a little bit behind rather than risk contaminating the RNA sample.

7. When precipitating RNA from small sample quantities RNase-free glycogen can be added at this stage. The glycogen will act as a carrier to the aqueous phase but will be co-precipitated with the RNA. It is important therefore to consider if this will have implications to downstream applications.

8. It is highly unlikely that a pellet will be visible; a little bit of faith is required at this stage. It helps to position the tubes in the centrifuge so that the general location of the pellet can be estimated. It is prudent to keep the supernatant, and store on ice, until you are sure the RNA extraction has been successful.

9. Keep the supernatant, and store on ice, until you are sure the RNA extraction has been successful.

10. Do not allow the RNA to dry completely as the pellet can lose solubility.

11. Make sure the pellet is completely dissolved (partially dissolved RNA samples have an A260/280 ratio <1.6) and then assess the quantity and quality using a NanoDrop or RiboGreen assay.

References

1. Domingo E, Sheldon J, Perales C (2012) Viral quasi species evolution. Microbiol Mol Biol Rev 76:159

2. Borucki MK, Allen JE, Chen-Harris H et al (2013) The role of viral population diversity in adaptation of bovine coronavirus to new host environments. PLoS One 8: e52752

3. Cotten M, Lam TT, Watson SJ et al (2013) Full-genome deep sequencing and phylogenetic analysis of novel human betacoronavirus. Emerg Infect Dis 19:736B–742B

4. Cotten M, Watson S, Zumla A et al (2014) Spread, circulation, and evolution of the Middle East respiratory syndrome coronavirus. MBio 5: pii: e01062–13. doi:10.1128/mBio.01062–13

Part III

Manipulating the Genomes of Coronaviruses

Chapter 12

Transient Dominant Selection for the Modification and Generation of Recombinant Infectious Bronchitis Coronaviruses

Sarah M. Keep, Erica Bickerton, and Paul Britton

Abstract

We have developed a reverse genetics system for the avian coronavirus infectious bronchitis virus (IBV) in which a full-length cDNA corresponding to the IBV genome is inserted into the vaccinia virus genome under the control of a T7 promoter sequence. Vaccinia virus as a vector for the full-length IBV cDNA has the advantage that modifications can be introduced into the IBV cDNA using homologous recombination, a method frequently used to insert and delete sequences from the vaccinia virus genome. Here, we describe the use of transient dominant selection as a method for introducing modifications into the IBV cDNA; this has been successfully used for the substitution of specific nucleotides, deletion of genomic regions, and the exchange of complete genes. Infectious recombinant IBVs are generated in situ following the transfection of vaccinia virus DNA, containing the modified IBV cDNA, into cells infected with a recombinant fowlpox virus expressing T7 DNA-dependent RNA polymerase.

Key words Transient dominant selection (TDS), Vaccinia virus, Infectious bronchitis virus (IBV), Coronavirus, Avian, Reverse genetics, Nidovirus, Fowlpox virus, T7 RNA polymerase

1 Introduction

Avian infectious bronchitis virus (IBV) is a gammacoronavirus that is the aetiological agent of infectious bronchitis (IB); an acute and high contagious disease of poultry. Coronaviruses are enveloped viruses which replicate in the cell cytoplasm. Coronavirus genomes consist of single stranded positive sense RNA, and are the largest of all the RNA viruses ranging from approximately 27–32 kb; the genome of IBV is 27.6 kb. Molecular analysis of the role of individual genes in the pathogenesis of RNA viruses has been advanced by the availability of full-length cDNAs, for the generation of infectious RNA transcripts that can replicate and result in infectious viruses. The assembly of full-length coronavirus cDNAs was hampered due to regions from the replicase gene being unstable in bacteria. We therefore devised a reverse genetics strategy for IBV

Helena Jane Maier et al. (eds.), *Coronaviruses: Methods and Protocols*, Methods in Molecular Biology, vol. 1282, DOI 10.1007/978-1-4939-2438-7_12, © Springer Science+Business Media New York 2015

involving the insertion of a full-length cDNA copy of the IBV genome, under the control of a T7 RNA promoter, into the vaccinia virus genome in place of the thymidine kinase (TK) gene. This was followed by the in situ recovery of infectious IBV in cells both transfected with vaccinia virus DNA and infected with a recombinant fowlpox virus expressing T7 RNA polymerase [1].

One of the main advantages of using vaccinia virus as a vector for IBV cDNA is its ability to accept large quantities of foreign DNA without loss of integrity and stability [2]. A second and equally important advantage is the ability to modify the IBV cDNA within the vaccinia virus vector through transient dominant selection (TDS), a method taking advantage of recombinant events between homologous sequences [3, 4]. The TDS method relies on a three-step procedure. In the first step, the modified IBV cDNA is inserted into a plasmid containing a selective marker under the control of a vaccinia virus promoter. In our case we use a plasmid, pGPTNEB193 (Fig. 1; [5]), which contains a dominant selective marker gene, *Escherichia coli guanine phosphoribosyltransferase* (*Ecogpt*; [6]), under the control of the vaccinia virus P7.5 K early/ late promoter.

In the second step, this complete plasmid sequence is integrated into the IBV sequence within the vaccinia virus genome (Fig. 2). This occurs as a result of a single cross-over event involving

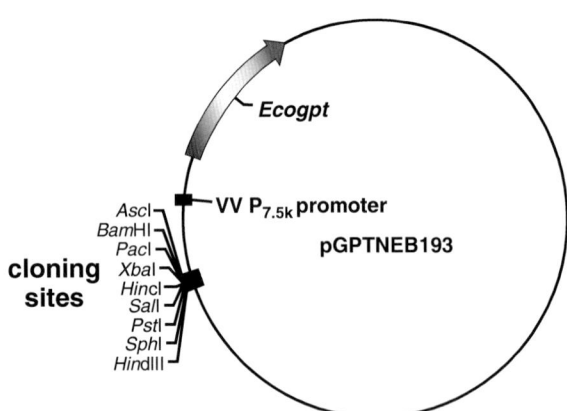

Fig. 1 Schematic diagram of the recombination vector for insertion of genes into a vaccinia virus genome using TDS. Plasmid pGPTNEB193 contains the *Ecogpt* selection gene under the control of the vaccinia virus early/late P$_{7.5K}$ promoter, a multiple cloning region for the insertion of the sequence to be incorporated into the vaccinia virus genome and the *bla* gene (not shown) for ampicillin selection of the plasmid in *E. coli*. For modification of the IBV genome, a sequence corresponding to the region being modified, plus flanking regions of 500–800 nucleotides for recombination purposes is inserted into the multiple cloning sites using an appropriate restriction endonuclease. The plasmid is purified from *E. coli* and transfected into Vero cells previously infected with a recombinant vaccinia virus containing a full-length cDNA copy of the IBV genome

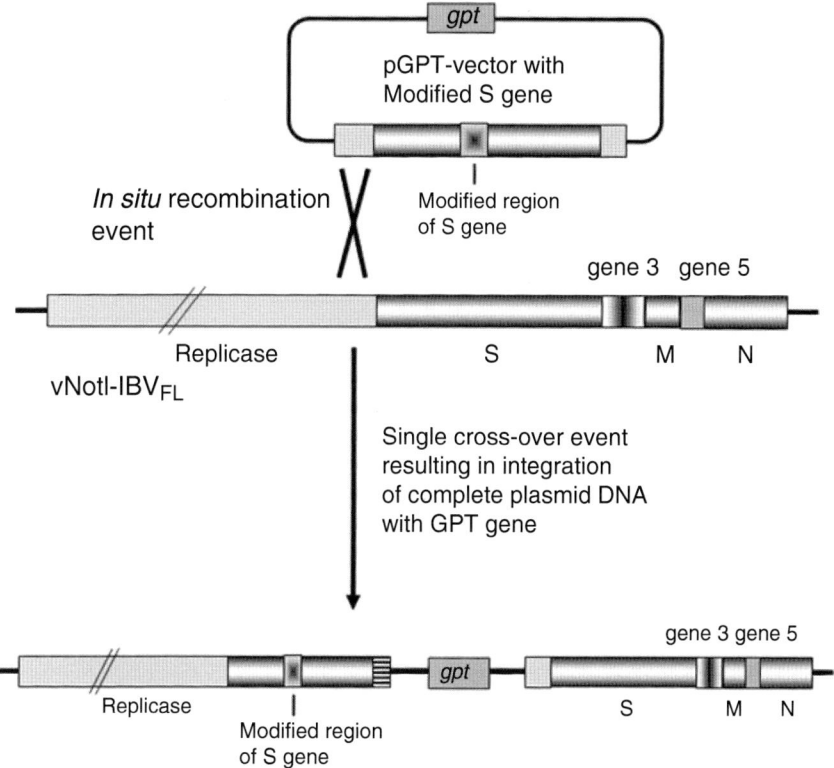

Fig. 2 Schematic diagram demonstrating the TDS method for integrating a modified IBV sequence into the full-length IBV cDNA within the genome of a recombinant vaccinia virus (vNotI-IBVFL). The diagram shows a potential first single-step recombination event between the modified IBV sequence within pGPTNEB193 and the IBV cDNA within vNotI-IBVFL. In order to guarantee a single-step recombination event any potential recombinant vaccinia viruses are selected in the presence of MPA; only vaccinia viruses expressing the *Ecogpt* gene are selected. The main IBV genes are indicated, the replicase, spike (S), membrane (M) and nucleocapsid (N) genes. The IBV gene 3 and 5 gene clusters that express three and two gene products, respectively, are also indicated. In the example shown a modified region of the S gene is being introduced into the IBV genome

homologous recombination between the IBV cDNA in the plasmid and the IBV cDNA sequence in the vaccinia virus genome. The resulting recombinant vaccinia viruses (rVV) are highly unstable due to the presence of duplicate sequences and are only maintained by the selective pressure of the *Ecogpt* gene, which confers resistance to mycophenolic acid (MPA) in the presence of xanthine and hypoxanthine [3]. In the third step, the MPA-resistant rVVs are grown in the absence of MPA selection, resulting in the loss of the *Ecogpt* gene due to a second single homologous recombination event between the duplicated sequences (Fig. 3). During this third step two recombination events can occur; one event will result in the generation of the original (unmodified) IBV sequence and the other in the generation of an IBV cDNA containing the desired modification (i.e., the modification within the

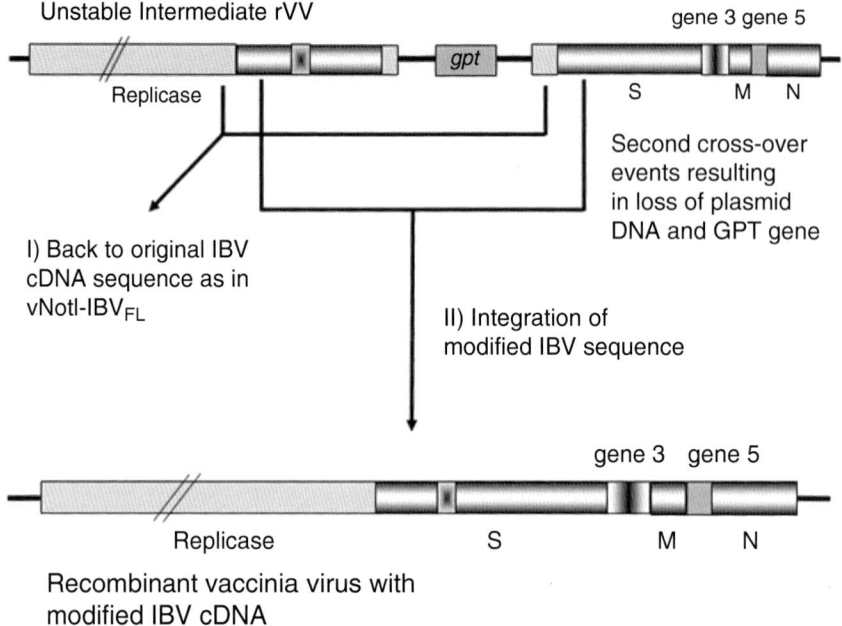

Recombinant vaccinia virus with
modified IBV cDNA

Fig. 3 Schematic diagram demonstrating the second step of the TDS method. Integration of the complete pGPTNEB193 plasmid into the vaccinia virus genome results in an unstable intermediate because of the presence of tandem repeat sequences, in this example the 3′ end of the replicase gene, the S gene and the 5′ end of gene 3. The second single-step recombination event is induced in the absence of MPA; loss of selection allows the unstable intermediate to lose one of the tandem repeat sequences including the *Ecogpt* gene. The second step recombination event can result in either (I) the original sequence of the input vaccinia virus IBV cDNA sequence, in this case shown as a recombination event between the two copies of the 3′ end of the replicase gene which results in loss of the modified S gene sequence along with *Ecogpt* gene; or (II) retention of the modified S gene sequence and loss of the original S gene sequence and *Ecogpt* gene as a result of a potential recombination event between the two copies of the 5′ end of the S gene sequence. This event results in a modified S gene sequence within the IBV cDNA in a recombinant vaccinia virus

plasmid sequence). In theory these two events will occur at equal frequency however in practice this is not necessarily the case.

To recover infectious rIBVs from the rVV vector, rVV DNA is transfected into primary chick kidney (CK) cells previously infected with a recombinant fowlpox virus expressing T7 RNA polymerase (rFPV-T7; [7]). In addition, a plasmid, pCi-Nuc [1, 8], expressing the IBV nucleoprotein (N), under the control of both the cytomegalovirus (CMV) RNA polymerase II promoter and the T7 RNA promoter, is co-transfected into the CK cells. Expression of T7 RNA polymerase in the presence of the IBV N protein and the rVV DNA, containing the full-length IBV cDNA under the control of a T7 promoter, results in the generation of infectious IBV RNA, which in turn results in the production of infectious rIBVs (Fig. 4).

The overall procedure is a multistep process which can be divided into two parts: the generation of an rVV containing the modified IBV cDNA (Fig. 5) and the recovery of infectious rIBV

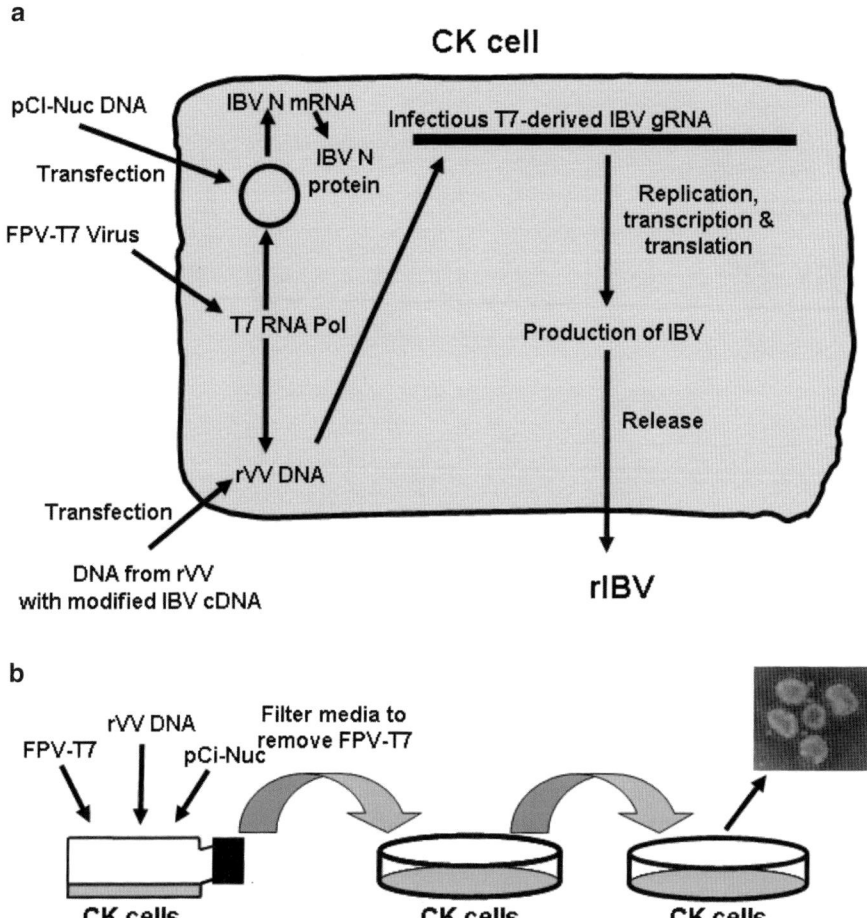

a

CK cell

pCI-Nuc DNA

Transfection

FPV-T7 Virus

IBV N mRNA

IBV N protein

Infectious T7-derived IBV gRNA

Replication, transcription & translation

T7 RNA Pol

Production of IBV

Release

rVV DNA

Transfection

DNA from rVV with modified IBV cDNA

rIBV

b

FPV-T7

rVV DNA

pCi-Nuc

Filter media to remove FPV-T7

CK cells
P₀

CK cells
P₁

CK cells
P₂

Fig. 4 A schematic representation of the recovery process for obtaining rIBV from DNA isolated from a recombinant vaccinia virus containing a full-length IBV cDNA under the control of a T7 promoter. (**a**) In addition to the vaccinia virus DNA containing the full-length IBV cDNA under the control of a T7 promoter a plasmid, pCI-Nuc, expressing the IBV nucleoprotein, required for successful rescue of IBV, is transfected into CK cells previously infected with a recombinant fowl pox virus, FPV-T7, expressing T7 RNA polymerase. The T7 RNA polymerase results in the synthesis of an infectious RNA from the vaccinia virus DNA that consequently leads to the generation of infectious IBV being released from the cell. (**b**) Any recovered rIBV present in the media of P_0 CK cells is used to infect P_1 CK cells. The media is filtered through a 0.22 μm filter to remove any FPV-T7 virus. IBV-induced CPE is normally observed in the P_1 CK cells following a successful recovery experiment. Any rIBV is passaged a further two times, P_2 and P_3, in CK cells. Total RNA is extracted from the P_1 to P_3 CK cells and the IBV-derived RNA analyzed by RT-PCR for the presence of the required modification

from the rVV vector (Fig. 4). The generation of the *Ecogpt* plasmids, based on pGPTNEB193, containing the modified IBV cDNA, is by standard *E. coli* cloning methods [9, 10] and is not described here. General methods for growing vaccinia virus have been published by Mackett et al. [11] and for using the TDS method for modifying the vaccinia virus genome by Smith [12].

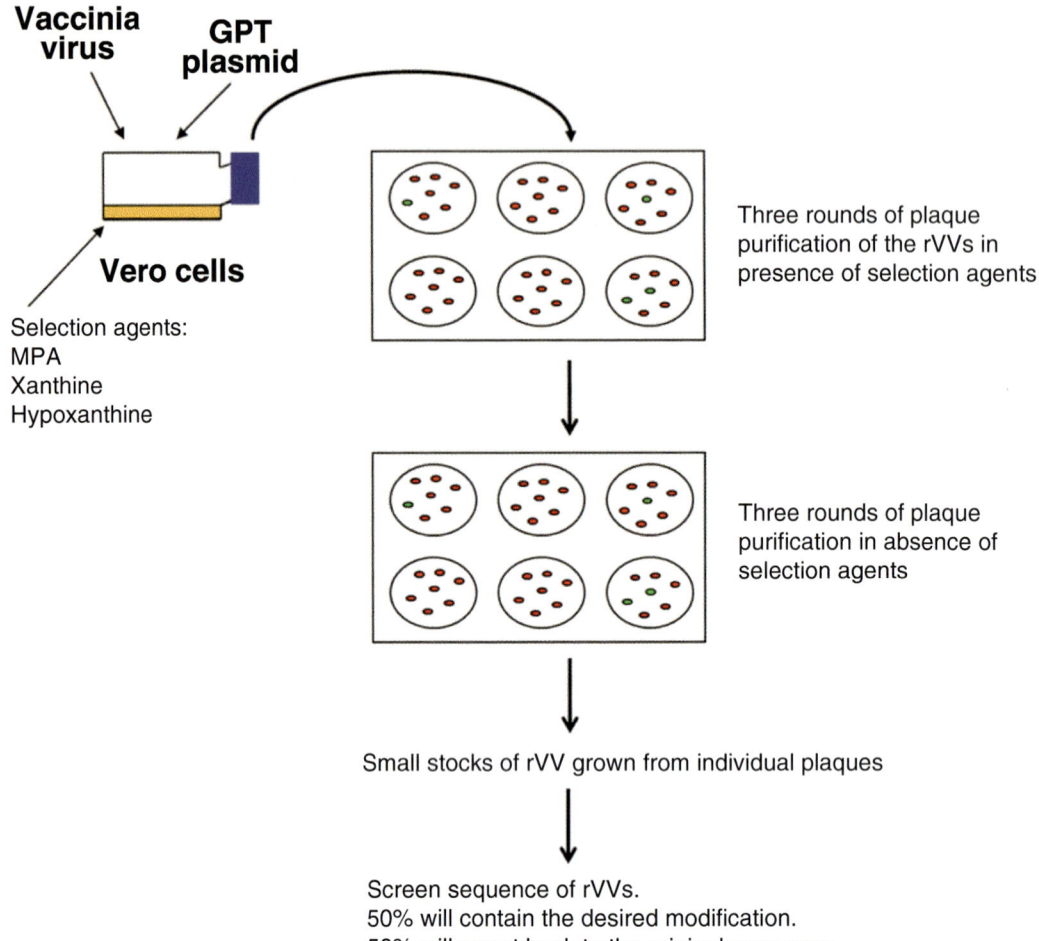

Fig. 5 Schematic detailing the multistep process of constructing a recombinant vaccinia virus. Vero cells are infected with rVV containing IBV cDNA and then transfected with a plasmid containing the IBV sequence to be inserted and the selective marker gene *Ecogpt*. Homologous recombination occurs and the complete plasmid sequence is inserted into the rVV. The *Ecogpt* gene allows positive selection of these rVV as it confers resistance to MPA in the presence of xanthine and hypoxanthine. The viruses are plaque purified three times in the presence of selection agents ensuring no wild type VV is present. The removal of the selection agents results in a second recombination event with the loss of the *Ecogpt* gene. Plaque purification in the absence of selection agents not only ensures the loss of the GPT gene but also ensures the maintenance of a single viral population. Small stocks of rVV are grown from individual plaques which are screened through PCR for the desired modification; this is found in theoretically 50 % of rVVs

2   Materials

2.1   Homologous Recombination and Transient Dominant Selection in Vero Cells

1. Vero cells.

2. PBSa: 172 mM NaCl, 3 mM KCl, 10 mM Na$_2$HPO$_4$, and 2 mM KH$_2$PO$_4$, adjusted to pH 7.2 with HCl.

3. 1× Eagle's Minimum Essential Medium (E-MEM) with Earle's salts, 2 mM L-glutamine, and 2.2 g/l sodium bicarbonate.

4. BES medium: 1× E-MEM, 0.3 % tryptose phosphate broth (TPB), 0.2 % bovine serum albumin (BSA), 20 mM N,N-Bis(2-hydroxyethyl)-2-aminoethanesulfonic acid (BES), 0.21 % sodium bicarbonate, 2 mM L-glutamine, 250 U/ml nystatin, 100 U/ml penicillin, and 100 U/ml streptomycin.

5. Opti-MEM 1 with GlutaMAX-1 (Life Technologies).

6. Lipofectin (Life Technologies).

7. Mycophenolic acid (MPA): 10 mg/ml in 0.1 M NaOH (30 mM); 400× concentrated.

8. Xanthine: 10 mg/ml in 0.1 M NaOH (66 mM); 40× concentrated. Heat at 37 °C to dissolve.

9. Hypoxanthine: 10 mg/ml in 0.1 M NaOH (73 mM); 667× concentrated.

10. Screw-top 1.5 ml microfuge tubes with gasket.

11. Cup form sonicator.

12. 2× E-MEM: 2× E-MEM, 10 % fetal calf serum, 0.35 % sodium bicarbonate, 4 mM L-glutamine, 1,000 U/ml nystatin, 200 U/ml penicillin, and 200 U/ml streptomycin.

13. 2 % agar.

14. *Ecogpt* selection medium: 1× E-MEM, 75 µM MPA, 1.65 mM xanthine, 109 µM hypoxanthine, 1 % agar (*see* **Note 7**).

15. Overlay medium: 1× E-MEM, 1 % agar.

16. 1 % Neutral red solution (H_2O).

2.2 Extraction of DNA from Recombinant Vaccinia Virus

1. 20 mg/ml Proteinase K.

2. 2× proteinase K buffer: 200 mM Tris/HCl pH 7.5, 10 mM EDTA, 0.4 % SDS, 400 mM NaCl.

3. Phenol–chloroform–isoamyl alcohol (25:24:1).

4. Chloroform.

5. Absolute ethanol.

6. 70 % ethanol.

7. QIAamp DNA mini kit (QIAGEN).

8. 3 M sodium acetate.

2.3 Production of Large Stocks of Vaccinia Virus

1. BHK-21 maintenance medium: Glasgow-Modified Eagle's Medium (G-MEM), 2 mM L-glutamine, 0.275 % sodium bicarbonate, 1 % fetal calf serum, 0.3 % TPB, 500 U/ml nystatin, 100 U/ml penicillin, and 100 U/ml streptomycin.

2. TE buffer: 10 mM Tris–HCl pH 9, 1 mM EDTA.

3. BHK-21 cells.

4. 50 ml Falcon tubes.

2.4 Vaccinia Virus Partial Purification

1. 30 % sucrose (w/v) in 1 mM Tris/HCl pH 9, filtered through 0.22 μm.

2. Superspin 630 rotor and Sorvall OTD65B ultracentrifuge or equivalent.

2.5 Analysis of Vaccinia Virus DNA by Pulse Field Agarose Gel Electrophoresis

1. 10× TBE buffer: 1 M Tris, 0.9 M Boric acid pH 8 and 10 mM EDTA.

2. Pulsed field certified ultrapure DNA grade agarose.

3. DNA markers (e.g., 8–48 kb markers, Bio-Rad).

4. 0.5 mg/ml ethidium bromide.

5. CHEF-DR® II pulsed field gel electrophoresis (PFGE) apparatus (Bio-Rad) or equivalent.

6. 6× sample loading buffer: 62.5 % glycerol, 62.5 mM Tris–HCl pH 8, 125 mM EDTA and 0.06 % bromophenol blue.

2.6 Preparation of rFPV-T7 Stock Virus

1. Chicken embryo fibroblast (CEF) cells.

2. CEF maintenance medium: 1× 199 Medium with Earle's Salts, 0.3 % TPB, 2 % new born calf serum (NBCS), 0.225 % sodium bicarbonate, 2 mM L-glutamine, 100 U/ml penicillin, 100 U/ml streptomycin, and 500 U/ml nystatin.

2.7 Recovery of rIBV and Serial Passage on CK Cells

1. Chick kidney (CK) cells.

2. Stock of rFPV-T7 virus.

3. The rVV DNA prepared from large partially purified stocks of rVV.

4. Plasmid pCi-Nuc which contains IBV nucleoprotein under the control of the CMV and T7 promoters.

5. 0.22 μm syringe driven filters.

6. 5 ml syringes.

---

3 Methods

3.1 Infection/ Transfection of Vero Cells with Vaccinia Virus

1. Freeze-thaw the vaccinia virus containing the full-length IBV cDNA genome to be modified three times (37 °C/dry ice) and sonicate for 2 min using a cup form sonicator, continuous pulse at 70 % duty cycle, seven output control (*see* **Notes 1–4**).

2. Infect six-well plates of 40 % confluent monolayers of Vero cells with the rVV at a multiplicity of infection (MOI) of 0.2. Use two independent wells per recombination (*see* **Notes 1–4**).

3. Incubate at 37 °C 5 % CO_2 for 2 h to allow the virus to infect the cells.

4. After 1 h of incubation, prepare the following solutions for transfection:

Solution A: For each transfection: Dilute 5 μg of modified pGPTNEB193 (containing the modified IBV cDNA) in 1.5 ml of Opti-MEM medium.

Solution B: Dilute 12 μl of Lipofectin in 1.5 ml of Opti-MEM for each transfection.

5. Incubate solutions A and B separately for 30 min at room temperature, then mix the two solutions together and incubate the mixture at room temperature for 15 min.

6. During the 15 min incubation, remove the inoculum from the vaccinia virus infected cells and wash the cells twice with Opti-MEM.

7. Add 3 ml of the transfection mixture (prepared in **step 5**) to each well.

8. Incubate for 60–90 min at 37 °C 5 % CO_2 (*see* **Note 5**).

9. Remove the transfection mixture from each well and replace it with 5 ml of BES medium.

10. Incubate the transfected cells overnight at 37 °C, 5 % CO_2.

11. The following morning add the MXH selection components, MPA 12.5 μl, xanthine 125 μl, and hypoxanthine 7.4 μl, directly to each well (*see* **Note 6**).

12. Incubate the cells at 37 °C 5 % CO_2 until they display extensive vaccinia virus induced cytopathic effect (CPE) (normally 2 days).

13. Harvest the infected/transfected cells into the cell medium of the wells and centrifuge for 3–4 min at $300 \times g$. Discard supernatant and resuspend the pellet in 400 μl 1× E-MEM and store at –20 °C.

3.2 Plaque Purification in the Presence of GPT Selection Agents: Selection of MPA Resistant Recombinant Vaccinia Viruses (GPT+ Phenotype)

1. Freeze-thaw the vaccinia virus produced from Subheading 3.1 three times and sonicate as described in the previous section (Subheading 3.1, **step 1**).

2. Remove the medium from confluent Vero cells in six-well plates and wash the cells once with PBSa.

3. Prepare 10^{-1} to 10^{-3} serial dilutions of the recombinant vaccinia virus in 1× E-MEM.

4. Remove the PBSa from the Vero cells and add 500 μl of the diluted virus per well.

5. Incubate for 1–2 h at 37 °C 5 % CO_2.

6. Remove the inoculum and add 3 ml of the *Ecogpt* selection medium (*see* **Note 7**).

7. Incubate for 3–4 days at 37 °C 5 % CO_2 and stain the cells by adding 2 ml of 1× E-MEM containing 1 % agar and 0.01 % neutral red.

8. Incubate the cells at 37 °C 5 % CO_2 for 6–24 h and pick two to three well isolated plaques for each recombinant, by taking a plug of agarose directly above the plaque. Place the plug of agar in 400 µl of 1× E-MEM.

9. Perform two further rounds of plaque purification for each selected recombinant vaccinia virus in the presence of *Ecogpt* selection medium, as described in **steps 1–8** (*see* **Note 8**).

3.3 Plaque Purification in the Absence of GPT Selection Agents: Selection of MPA Sensitive Recombinant Vaccinia Viruses (Loss of GPT⁺ Phenotype)

1. Take the MPA resistant plaque-purified rVVs which have been plaque purified a total of three times as described in Subheading 3.2 and freeze-thaw and sonicate as described in Subheading 3.1, **step 1**.

2. Remove the medium from confluent Vero cells in six-well plates and wash the cells with PBSa.

3. Prepare 10^{-1} to 10^{-3} serial dilutions of the recombinant vaccinia virus in 1× E-MEM.

4. Remove the PBSa from the Vero cells and add 500 µl of the diluted virus per well.

5. Incubate for 1–2 h at 37 °C 5 % CO_2.

6. Remove the inoculum and add 3 ml of the overlay medium (*see* **Note 9**).

7. Incubate for 3–4 days at 37 °C 5 % CO_2 and stain the cells by adding 2 ml 1× E-MEM containing 1 % agar and 0.01 % neutral red.

8. Incubate the cells at 37 °C 5 % CO_2 for 6–24 h and pick three to six well-isolated plaques for each recombinant, by taking a plug of agar directly above the plaque. Place the plug of agar in 400 µl of 1× E-MEM (*see* **Note 8**).

9. Perform two further rounds of plaque purification for each selected recombinant vaccinia virus in the presence of selection medium, as described in **steps 1–8**.

3.4 Production of Small Stocks of Recombinant Vaccinia Viruses

1. Take the MPA sensitive plaque-purified rVVs which have been plaque purified a total of three times as described in Subheading 3.3 and freeze-thaw and sonicate as described in Subheading 3.1, **step 1**.

2. Remove the medium from confluent Vero cells in six-well plates and wash the cells with PBSa.

3. Dilute 150 µl of the sonicated rVVs in 350 µl of BES medium.

4. Remove the PBSa from the Vero cells and add 500 µl of the diluted rVVs per well.

5. Incubate at 37 °C and 5 % CO_2 for 1–2 h.

6. Add 2.5 ml per well of BES medium.

7. Incubate the infected Vero cells at 37 °C and 5 % CO_2 until the cells show signs of extensive vaccinia virus-induced CPE (approx. 4 days).

8. Scrape the Vero cells into the medium, and harvest into 1.5 ml screw cap tubes with gaskets.

9. Centrifuge for 3 min at 16,000×g in a bench top centrifuge.

10. Discard the supernatants and resuspend the cells in a total of 400 μl of BES cell culture medium and store at –20 °C.

3.5   DNA Extraction from Small Stocks of Recombinant Vaccinia Virus for Screening by PCR

There are two methods for DNA extraction:

3.5.1   DNA Extraction Using Phenol–Chloroform–Isoamyl Alcohol

1. To 100 μl of rVV stock produced in Subheading 3.4, add 100 μl 2× proteinase K buffer and 2 μl of the proteinase K stock. Gently mix and incubate at 50 °C for 2 h.

2. Add 200 μl of phenol–chloroform–isoamyl alcohol to the proteinase K-treated samples and mix by inverting the tube five to ten times and centrifuge at 16,000×g for 5 min (*see* **Note 10**).

3. Take the upper aqueous phase and repeat **step 2** twice more.

4. Add 200 μl of chloroform to the upper phase and mix and centrifuge as in **step 2**.

5. Take the upper phase and precipitate the vaccinia virus DNA by adding 2.5 volumes of absolute ethanol; the precipitated DNA should be visible. Centrifuge the precipitated DNA at 16,000×g for 20 min. Discard the supernatant.

6. Wash the pelleted DNA with 400 μl 70 % ethanol and centrifuge at 16,000×g for 10 min. Discard the supernatant, carefully, and remove the last drops of 70 % ethanol using a capillary tip.

7. Resuspend the DNA in 30 μl water and store at 4 °C (*see* **Note 11**).

3.5.2   Extraction of rVV DNA Using the Qiagen QIAamp DNA Mini Kit

1. Follow the blood/bodily fluids spin protocol and start with 200 μl of rVV stock produced in Subheading 3.4.

2. Elute the rVV DNA in 200 μl buffer AE (provided in the kit) and store at 4 °C.

At this stage the extracted rVV DNA is analyzed by PCR and/or sequence analysis for the presence/absence of the *Ecogpt* gene and for the modifications within the IBV cDNA sequence.

Once an rVV is identified that has both lost the *Ecogpt* gene and also contains the desired IBV modification, large stocks are produced. Typically two rVVs will be taken forward at this stage, which ideally have been generated from different wells of the infection/transfection of Vero cells stage previously described in Subheading 3.1. Once the large stocks of the chosen rVVs have been produced, rVV DNA will be extracted and prepared for the recovery of rIBV.

3.6 Production of Large Stocks of Vaccinia Virus

1. Freeze-thaw and sonicate the chosen rVV stocks from Subheading 3.4 as described in Subheading 3.1, **step 1**.

2. Dilute the sonicated virus in BHK-21 maintenance medium and infect $11 \times T150$ flasks of confluent monolayers of BHK-21 cells using 2 ml of the diluted vaccinia virus per flask at a MOI of 0.1–1.

3. Incubate the infected cells for 1 h at 37 °C and 5 % CO_2.

4. Add 18 ml of pre-warmed (37 °C) BHK-21 maintenance medium and incubate the infected cells at 37 °C and 5 % CO_2 until the cells show an advanced CPE (normally about 2–3 days post-infection). At this stage the cells should easily detach from the plastic.

5. Either continue to **step 6** or freeze the flasks in plastic boxes lined with absorbent material and labeled with biohazard tape at –20 °C until further use.

6. If prepared from frozen, the flasks need to be defrosted by leaving them at room temperature for 15 min and then at 37 °C until the medium over the cells has thawed.

7. Tap the flasks to detach the cells from the plastic, if necessary use a cell scraper.

8. Transfer the medium containing the cells to 50 ml Falcon tubes and centrifuge at $750 \times g$ for 15 min at 4 °C to pellet the cells.

9. Discard the supernatant (99 % of vaccinia virus is cell-associated) and resuspend the cells in 1 ml of TE buffer per flask.

10. Pool the resuspended cells then aliquot into screw top microfuge tubes with gasket and store at –70 °C.

11. Use one 1 ml aliquot of the resuspended cells as a virus stock. Use the resuspended cells from the remaining ten flasks for partial purification.

3.7 Vaccinia Virus Partial Purification

1. Freeze-thaw and sonicate the resuspended cells generated from Subheading 3.6 as described in Subheading 3.1, **step 1**.

2. Centrifuge at $750 \times g$ for 10 min at 4 °C to remove the cell nuclei.

3. Keep the supernatant and add TE buffer to give a final volume of 13 ml.

4. Add 16 ml of the 30 % sucrose solution into a Beckman ultra-clear (25 × 89 mm) ultracentrifuge tube and carefully layer 13 ml of the cell lysate from **step 3** on to the sucrose cushion.

5. Centrifuge the samples using an ultracentrifuge at 36,000 ×*g*, 4 °C for 60 min.

6. The partially purified vaccinia virus particles form a pellet under the sucrose cushion. After centrifugation, carefully remove the top layer (usually pink) and the sucrose layer with a pipette. Wipe the sides of the tube carefully with a tissue to remove any sucrose solution.

7. Resuspend each pellet in 5 ml TE buffer and store at −70 °C.

3.8 Extraction of Vaccinia Virus DNA from Large Partially Purified rVV Stocks

1. Defrost the partially purified vaccinia virus from Subheading 3.7 at 37 °C.

2. Add 5 ml of pre-warmed 2× proteinase K buffer and 100 μl of 20 mg/ml proteinase K to the partially purified vaccinia virus in a 50 ml Falcon tube. Incubate at 50 °C for 2.5 h (*see* **Notes 1–4**).

3. Transfer into a clean 50 ml Falcon tube.

4. Add 5 ml of phenol–chloroform–isoamyl alcohol, mix by inverting the tube five to ten times, and centrifuge at 1,100 ×*g* in a benchtop centrifuge for 15 min at 4 °C. Transfer the upper phase to a clean 50 ml Falcon tube using wide-bore pipette tips (*see* **Notes 10** and **11**).

5. Repeat **step 3**.

6. Add 5 ml chloroform, mix by inverting the tube five to ten times, and centrifuge at 1,100 ×*g* for 15 min at 4 °C. Transfer the upper phase into a clean 50 ml Falcon tube.

7. Precipitate the vaccinia virus DNA by adding 2.5 volumes of −20 °C absolute ethanol and 0.1 volumes of 3 M sodium acetate. Centrifuge at 1,200 ×*g*, 4 °C for 60–90 min. A glassy pellet should be visible.

8. Discard the supernatant and wash the DNA using 10 ml −20 °C 70 % ethanol. Leave on ice for 5 min and centrifuge at 1,200 ×*g*, 4 °C for 30–45 min. Discard the supernatant and remove the last drops of ethanol using a capillary tip. Dry the inside of the tube using a tissue to remove any ethanol.

9. Air-dry the pellet for 5–10 min.

10. Resuspend the vaccinia DNA in 100 μl of water. Do not pipette to resuspend as shearing of the DNA will occur.

11. Leave the tubes at 4 °C overnight. If the pellet has not dissolved totally, add more water.

12. Measure the concentration of the extracted DNA using a NanoDrop or equivalent.

13. Store the vaccinia virus DNA at 4 °C. DO NOT FREEZE (*see* **Note 6**).

3.9 Analysis of Vaccinia Virus DNA by Pulsed Field Agarose Gel Electrophoresis (PFGE)

1. Prepare 2 l of 0.5×TBE buffer for preparation of the agarose gel and as electrophoresis running buffer; 100 ml is required for a 12.7×14 cm agarose gel and the remainder is required as running buffer.

2. Calculate the concentration of agarose that is needed to analyze the range of DNA fragments to be analyzed. Increasing the agarose concentration decreases the DNA mobility within the gel, requiring a longer run time or a higher voltage. However, a higher voltage can increase DNA degradation and reduce resolution. A 0.8 % agarose gel is suitable for separating DNA ranging between 50 and 95 kb. A 1 % agarose gel is suitable for separating DNA ranging between 20 and 300 kb.

3. Place the required amount of agarose in 100 ml 0.5× TBE buffer and microwave until the agarose is dissolved. Cool to approximately 50–60 °C.

4. Clean the gel frame and comb with MQ water followed by 70 % ethanol. Place the gel frame on a level surface, assemble the comb and pour the cooled agarose into the gel frame. Remove any bubbles using a pipette tip and allow the agarose to set (approx. 30–40 min) and store in the fridge until required.

5. Place the remaining 0.5× TBE buffer into the CHEF-DR® II PFGE electrophoresis tank and switch the cooling unit on. Leave the buffer circulating to cool.

6. Digest 1 μg of the DNA with a suitable restriction enzyme such as Sal I in a 20 μl reaction.

7. Add the sample loading dye to the digested vaccinia virus DNA samples and incubate at 65 °C for 10 min.

8. Place the agarose gel in the electrophoresis chamber; load the samples using wide bore tips and appropriate DNA markers (*see* **Note 11**).

9. The DNA samples are analyzed by PFGE at 14 °C in gels run with a 0.1–1.0 s switch time for 16 h at 6 V/cm at an angle of 120° or with a 3.0–30.0 s switch time for 16 h at 6 V/cm depending on the concentration of agarose used.

10. Following PFGE, place the agarose gel in a sealable container containing 400 ml 0.1 μg/ml ethidium bromide and gently shake for 30 min at room temperature.

11. Wash the ethidium bromide-stained agarose gel in 400 ml MQ water by gently shaking for 30 min.

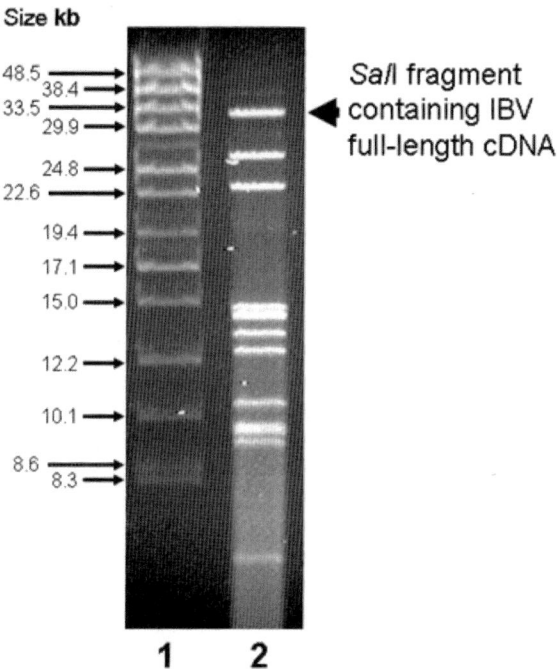

Size kb

Fig. 6 Analysis of Sal I digested vaccinia virus DNA by PFGE. *Lane 1* shows DNA markers and *Lane 2* the digested vaccinia virus DNA. The IBV cDNA used does not contain a Sal I restriction site; therefore the largest DNA fragment (~31 kb) generated from the recombinant vaccinia virus DNA represents the IBV cDNA with some vaccinia virus-derived DNA at both ends

12. Visualize DNA bands using a suitable UV system for analyzing agarose gels. An example of recombinant vaccinia virus DNA digested with the restriction enzyme *Sal I* and analyzed by PFGE is shown in Fig. 6.

3.10 Preparation of rFPV-T7 Stock

Infectious recombinant IBVs are generated in situ by co-transfection of vaccinia virus DNA, containing the modified IBV cDNA, and pCi-Nuc (a plasmid containing the IBV N gene) into CK cells previously infected with a recombinant fowlpox virus expressing the bacteriophage T7 DNA-dependent RNA polymerase under the direction of the vaccinia virus P7.5 early-late promoter 8 (rFPV-T7). This protocol covers the procedure for preparing a stock of rFPV/T7 by infecting primary avian chicken embryo fibroblasts (CEFs).

Preparation of a 200 ml stock of rFPV-T7 uses ten T150 flasks containing confluent monolayers of CEFs.

1. Remove the culture growth medium from the cells and infect with 2 ml rFPV/T7 at a MOI of 0.1, previously diluted in CEF maintenance medium.

2. Incubate the infected cells for 1 h at 37 °C 5 % CO_2 then without removing the inoculum add 20 ml of CEF maintenance medium.

3. After 4 days post infection check for CPE (90 % of the cells should show CPE). Tap the flasks to detach the cells from the plastic and disperse the cells into the medium by pipetting them up and down.

4. Harvest into 50 ml Falcon tubes and freeze-thaw the cells three times as described in Subheading 3.1, **step 1**.

5. Centrifuge at $750 \times g$, 4 °C for 5 min to remove the cell debris. Take the supernatant containing the virus stock and store at −70 °C until required.

6. Determine the titre of the virus stock using CEF cells. The titre should be in the order of 10^6–10^7 PFU/ml.

3.11 Infection and Transfection of CK Cells for the Recovery of rIBV

1. Wash 40 % confluent CK cells in six-well plates once with PBSa.

2. Infect the cells with rFPV-T7 at a MOI of 10 in 1 ml of CK cell culture medium. Typically we carry out ten replicates per recovery experiment.

3. Incubate for 1 h at 37 °C 5 % CO_2.

4. During this infection period prepare the transfection reaction solutions.

   Solution A: 1.5 ml Opti-MEM, 10 μg rVV DNA, and 5 μg pCi-Nuc per replicate.

   Solution B: 1.5 ml Opti-MEM and 30 μl Lipofectin per replicate.

5. Incubate solutions A and B at room temperature for 30 min.

6. Mix solutions A and B together producing solution AB, and incubate for a further 15 min at room temperature.

7. Remove the rFPV-T7 from each well and wash the CK cells twice with Opti-MEM and carefully add 3 ml of solution AB per well.

8. Incubate the transfected cells at 37 °C 5 % CO_2 for 16–24 h.

9. Remove the transfection medium from each well and replace with 5 ml of BES medium and incubate at 37 °C 5 % CO_2.

10. Two days after changing the transfection media, when FPV/IBV-induced CPE is extensive, harvest the cell supernatant from each well and using a 5 ml syringe, filter through 0.22 μm to remove any rFPV-T7 virus present.

11. Store the filtered supernatant, referred to as passage 0 (P_0 CKC) supernatant at −70 °C.

3.12 Serial Passage of rIBVs in CK Cells

To check for the presence of any recovered rIBVs the P_0 CKC supernatant is passaged three times, P_1 to P_3, in CK cells (Fig. 4b). At each passage the cells are checked for any IBV-associated CPE and for further confirmation RNA is extracted from P_3 CKC supernatant and is analyzed by RT-PCR (*see* **Note 12**).

For passage 1 (P_1):

1. Wash the confluent CK cells in six-well plates once with PBSa.

2. Add 1 ml of the P_0 CKC supernatant per well and incubate at 37 °C 5 % CO_2 for 1 h.

3. Without removing the inoculum add 2 ml of BES medium per well.

4. Check cells for IBV-associated CPE over the next 2–3 days using a bright-field microscope.

5. Harvest the supernatant from each well and store at –70 °C.

6. Repeat **steps 1–6** for passages P_2 and P_3 in CK cells.

7. At P_3 any recovered virus is used to prepare a large stock for analysis of the virus genotype and phenotype.

4   Notes

1. Vaccinia virus is classified as a category 2 human pathogen, and its use is therefore subject to local regulations and rules that have to be followed.

2. Always discard any medium of solution containing vaccinia virus into a 1 % solution of Virkon, leave at least 12 h before discarding.

3. Flasks of cells infected with vaccinia virus should be kept in large plastic boxes, which should be labeled with the word vaccinia and biohazard tape. A paper towel should be put on the bottom of the boxes to absorb any possible spillages.

4. During centrifugation of vaccinia virus infected cells use sealed buckets for the centrifugation to avoid possible spillages.

5. After 2 h of incubation with the transfection mixture, the cells begin to die. It is best therefore not to exceed 90 min incubation.

6. It is important that after the addition of each selection agent, the medium is mixed to ensure the selection agents are evenly distributed. This can be achieved by *gently* rocking/swirling the plate.

7. Add an equal volume of 2 % agar to the 2× EMEM containing MPA, xanthine and hypoxanthine and mix well before adding it to vaccinia virus infected cells. There is skill to making the overlay medium and adding it to the cells before the agar sets. There are a number of methods including adding hot agar to cold medium, or pre warming the medium to 37 °C and adding agar which has been incubated at 50 °C. Despite the method chosen it is important that all components of the overlay medium are mixed well, and the medium is not too hot when it is added to the cells. If there are problems, 1 % agar can be substituted with 1 % low melting agarose.

8. The first recombinant event in the TDS system will not necessarily occur in the same place in every rVV. It is therefore important to pick a number of plaques from the first round of plaque purification in presence of GPT selection agents and take a variety of them forward. The following two rounds of plaque purification in the presence of GPT selection agents ensure a single virus population and also that no carry through of the input receiver/wild type vaccinia virus has occurred.

9. Previous chapters and protocols have instructed during plaque purification in the absence of GPT selection agents to plate 10^{-1} rVV dilution in the presence of GPT selection medium and rVV dilutions 10^{-2} and 10^{-3} in the absence. When there are no plaques in the 10^{-1} dilution, it means that the rVV has lost the GPT gene and the plaques are ready to amplify and check for the presence of mutations.

10. There are risks associated with working with phenol–chloroform–isoamyl alcohol and chloroform. It is important to check the local COSHH guidelines and code of practices.

11. Vaccinia virus DNA is a very large molecule that is very easy to shear, therefore when working with the DNA be gentle and use wide bore tips or cut the ends off ordinary pipette tips. In addition always store vaccinia virus DNA at 4 °C; do not freeze as this leads to degradation. However, there is an exception to this if the vaccinia virus DNA has been extracted using the Qiagen QIAamp DNA mini kit, as this DNA will have already been sheared (the kit only purifies intact DNA fragments up to 50 bp). This DNA can be stored at −20 °C but it is only suitable for analysis of the rVV genome by PCR and is *not* suitable for the infection and transfection of CK cells for the recovery of rIBV.

12. There is always the possibility that the recovered rIBV is not cytopathic. In this case, check for the presence of viral RNA by RT-PCR at passage 3 (P_3). It is quite common even with a cytopathic rIBV not to see easily definable IBV induced CPE at P_1 and P_2. The recovery process is a low probability event and the serial passage of rIBVs in CK cells acts as an amplification step.

References

1. Casais R, Thiel V, Siddell SG et al (2001) Reverse genetics system for the avian coronavirus infectious bronchitis virus. J Virol 75:12359–12369

2. Thiel V, Siddell SG (2005) Reverse genetics of coronaviruses using vaccinia virus vectors. Curr Top Microbiol Immunol 287:199–227

3. Falkner FG, Moss B (1990) Transient dominant selection of recombinant vaccinia viruses. J Virol 64:3108–3111

4. Britton P, Evans S, Dove B et al (2005) Generation of a recombinant avian coronavirus infectious bronchitis virus using transient dominant selection. J Virol Meth 123:203–211

5. Boulanger D, Green P, Smith T et al (1998) The 131-amino-acid repeat region of the essential 39-kilodalton core protein of fowlpox virus FP9, equivalent to vaccinia virus A4L protein, is nonessential and highly immunogenic. J Virol 72:170–179

6. Mulligan R, Berg P (1981) Selection for animal cells that express the E. coli gene coding for xanthine-guanine phosphoribosyl transferase. Proc Natl Acad Sci U S A 78: 2072–2076

7. Britton P, Green P, Kottier S et al (1996) Expression of bacteriophage T7 RNA polymerase in avian and mammalian cells by a recombinant fowlpox virus. J Gen Virol 77: 963–967

8. Hiscox JA, Wurm T, L W et al (2001) The coronavirus infectious bronchitis virus nucleo-protein localizes to the nucleolus. J Virol 75: 506–512

9. Ausubel FM, Brent R, Kingston RE et al (1987) Current protocols in molecular biology. Wiley, New York

10. Sambrook J, Fritsch EF, Maniatis T (1989) Molecular cloning: a laboratory manual, 2nd edn. Cold Spring Harbor Laboratory, New York

11. Mackett M, Smith GL, Moss B (1985) The construction and characterisation of vaccinia virus recombinants expressing foreign genes. In: Glover DM (ed) DNA cloning, a practical approach. IRL Press, Oxford, pp 191–211

12. Smith GL (1993) Expression of genes by vaccinia virus vectors. In: Davison MJ, Elliot RM (eds) Molecular virology, a practical approach. IRL Press, Oxford, pp 257–283

Chapter 13

Engineering Infectious cDNAs of Coronavirus as Bacterial Artificial Chromosomes

Fernando Almazán, Silvia Márquez-Jurado, Aitor Nogales, and Luis Enjuanes

Abstract

The large size of the coronavirus (CoV) genome (around 30 kb) and the instability in bacteria of plasmids carrying CoV replicase sequences represent serious restrictions for the development of CoV infectious clones using reverse genetic systems similar to those used for smaller positive sense RNA viruses. To overcome these problems, several approaches have been established in the last 13 years. Here we describe the engineering of CoV full-length cDNA clones as bacterial artificial chromosomes (BACs), using the Middle East respiratory syndrome CoV (MERS-CoV) as a model.

Key words Coronavirus, MERS, Reverse genetics, Infectious clones, Bacterial artificial chromosomes

1 Introduction

Coronaviruses (CoV) are enveloped, single-stranded, positive-sense RNA viruses relevant in animal and human health [1, 2]. Historically, CoV infection in humans has been associated with mild upper respiratory tract diseases [1]. However, the identification of the severe acute respiratory syndrome CoV (SARS-CoV) in 2003 [3] and the recently emerged (April 2012) Middle East respiratory syndrome CoV (MERS-CoV) [4], which has been associated with acute pneumonia, redefined historic perceptions and potentiated the relevance of CoVs as important human pathogens. In this sense, the development of CoV infectious clones provides a valuable molecular tool to study fundamental viral processes, to develop genetically defined vaccines, and to test antiviral drugs. However, the generation of CoV infectious clones has been hampered for a long time due to the huge size of the CoV genome (around 30 kb) and the toxicity of some CoV replicase gene sequences during its propagation in bacteria. Recently, these problems were overcome using nontraditional approaches based

Helena Jane Maier et al. (eds.), *Coronaviruses: Methods and Protocols*, Methods in Molecular Biology, vol. 1282, DOI 10.1007/978-1-4939-2438-7_13, © Springer Science+Business Media New York 2015

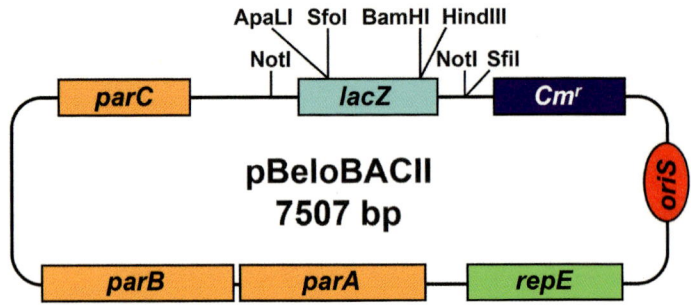

Fig. 1 Schematic of plasmid pBeloBAC11. The regulatory genes *parA, parB, parC,* and *repE*, the F-factor replication origin (*OriS*), the chloramphenicol resistance gene (*Cmr*), the *lacZ* gene, and the restriction sites that can be used to clone foreign DNAs are indicated

on the use of bacterial artificial chromosomes (BACs) [5], in vitro ligation of cDNA fragments [6], and vaccinia virus as a vector for the propagation of CoV full-length cDNAs [7, 8].

In this chapter we describe the protocol to assemble CoV full-length cDNAs in BACs using the MERS-CoV EMC12 strain [9] as an example. In this system, the full-length cDNA copy of the viral genome is assembled in the BAC plasmid pBeloBAC11 [10] (Fig. 1), a low-copy-number plasmid based on the *Escherichia coli* (*E. coli*) F-factor [11] that presents a strictly controlled replication leading to one or two plasmid copies per cell. This plasmid minimizes the instability problem of several CoV sequences when amplified in high-copy-number plasmids, allows the stable maintenance of large DNA fragments in bacteria [11], and its manipulation is similar to that of conventional plasmids. The cDNA of the CoV genome is assembled in the BAC under the control of the cytomegalovirus (CMV) immediate-early promoter and it is flanked at the 3′-end by a 25-bp synthetic poly(A) followed by the sequences of the hepatitis delta virus (HDV) ribozyme and the bovine growth hormone (BGH) termination and polyadenylation signals to produce synthetic RNAs bearing authentic 5′- and 3′-ends of the viral genome. This DNA-launched system couples expression of the viral RNA in the nucleus from the CMV promoter [12] with a second amplification step in the cytoplasm driven by the viral polymerase, allowing the recovery of infectious virus from the cDNA clone without the need for in vitro ligation and transcription steps. Although some splicing events could occur during the nuclear expression of the viral genome, the efficiency of this phenomenon is very low and does not affect the recovery of infectious virus [5].

The BAC approach, originally applied to the transmissible gastroenteritis coronavirus (TGEV) [5], has been successfully used to engineer the infectious clones of the feline infectious peritonitis virus (FIPV) [13] and the human CoVs (HCoVs): HCoV-OC43 [14],

SARS-CoV [15], and MERS-CoV [16], and it is potentially applicable to the cloning of other CoV cDNAs, other viral genomes, and large-size RNAs of biological relevance.

2 Materials

To reach optimal results, all solutions should be prepared using pure Milli-Q grade water (resistivity of 18.2 MΩ/cm at 25 °C) and analytical grade reagents.

2.1 Assembly and Manipulation of BAC Clones

2.1.1 Plasmids and Bacterial Strains

1. Plasmid pBeloBAC11 [10]. This plasmid contains genes *parA*, *parB*, and *parC* derived from the *E. coli* F-factor to ensure the accurate partitioning of plasmids to daughter cells, avoiding the possibility of coexistence of multiple BACs in a single cell. In addition, this plasmid carries gene *repE* and the element *oriS* involved in initiation and orientation of DNA replication, the chloramphenicol resistance gene (*Cmr*), the *lacZ* gene to allow color-based identification of recombinants by α-complementation, and the restriction sites ApaLI, SfoI, BamHI, HindIII, and SfiI to clone large DNA fragments (Fig. 1).

2. pBeloBAC11^{-StuI}, a pBeloBAC without the StuI restriction site.

3. *E. coli* DH10B strain [F$^-$ *mcr*A Δ (*mrr-hsd*RMS-*mcr*BC) Ø80d*lacZ* ΔM15 Δ*lac*X74 *deo*R *rec*A1 *end*A1 *ara*D139 (*ara, leu*)7697 *gal*U *gal*K λ$^-$ *rps*L *nup*G] (*see* **Note 1**).

4. DH10B electrocompetent cells. These bacterial cells could be purchased or prepared following the procedure described in Subheading 3.2.3.

2.1.2 Culture Media for E. coli

1. LB medium: 1 % (w/v) tryptone, 0.5 % (w/v) yeast extract, 1 % (w/v) NaCl. Adjust the pH to 7.0 with 5 N NaOH. Sterilize by autoclaving on liquid cycle.

2. LB agar plates: LB medium containing 15 g/l of Bacto Agar. Prepare LB medium and just before autoclaving add 15 g/l of Bacto Agar. Sterilize by autoclaving on liquid cycle and dispense in 90-mm petri plates.

3. LB agar plates containing 12.5 μg/ml chloramphenicol. After autoclaving the LB agar medium, allow the medium to cool to 45 °C, add the chloramphenicol to a final concentration of 12.5 μg/ml from a stock solution of 34 mg/ml, and dispense in 90-mm petri plates.

4. SOB medium: 2 % (w/v) tryptone, 0.5 % (w/v) yeast extract, 0.05 % (w/v) NaCl, 2.5 mM KCl. Adjust the pH to 7.0 with 5 N NaOH and sterilize by autoclaving on liquid cycle (*see* **Note 2**).

5. SOC medium: SOB medium containing 10 mM MgCl$_2$, 10 mM MgSO$_4$, 20 mM glucose. After autoclaving the SOB medium, cool to 45 °C and add the MgCl$_2$, MgSO$_4$ and glucose from filter sterilized 1 M stock solutions.

2.1.3 Enzymes and Buffers

1. Restriction endonucleases.
2. Shrimp alkaline phosphatase.
3. T4 DNA ligase.
4. Taq DNA polymerase.
5. High-fidelity thermostable DNA polymerase.
6. Reverse transcriptase.
7. dNTPs.
8. Enzyme reaction buffers. Use the buffer supplied with the enzyme by the manufacturer.

2.1.4 Special Buffers and Solutions

1. LB freezing buffer: 40 % (v/v) glycerol in LB medium. Sterilize by passing it through a 0.45-μm disposable filter.
2. Chloramphenicol stock (34 mg/ml). Dissolve solid chloramphenicol in ethanol to a final concentration of 34 mg/ml and store the solution in a light-tight container at −20 °C. This solution does not have to be sterilized.
3. Ice-cold 10 % glycerol in sterile water.

2.1.5 Reagents

1. Qiagen QIAprep Miniprep Kit.
2. Qiagen Large-Construct Kit.
3. Qiagen QIAEX II Kit.

2.1.6 Special Equipment

1. Equipment for electroporation.
2. Cuvettes fitted with electrodes spaced 0.2 cm.

2.2 Rescue of Recombinant Viruses

1. Baby hamster kidney cells (BHK-21).
2. Human liver-derived Huh-7 cells (*see* **Note 3**).

2.2.1 Cells

2.2.2 Cell Culture Medium, Solutions and Reagents

1. Cell growth medium: Dulbecco's Modified Eagle Medium (DMEM) supplemented with 1 % nonessential amino acids, gentamicin (50 mg/ml), and 10 % fetal calf serum (FCS).
2. Opti-MEM I Reduced Serum Medium.
3. Trypsin–EDTA solution: 0.25 % (w/v) trypsin, 0.02 % (w/v) EDTA.
4. Lipofectamine 2000.

3   Methods

3.1   Assembly of Full-Length CoV cDNAs in BACs

The basic strategy for the generation of CoV infectious clones using BACs is described for the MERS-CoV EMC12 strain (GenBank accession number JX869059) [9] as a model (Fig. 2).

3.1.1   Selection of Restriction Endonuclease Sites in the Viral Genome

1. The first step for the assembly of the full-length cDNA clone is the selection of appropriate restriction endonuclease sites in the viral genome. These restriction sites must be absent in the BAC plasmid (*see* **Note 4**). In the case of MERS-CoV, the restriction sites BamHI (genomic position 806), StuI (genomic positions 7,620 and 9,072), SwaI (genomic position 20,898), and PacI (genomic position 25,836) were selected to assemble the infectious clone (Fig. 2).

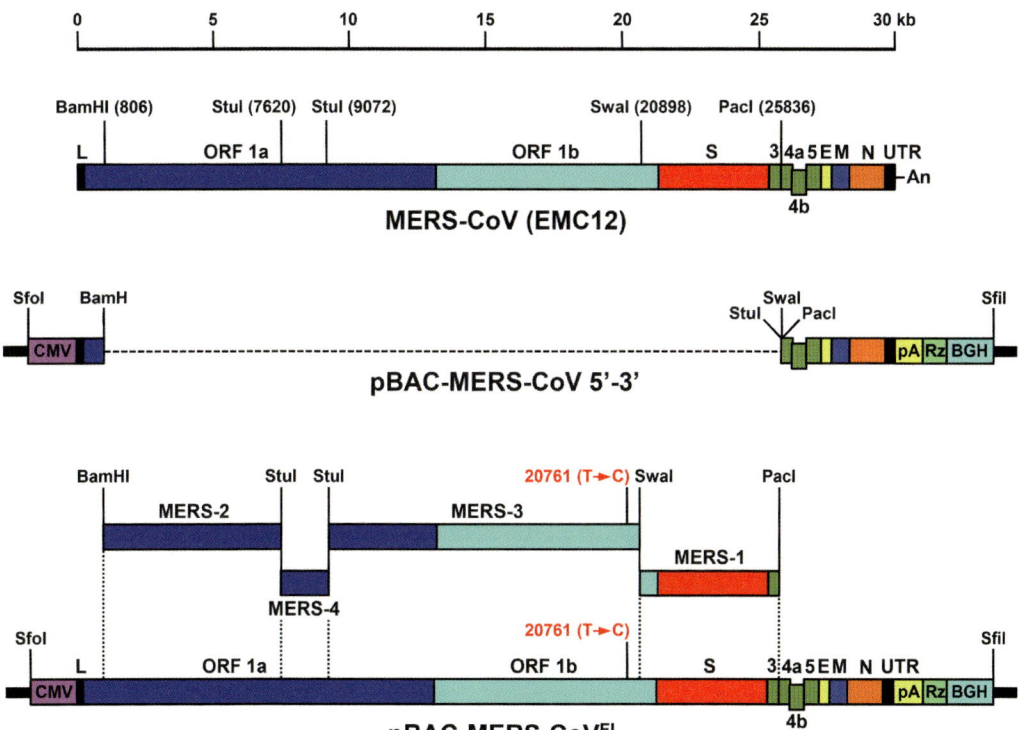

Fig. 2 Strategy to assemble a MERS-CoV infectious cDNA clone as a BAC. After selection of appropriate restriction sites in the genome of the MERS-CoV EMC12 strain (*top* of the figure), the intermediate plasmid pBAC-MERS-CoV 5′–3′ was generated and used as the backbone to assemble the full-length cDNA clone (pBAC-MERS-CoVFL) by sequential cloning of four overlapping cDNA fragments (MERS-1 to MERS-4) covering the entire viral genome. The full-length clone is assembled in BAC under the control of the CMV promoter and it is flanked at the 3′-end by a 25-bp poly(A) tail (pA) followed by the HDV ribozyme (Rz) and the BGH termination and polyadenylation sequences (BGH). The viral genes (ORF 1a, ORF 1b, S, 3, 4a, 4b, 5, E, M, and N), relevant restriction sites (genomic positions in *brackets*) and the genetic marker (T to C) introduced at position 20,761 to abrogate the SwaI restriction site at position 20,760 are indicated. *L* leader sequence, *UTR* untranslated region. Figure adapted from ref. [16]

2. In case that no adequate restriction sites were available in the viral genome, new restriction sites, appropriately spaced in the viral genome, could be generated by the introduction of silent mutations. In addition, natural restriction sites could be eliminated following the same approach to facilitate the assembly of the infectious clone (*see* **Note 5**).

3.1.2 Construction of an Intermediate BAC Plasmid as the Backbone to Assemble the Full-Length cDNA Clone

The assembly of the infectious clone in a BAC is facilitated by the construction of an intermediate BAC plasmid containing the 5′-end of the genome (until the first restriction site selected) under the control of the CMV promoter, a multicloning site containing the restriction sites selected in the first step, and the 3′-end of the genome (from the last restriction site selected to the end of the genome) followed by a 25-nt poly(A) tail, the HDV ribozyme, and the BGH termination and polyadenylation sequences. All these elements have to be precisely assembled to produce synthetic RNAs bearing authentic 5′- and 3′-ends of the viral genome. A detailed protocol for the generation of the MERS-CoV intermediate plasmid, pBAC-MERS-CoV 5′-3′, is described next (Fig. 2).

1. Generate by chemical synthesis a DNA fragment containing the CMV promoter [12] precisely fused to the first 811 nt of the viral genome (from the first nucleotide to the restriction site BamHI), flanked at the 5′- and 3′-ends by restriction sites SfoI and BamHI, respectively. Alternatively, this DNA fragment could be generated by PCR using two overlapping PCR fragments as template (*see* **Notes 6** and **7**). One of these fragments should contain the CMV promoter flanked at the 5′-end by the restriction site SfoI and at the 3′-end by the 20 first nucleotides of the genome as overlapping sequence. The second overlapping PCR fragment should expand from the first nucleotide to the restriction site BamHI.

2. The generated DNA fragment is digested with SfoI and BamHI, and cloned into pBeloBAC11[-StuI] digested with the same restriction enzymes to generate the plasmid pBAC-MERS-CoV 5′.

3. Generate a second DNA fragment, containing the last 4,272 nt of the viral genome (from the restriction site PacI at genomic position 25,836 to the end of the genome) precisely joined to a 25-nt poly(A) tail, the HDV ribozyme, and the BGH termination and polyadenylation sequences, flanked at the 5′-end by a multicloning site with the restriction sites selected before (BamHI, StuI, SwaI, and PacI) and at the 3′-end by the restriction site SfiI. This DNA fragment could be generated by chemical synthesis or by overlapping PCR as described before (*see* **Notes 6** and **7**).

4. Digest with BamHI and SfiI the second DNA fragment (containing the multicloning site, the viral 3′-end followed by the

poly(A) tail, and the HDV-BGH sequences) and clone it into the plasmid pBAC-MERS-CoV 5′ digested with the same restriction enzymes to generate the intermediate plasmid pBAC-MERS-CoV 5′-3′ (Fig. 2).

5. After each cloning step, the integrity of the cloned DNA fragments is verified by restriction analysis and sequencing.

3.1.3 Assembly of the Full-Length cDNA Clone

1. The full-length cDNA clone (pBAC-MERS-CoV^{FL}) is assembled by sequential cloning of overlapping DNA fragments (MERS-1 to MERS-4), covering the entire viral genome, into the intermediate plasmid pBAC-MERS-CoV 5′-3′ using the restriction sites selected in the first step of the cloning strategy (Fig. 2) (*see* **Note 8**).

2. The overlapping DNAs flanked by the appropriated restriction sites are generated either by chemical synthesis or by standard reverse transcriptase PCR (RT-PCR) (*see* **Note 6**) using specific oligonucleotides and total RNA from infected cells as template. In the case of fragment MERS-3, a silent mutation (T to C) was introduced at position 20,761 to abrogate the SwaI restriction site at position 20,760. This mutation facilitates the cloning process and can be used as a genetic marker to identify the virus recovered from the full-length cDNA clone.

3. The genetic integrity of the cloned DNAs is verified throughout the subcloning and assembly process by extensive restriction analysis and sequencing.

3.2 Generation and Manipulation of BAC Clones

One of the major advantages of using BAC vectors to generate infectious clones is that the manipulation of BAC clones is relatively easy and essentially the same as that of a conventional plasmid with slight modifications due to the huge size of the BAC clones and the presence of this plasmid in only one or two copies per cell [11]. The amplification and isolation of BAC plasmids is performed using standard procedures described for conventional plasmids but using large volumes of bacterial cultures.

3.2.1 Isolation of BAC Plasmids from Small-Scale Cultures

Small amounts of BAC DNAs are prepared from 5 ml cultures of BAC transformed DH10B cells by the alkaline lysis method. Any commercial kit could be used, but we suggest the QIAprep Miniprep Kit (Qiagen) following the recommendations for purification of large low-copy plasmids.

1. Streak the bacterial stock containing the BAC plasmid onto a LB agar plate containing 12.5 µg/ml chloramphenicol and incubate for 16 h at 37 °C (*see* **Note 9**).

2. Inoculate a single colony in 5 ml of LB medium plus 12.5 µg/ml chloramphenicol in a flask with a volume of at least four times the volume of the culture and incubate for 16 h at 37 °C with vigorous shaking (250 rpm) (*see* **Note 10**).

3. Harvest the bacterial cells in 15 ml centrifuge tubes by centrifugation at $6,000 \times g$ for 10 min at 4 °C and pour off the supernatant fluid.

4. Purify the BAC plasmid following the manufacturer's instructions. Owing to the size of BAC DNAs and the need to use large culture volumes, we recommend duplicating the volume of buffers P1, P2, and N3, performing the optional wash step with buffer PB, and eluting the DNA from the QIAprep membrane using buffer EB preheated at 70 °C (*see* **Note 11**).

5. Depending of the BAC size, yields of 0.1–0.4 µg can be obtained. Although the BAC DNA prepared by this method is contaminated with up to 30 % of bacterial genomic DNA, it is suitable for analysis by restriction enzyme digestion or PCR.

3.2.2 Isolation of Ultrapure BAC Plasmids from Large-Scale Cultures

Large-scale preparation of ultrapure BAC DNA suitable for all critical applications, including subcloning, DNA sequencing or transfection experiments, is performed by alkaline lysis with the Qiagen Large-Construct Kit, which has been specifically developed and adapted for BAC purification. This kit integrates an ATP-dependent exonuclease digestion step that enables efficient removal of bacterial genomic DNA contamination to yield ultrapure BAC DNA.

1. Inoculate a single colony from a freshly streaked plate (LB agar plate containing 12.5 µg/ml chloramphenicol) (*see* **Note 9**) in 5 ml of LB medium containing 12.5 µg/ml chloramphenicol and incubate for 8 h at 37 °C with vigorous shaking (250 rpm).

2. Dilute 1 ml of the culture into 500 ml of selective LB medium (*see* **Note 10**) pre-warmed to 37 °C and grow the cells with vigorous shaking (250 rpm) in a 2 l flask at 37 °C for 12–16 h, to an OD at 550 nm between 1.2 and 1.5. This cell density typically corresponds with the transition from a logarithmic to a stationary growth phase (*see* **Note 12**).

3. Harvest the bacterial cells by centrifugation at $6,000 \times g$ for 15 min at 4 °C and purify the BAC DNA with the Qiagen Large-Construct Kit according to the manufacturer's specifications (*see* **Note 13**). Depending of the BAC size, yields of 20–35 µg of ultrapure BAC DNA can be obtained.

3.2.3 Preparation of DH10B Competent Cells for Electroporation

Owing to the large size of BAC plasmids, the cloning of DNA fragments in BACs requires the use of DH10B competent cells with transformation efficiencies higher than 1×10^8 transformant colonies per µg of DNA. These efficiencies are easily obtained by the electroporation method, which is more reproducible and efficient than the chemical methods. Here we described the protocol for preparing electrocompetent DH10B cells from 1 l of bacterial culture. All the steps of this protocol should be carried out under sterile conditions.

1. Inoculate a single colony of DH10B cells from a freshly streaked LB agar plate into a flask containing 10 ml of SOB medium and incubate the culture overnight at 37 °C with vigorous shaking (250 rpm).

2. Dilute 1 ml of the overnight culture into 1 l of SOB medium pre-warmed at 37 °C and grow the cells with vigorous shaking (250 rpm) in a 2 l flask at 37 °C until the OD at 550 nm reaches 0.7 (this can take 4–5 h) (*see* **Note 14**).

3. Transfer the flask to an ice-water bath for about 20 min. Swirl the culture occasionally to ensure that cooling occurs evenly. From this point on, it is crucial that the temperature of the bacteria not rise above 4 °C.

4. Divide the bacteria culture in two ice-cold 500 ml centrifuge bottles and harvest the cells by centrifugation at 6,000×*g* for 10 min at 4 °C. Discard the supernatant and resuspend each cell pellet in 500 ml of ice-cold 10 % glycerol in sterile water.

5. Harvest the cells by centrifugation at 6,000×*g* for 15 min at 4 °C. Carefully pour off the supernatant and resuspend each cell pellet in 250 ml of ice-cold 10 % glycerol (*see* **Note 15**).

6. Repeat **step 5** reducing the resuspension volume to 125 ml from each cell pellet.

7. Harvest the cells by centrifugation at 6,000×*g* for 15 min at 4 °C. Carefully pour off the supernatant (*see* **Note 15**) and remove any remaining drops of buffer using a Pasteur pipette attached to a vacuum line.

8. Resuspend the cells in a final volume of 3 ml ice-cold 10 % glycerol, avoiding the generation of bubbles. This volume has been calculated to reach an optimal cell concentration of 2–4×10^{10} cells/ml.

9. Transfer 50 µl of the suspension to an ice-cold electroporation cuvette (0.2-cm gap) and test whether arcing occurs when an electrical discharge is applied with the electroporation apparatus using the conditions described in Subheading 3.2.6, **step 4**. Arcing is usually manifested by the generation of a popping sound in the cuvette during the electrical pulse. If arcing occurs, wash the cell suspension once more with 100 ml 10 % glycerol and repeat **steps 7** and **8**.

10. Dispense 100 µl aliquots of the final cell suspension into sterile, ice-cold 1.5 ml microfuge tubes, freeze quickly in a dry ice–methanol bath, and transfer to a –70 °C freezer. Electrocompetent DH10B cells could be stored at –70 °C for up to 6 months without loss of transforming efficiency.

3.2.4 Cloning of DNA Fragments in BACs: Preparation of BAC Vectors and DNA Inserts

The same standard techniques used for the cloning of DNA in conventional plasmids are applied to BACs with special considerations owing to the large size of BAC plasmids.

1. Digest the BAC vector and foreign DNA with a two- to three-fold excess of the desired restriction enzymes for 3 h using the buffers supplied with the enzymes and check a small aliquot of the digestions by agarose gel electrophoresis to ensure that the entire DNA has been cleaved. Use an amount of target DNA sufficient to yield 2 µg of the BAC vector and 0.25–0.5 µg of the desired DNA insert.

2. When two enzymes requiring different buffers are used to digest the DNA, carry out the digestion sequentially with both enzymes. Clean the DNA after the first digestion by extraction with phenol–chloroform and standard ethanol precipitation or by using the Qiagen QIAEX II Gel Extraction Kit following the manufacturer's instructions for purifying DNA fragments from aqueous solutions (*see* **Note 16**).

3. Purify the digested BAC vector and the DNA insert by agarose gel electrophoresis using the Qiagen QIAEX II Gel Extraction Kit following the manufacturer's instructions (*see* **Note 16**).

4. Determine the concentration of the BAC vector and the insert by UV spectrophotometry or by quantitative analysis on an agarose gel.

5. If the BAC vector was digested with only one restriction enzyme or with restriction enzymes leaving compatible or blunt ends, the digested BAC vector has to be dephosphorylated prior to its purification by agarose gel electrophoresis to suppress self-ligation of the BAC vector. We recommend cleaning the DNA before the dephosphorylation reaction as described in **step 2** and using shrimp alkaline phosphatase following the manufacturer's specifications.

3.2.5 Ligation Reaction

1. For protruding-ended DNA ligation, mix in a sterile microfuge tube 150 ng of purified digested BAC vector, an amount of the purified insert equivalent to a molar ratio of insert to vector of 3:1, 1.5 µl of 10× T4 DNA ligase buffer containing 10 mM ATP, 3 Weiss unit of T4 DNA ligase, and water to a final volume of 15 µl. In separate tubes, set up two additional ligations as controls, one containing only the vector and the other containing only the insert. Incubate the reaction mixtures for 16 h at 16 °C (*see* **Note 17**).

2. In the case of blunt-ended DNAs, to improve the ligation efficiency use 225 ng of vector, the corresponding amount of insert, 6 Weiss unit of T4 DNA ligase, and incubate the reaction mixtures for 20 h at 14 °C.

3.2.6 Transformation of DH10B Competent Cells by Electroporation

1. Thaw the electrocompetent DH10B cells at room temperature and transfer them to an ice bath.

2. For each transformation, pipette 50 µl of electrocompetent cells into an ice-cold sterile 1.5 ml microfuge tube and place it on ice together with the electroporation cuvettes.

3. Dilute 2.5 µl of the ligation reaction (about 25 ng of DNA) in 47.5 µl of sterile water, mix with the competent cells and incubate the mixture on ice for 1 min. For routine transformation with supercoiled BACs, add 0.1 ng of DNA in a final volume of 2 µl. Include all the appropriate positive and negative controls.

4. Set the electroporation machine to deliver an electrical pulse of 25 µF capacitance, 2.5 kV, and 100 Ω resistance (*see* **Note 18**).

5. Add the DNA–cells mixture into the cold electroporation cuvette avoiding bubbles formation and ensuring that the DNA–cells mixture sits at the bottom of the cuvette. Dry the outside of the cuvette with filter paper and place the cuvette in the electroporation device.

6. Deliver an electrical pulse at the settings indicated above. A time constant of 4–5 ms should be registered on the machine (*see* **Note 19**).

7. Immediately after the electrical pulse, remove the cuvette and add 1 ml of SOC medium pre-warmed at room temperature.

8. Transfer the cells to a 17×100-mm polypropylene tube and incubate the electroporated cells for 50 min at 37 °C with gentle shaking (250 rpm).

9. Plate different volumes of the electroporated cells (2.5, 20, and 200 µl) onto LB agar plates containing 12.5 µg/ml chloramphenicol and incubate them at 37 °C for 16–24 h (*see* **Note 20**).

3.2.7 Screening of Bacterial Colonies by PCR

The recombinant colonies containing the insert in the correct orientation are identified by direct PCR analysis using specific oligonucleotides and conventional Taq DNA polymerase (*see* **Note 21**).

1. For each bacterial colony prepare a PCR tube with 25 µl of sterile water.

2. Using sterile yellow tips, pick the bacterial colonies, make small streaks (2–3 mm) on a fresh LB agar plate containing 12.5 µg/ml chloramphenicol to make a replica, and transfer the tips to the PCR tubes containing the water (*see* **Note 22**). In separate tubes, set up positive and negative controls. Leave the tips inside the PCR tubes for 5 min at room temperature.

3. During this incubation time, prepare a 2× master mix containing 2× PCR buffer, 3 mM $MgCl_2$ (it has to be added only in the case that the PCR buffer does not contain $MgCl_2$), 0.4 mM dNTPs, 2 µM of each primer, and 2.5 U of Taq DNA

polymerase per each 25 μl of master mix. Prepare the appropriate amount of 2× master mix taking into consideration that the analysis of each colony requires 25 μl of this master mix.

4. Remove the yellow tip and add 25 μl of 2× master mix to each PCR tube.

5. Transfer the PCR tubes to the thermocycler and run a standard PCR, including an initial denaturation step at 95 °C for 5 min to liberate and denature the DNA templates and to inactivate proteases and nucleases.

6. Analyze the PCR products by electrophoresis through an agarose gel.

7. Pick the positive colonies from the replica plate and isolate the BAC DNA as described in Subheadings 3.2.1 and 3.2.2 for further analysis.

3.2.8 Storage of Bacterial Cultures

1. Mix 0.5 ml of LB freezing buffer with 0.5 ml of an overnight bacterial culture in a cryotube with a screw cap.

2. Vortex the culture to ensure that the glycerol is evenly dispersed, freeze in ethanol–dry ice, and transfer to –70 °C for long-term storage.

3. Alternatively, a bacterial colony can be stored directly from the agar plate without being grown in liquid media. Using a sterile yellow tip, scrape the bacteria from the agar plate and resuspend the cells into 200 μl of LB medium in a cryotube with a screw cap. Add an equal volume of LB freezing buffer, vortex the mixture, and freeze the bacteria as described in **step 2** (*see* **Note 23**).

3.2.9 Modification of BAC Clones

The modification of BAC clones is relatively easy and it is performed using the same techniques as for conventional plasmids with the modifications described in this chapter. We recommend introducing the desired modifications into intermediate BAC plasmids containing the different viral cDNA fragments used during the assembly of the full-length cDNA clone, and then inserting the modified cDNA into the infectious clone by restriction fragment exchange. Besides standard protocols, the BAC clones can be easily and efficiently modified in *E. coli* by homologous recombination using a two-step procedure that combines the Red recombination system and counterselection with the homing endonuclease I-SceI [17–20] (*see* **Note 24**).

3.3 Rescue of Recombinant Viruses

Infectious virus is recovered by transfection of susceptible cells with the full-length cDNA clone using the cationic lipid Lipofectamine 2000 as transfection reagent (*see* **Note 25**). When the transfection efficiency of the susceptible cells is very low, we recommend first transfecting BHK-21 cells and then plating these cells over a monolayer of susceptible cells to allow virus propagation. BHK-21 cells are selected because they present good transfection efficiencies and support the replication of most known CoVs after transfection of the viral genome. The following protocol is indicated for a 35-mm-diameter dish and can be upscaled or downscaled if desired (*see* **Note 26**).

1. One day before transfection, plate 4×10^5 BHK-21 cells in 2 ml of growth medium without antibiotics to obtain 90–95 % confluent cell monolayers by the time of transfection (*see* **Note 27**). Also plate susceptible cells (Huh-7 cells in the case of MERS-CoV) at the required confluence for the amplification of the recombinant virus after transfection.

2. Before transfection, equilibrate the Opti-MEM I Reduced Serum Medium at room temperature and put the DNA (*see* **Note 28**) and the Lipofectamine 2000 reagent on ice. For each transfection sample, prepare transfection mixtures in sterile microfuge tubes as follows:

   (a) Dilute 5 μg of the BAC clone in 250 μl of Opti-MEM medium. Mix carefully, avoiding prolonged vortexing or pipetting to prevent plasmid shearing.

   (b) Mix Lipofectamine 2000 gently before use. Dilute 12 μl of Lipofectamine 2000 in 250 μl of Opti-MEM medium (*see* **Note 29**), mix by vortexing, and incubate the diluted Lipofectamine 2000 at room temperature for 5 min.

   (c) Combine the diluted DNA with diluted Lipofectamine 2000, mix carefully, and incubate for 20 min at room temperature.

3. During this incubation period, wash the BHK-21 cells once with growth medium without antibiotics and leave the cells in 1 ml of the same medium per dish.

4. Add the 500 μl of the DNA–Lipofectamine 2000 mixture onto the washed cells and mix by rocking the plate back and forth. Incubate the cells at 37 °C for 6 h (*see* **Note 30**).

5. Remove the transfection medium, wash the cells with trypsin–EDTA solution, and detach the cells using 300 μl of trypsin–EDTA solution.

6. Add 700 μl of growth media to collect the cells and reseed them over a confluent monolayer of susceptible cells containing 1 ml of normal growth medium.

7. Incubate at 37 °C until a clear cytopathic effect was observed.

8. Analyze the presence of virus in the supernatant by titration.

9. Clone the virus by three rounds of plaque purification and analyze the genotypic and phenotypic properties of the recovered virus.

4   Notes

1. *E. coli* DH10B strain is a recombination-defective strain used for the propagation of BACs to avoid unwanted rearrangements.

2. SOB medium should be Mg^{2+}-free to avoid arcing during the electroporation step.

3. The Huh-7 cell line has never been deposited at ATCC but it can be purchased from the Japanese Collection of Research Bioresources (JCRB) Cell Bank.

4. In case that a restriction site present in the BAC plasmid was selected, it must be removed in the plasmid by the introduction of silent mutations.

5. The silent mutations introduced in the viral genome to generate new restriction sites or to abrogate preexisting ones can be used as genetic markers to identify the virus recovered from the infectious clone.

6. To reduce the number of undesired mutations, perform all PCR reactions with a high-fidelity polymerase, according to the manufacturer's instructions.

7. The CMV promoter and the BGH termination and polyadenylation sequences can be amplified from pcDNA3.1. Alternatively, these sequences together with the HDV ribozyme could be amplified from plasmid pBAC-TGEV 5′-3′ that is available from the authors upon request.

8. In general, the cloning of CoV full-length cDNAs in BACs allows the stable propagation of the infectious clone in *E. coli* DH10B cells. If a residual toxicity, characterized by a small colony phenotype and a delay in the bacterial growth, is observed during the assembly of the infectious clone, we recommend inserting the DNA fragment responsible for this toxicity in the last cloning step to minimize the toxicity problem.

9. Cultures of BAC transformed bacteria should be grown from a single colony isolated from a freshly streaked selective plate. Subculturing directly from glycerol stocks or plates that have been stored for a long time may lead to loss of the construct.

10. LB broth is the recommended culture medium, since richer broths such as TB (Terrific Broth) lead to extremely high cell densities, which can overload the purification system, resulting in lower yield and less purity of the BAC DNA.

11. When other kits are used instead of the Qiagen QIAprep Miniprep Kit, equivalent modifications have to be included to optimize the recovery of BAC DNA.

12. To avoid DNA degradation and unwanted rearrangements owing to culture overaging, it is important to prevent growing the culture up to the late stationary growth phase.

13. The use of a swinging bucket rotor is recommended for the last isopropanol precipitation step to facilitate the further resuspension of the BAC DNA. After washing with 70 % ethanol, air-dry the pellet for only 5 min. Never use vacuum, as overdrying the pellet will make the BAC DNA difficult to dissolve. Carefully remove any additional liquid drop, add 250 µl of 10 mM Tris–HCl (pH 8.5) (DNA dissolves better under slightly alkaline conditions) and resuspend the DNA overnight at 4 °C. To prevent plasmid shearing, avoid vortexing or pipetting to promote resuspension of the BAC DNA. Transfer the DNA to a clean 1.5 ml microfuge tube, remove any possible resin traces by centrifugation for 1 min in a tabletop microfuge, and keep the supernatant in a clean tube at 4 °C. If the purified BAC DNA is not going to be used for a long period of time we recommend storage at –20 °C. Avoid repeated freeze-thaw cycles to prevent plasmid shearing.

14. For efficient cell transformation, bacterial culture OD at 550 nm should not exceed 0.8. To ensure that the culture does not grow to a higher density, OD measurement every 20 min after 3 h of growth is highly recommended.

15. Take care when decanting the supernatant as the bacterial pellets lose adherence in 10 % glycerol.

16. The Qiagen QIAEX II resin can be used to efficiently purify DNA fragments from 40 bp to 50 kb from aqueous solutions and from standard or low-melt agarose gels in TAE or TBE buffers. Other commercial kits are available, but check whether they have been optimized for purification of DNA fragments larger than 10 kb, as most BAC constructs used during the assembly of the infectious clone are larger than 10 kb.

17. The large size of the BAC vectors reduces the ligation efficiency. To increase this efficiency, it is essential to use larger amounts of vector, insert, and T4 DNA ligase than when using conventional plasmids.

18. Most electroporation machines contain programs with defined parameters for transforming specific cell types. In this case, choose the program containing the conditions closest to those described in this protocol.

19. The presence of salt increases the conductivity of the solution and could cause arcing during the electrical pulse, drastically reducing the transformation efficiency. If arcing occurs, use a smaller amount of the ligation reaction in the electroporation or remove salt from the DNA using any commercial kit or by extraction with phenol–chloroform followed by precipitation with ethanol and 2 M ammonium acetate.

20. Plating volumes higher than 200 µl of electroporated cells on a single plate may inhibit the growth of transformants owing to the large number of dead cells resulting from electroporation.

If only small numbers of transformant colonies are expected, it is recommended to spread 200 μl-aliquots of the electroporated cells on different plates.

21. A mix of small and large colonies indicates that the cloned DNA fragment presents some toxicity when amplified in *E. coli*. Choose the small colonies, which may contain the correct insert, and always grow the bacteria containing this recombinant BAC plasmid at 30 °C to minimize the toxicity problem. In this case, we strongly recommend inserting this toxic DNA fragment into the infectious clone in the last cloning step, in order to reduce the manipulation and minimize the possibility of introducing unwanted mutations. Infectious BAC cDNA clones presenting a residual toxicity should be grown at 30 °C.

22. It is important to avoid overloading the reaction by adding too much bacteria, which may alter the ionic balance of the reaction and inhibit the amplification by the Taq polymerase.

23. We recommend to use this storage method for BAC clones that present a residual toxicity and are not fully stable when amplified in *E. coli*.

24. The Red recombination system combined with counterselection with I-SceI endonuclease results in an accurate and highly efficient method to introduce insertions, deletions, or point mutations in BAC clones without retention of unwanted foreign sequences.

25. The transfection of BACs containing large inserts into mammalian cells has been optimized in our laboratory and the best transfection efficiencies were provided using Lipofectamine 2000 as transfection reagent.

26. All work involving MERS-CoV has to be performed in a Biosafety Level 3 (BSL3) laboratory, following the guidelines of the European Commission and the National Institutes of Health (NIH) of the USA.

27. Do not add antibiotics to media during transfection as this may decrease transfection activity. A healthy cell culture is critical for an efficient transfection. The use of low passage-number cells is recommended.

28. Use a BAC DNA isolated with the Qiagen Large-Construct Kit since a DNA preparation of high purity is required in the transfection step.

29. Opti-MEM I Reduced Serum Medium is recommended for dilution of the cationic lipid Lipofectamine 2000 reagent prior to complexing with DNA, although other media without serum may also be used. However, owing that some serum-free media formulations can inhibit cationic lipid-mediated transfection, test any new serum-free medium for compatibility

with the transfection reagent prior to use. Some media formulations that have been found to inhibit cationic lipid-mediated transfection are CD 293 Medium, 293 SFM II, and VP-SFM.

30. If susceptible cells are directly transfected, incubate them at 37 °C until the cytopathic effect is observed and proceed to clone and characterize the recovered virus. In this case, optimization of the transfection of the desired cells with the BAC clone using Lipofectamine 2000 should be required. For transfection optimization, use a similar size plasmid expressing GFP. This plasmid is available from the authors upon request.

Acknowledgements

This work was supported by grants from the Ministry of Science and Innovation of Spain (MCINN) (BIO2010-16705), the European Community's Seventh Framework Programme (FP7/2007–2013) under the project "EMPERIE" (HEALTH-F3-2009-223498), and the National Institute of Health (NIH) of the USA (2P01AI060699-06A1). S. M. received a predoctoral fellowship from the National Institute of Health (ISCIII) of Spain.

References

1. Masters PS (2006) The molecular biology of coronaviruses. Adv Virus Res 66:193–292

2. Lai MMC, Perlman S, Anderson L (2007) Coronaviridae. In: Knipe DM, Howley PM, Griffin DE, Lamb RA, Martin MA, Roizman B, Straus SE (eds) Fields virology, vol 1, 5th edn. Lippincott Williams and Wilkins, Philadelphia, pp 1305–1335

3. Stadler K, Masignani V, Eickmann M et al (2003) SARS-beginning to understand a new virus. Nat Rev Microbiol 1:209–218

4. Zaki AM, van Boehmen S, Bestebroer TM et al (2012) Isolation of a novel coronavirus from a man with pneumonia in Saudi Arabia. N Engl J Med 367:1814–1820

5. Almazán F, González JM, Pénzes Z et al (2000) Engineering the largest RNA virus genome as an infectious bacterial artificial chromosome. Proc Natl Acad Sci U S A 97:5516–5521

6. Yount B, Curtis KM, Baric RS (2000) Strategy for systematic assembly of large RNA and DNA genomes: transmissible gastroenteritis virus model. J Virol 74:10600–10611

7. Thiel V, Herold J, Schelle B et al (2001) Infectious RNA transcribed in vitro from a cDNA copy of the human coronavirus genome cloned in vaccinia virus. J Gen Virol 82:1273–1281

8. Casais R, Thiel V, Siddell SG et al (2001) Reverse genetics system for the avian coronavirus infectious bronchitis virus. J Virol 75:12359–12369

9. van Boehmen S, de Graaf M, Lauber C et al (2012) Genomic characterization of a newly discovered coronavirus associated with acute respiratory distress syndrome in humans. mBio 3, e00473-12

10. Wang K, Boysen C, Shizuya H et al (1997) Complete nucleotide sequence of two generations of a bacterial artificial chromosome cloning vector. Biotechniques 23:992–994

11. Shizuya H, Birren B, Kim UJ et al (1992) Cloning and stable maintenance of 300-kilobase-pair fragments of human DNA in *Escherichia coli* using an F-factor-based vector. Proc Natl Acad Sci U S A 89: 8794–8797

12. Dubensky TW, Driver DA, Polo JM et al (1996) Sindbis virus DNA-based expression vectors: utility for in vitro and in vivo gene transfer. J Virol 70:508–519

13. Bálint A, Farsang A, Zádori Z et al (2012) Molecular characterization of feline infectious peritonitis virus strain DF-2 and studies of the role of ORF3abc in viral cell tropism. J Virol 86:6258–6267

14. St-Jean JR, Desforges M, Almazán F et al (2006) Recovery of a neurovirulent human coronavirus OC43 from an infectious cDNA clone. J Virol 80:3670–3674

15. Almazán F, DeDiego ML, Galán C et al (2006) Construction of a severe acute respiratory syndrome coronavirus infectious cDNA clone and a replicon to study coronavirus RNA synthesis. J Virol 80:10900–10906

16. Almazán F, DeDiego ML, Sola I et al (2013) Engineering a replication-competent, propagation-defective Middle East respiratory syndrome coronavirus as a vaccine candidate. mBio 4, e00650-00613

17. Zhang Y, Buchholz F, Muyrers JPP et al (1998) A new logic for DNA engineering using recombination in *Escherichia coli*. Nat Genet 20:123–128

18. Lee EC, Yu D, Martinez de Velasco J et al (2001) A highly efficient Escherichia coli-based chromosome engineering system adapted for recombinogenic targeting and sub-cloning of BAC DNA. Genomics 73:56–65

19. Jamsai D, Orford M, Nefedov M et al (2003) Targeted modification of a human beta-globin locus BAC clone using GET recombination and an I-SceI counterselection cassette. Genomics 82:68–77

20. Tischer BK, von Einem J, Kaufer B et al (2006) Two-step red-mediated recombination for versatile high-efficiency markerless DNA manipulation in *Escherichia coli*. Biotechniques 40:191–197

Part IV

Coronavirus Attachment and Entry

Chapter 14

Protein Histochemistry Using Coronaviral Spike Proteins: Studying Binding Profiles and Sialic Acid Requirements for Attachment to Tissues

Iresha N. Ambepitiya Wickramasinghe and M. Hélène Verheije

Abstract

Protein histochemistry is a tissue-based technique that enables the analysis of viral attachment patterns as well as the identification of specific viral and host determinants involved in the first step in the infection of a host cell by a virus. Applying recombinantly expressed spike proteins of infectious bronchitis virus onto formalin-fixed tissues allows us to profile the binding characteristics of these viral attachment proteins to tissues of various avian species. In particular, sialic acid-mediated tissue binding of spike proteins can be analyzed by pretreating tissues with various neuraminidases or by blocking the binding of the viral proteins with specific lectins. Our assay is particularly convenient to elucidate critical virus–host interactions for viruses for which infection models are limited.

Key words Protein histochemistry, Spike protein, Neuraminidase, Lectin, Formalin-fixed tissues, Infectious bronchitis virus (IBV), Attachment, Glycan

1 Introduction

Infectious bronchitis virus (IBV), an avian coronavirus belonging to the genus *Gammacoronavirus*, is the major cause of contagious respiratory disease or infectious bronchitis in poultry. Many IBV serotypes have been isolated so far and some serotypes induce pathological changes in organs other than the respiratory tissues [1]. This variable tissue tropism is likely due to tissue-specific factors resulting in differences in binding or entry of the virus. Although a specific protein receptor for IBV is yet to be revealed it has been shown by removing sialic acids from the susceptible cell surface, that $\alpha2, 3$-linked sialic acids are a determinant of cell attachment and entry of IBV [2, 3]. Further elucidation of host–virus interactions is, however, hampered due to limitations in in vitro infection model systems for pathogenic IBV strains.

Helena Jane Maier et al. (eds.), *Coronaviruses: Methods and Protocols*, Methods in Molecular Biology, vol. 1282, DOI 10.1007/978-1-4939-2438-7_14, © Springer Science+Business Media New York 2015

For IBV the initial cell attachment and entry is mediated by a glycoprotein called spike protein residing in the viral envelope. By swapping the gene encoding for spike protein between different IBV serotypes it has been shown that the spike determines the tissue tropism [4]. The spike protein is cleaved into an S1 and an S2 subunit [5, 6]; while S1 mediates the first step in infection via the initial virus–cell binding, S2 is responsible for cell entry [7]. Analyzing the binding of S1 to tissues with our protein histochemistry protocol enables us not only to profile the attachment of avian coronavirus S1 proteins to various avian tissues but also to elucidate glycan binding specificities of IBV S1 [8] as well as determinants within S1 for tissue attachment [9]. Thereby, this method aids to understand the in vivo tissue tropism of avian coronaviruses.

2 Materials

The amounts of buffers or chemicals prepared are described such to result in a convenient volume. Any other required volume can be calculated from this.

2.1 Components for Expression of Spike Protein (S1) of IBV

1. Expression plasmid harboring a CMV promoter, signal sequence, GCN4 trimerization domain and *Strep*-tag for purification and detection: Use codon-optimized IBV S1 sequence of the serotype of interest (*see* **Note 1**) and clone S1 into for example pCD5 expression plasmid (*see* **Note 2**) in frame with CMV, GCN4, and *Strep*-tag (Fig. 1).

2. Polyethylenimine (PEI): Dissolve the powder at a concentration of 1 mg/ml at 50–60 °C (*see* **Note 3**). Test the efficiency by transfecting human embryo kidney cells (HEK) 293T with pCMV-EGFP-N1 or any other vector expressing a fluorescent protein (*see* **Note 3**).

3. Supplemented Dulbecco's modified eagle medium (DMEM): DMEM, 2 % glutamine, 10 % fetal calf serum, 0.1 mg/ml gentamicin.

4. Supplemented 293 SFM II expression medium: 293 SFM II, 3.7 g/l sodium bicarbonate, 2.0 g/l glucose, 3.0 g/l Primatone RL-UF, 0.1 mg/ml gentamicin, 1× GlutaMAX, 1.5 % dimethyl sulfoxide. Sterilize the medium by filtering.

5. T175 culture flasks.

CMV   SS   S1   GCN4 ST2

Fig. 1 Diagrammatic representation of S1 expression cassette. S1 was cloned into pCD5 expression plasmid in frame with signal sequence (SS), trimerization motif (GCN4), and *Strep*-tag (ST2). The promoter sequence was from Cytomegalovirus (CMV)

6. *Strep*-Tactin sepharose 50 % suspension or *Strep*-Tactin gravity flow columns

7. Elution buffer: Biotin elution buffer 10×, dilute 10× concentrated to 1× in distilled water (working solution).

8. Vivaspin 10 or 50 MWCO 3000.

9. Tube roller.

2.2 Components for Protein Histochemistry

1. Xylene.

2. Ethanol at 100, 96, and 70 %.

3. Tissue section slides of 3–4 μm on Superfrost Plus or KP plus glasses (*see* **Note 4**).

4. Citrate buffer (pH 6.0): Add 2.1 g of citrate buffer monohydrate to 800 ml of distilled water and while stirring adjust the pH to 6 at room temperature by adding 10 N NaOH drop wise. Then add up the total volume to 1,000 ml by adding distilled water.

5. *Strep*-Tactin HRP conjugated.

6. PBS 10×: Add 35.6 g of $Na_2HPO_4 \cdot 2H_2O$ and 6.24 g of $NaH_2PO_4 \cdot 2H_2O$ into 2.4 l of distilled water. Check if the pH is 7.4–7.5 and add 216 g of NaCl.

7. PBS: Dilute stock PBS 10× in distilled water to prepare 1× working solution.

8. PBS–Tween 0.1 %: Dilute 500 ml 10× stock PBS into 4,500 ml of distilled water. Add 5 ml of Tween 20.

9. PBS (pH 5.0): Adjust the pH to 5.0 by adding 6 N HCl into PBS.

10. 1 % hydrogen peroxide: Add 2.85 ml of 35 % hydrogen peroxide to 97.15 ml of absolute methanol.

11. VECTASTAIN ABC Kit (Vector Laboratories Inc.): Add 10 μl of solution A to 240 μl of PBS and add 10 μl of solution B to 240 μl of PBS. Mix and incubate for 30 min at RT.

12. 3-Amino-9-ethylcarbazole (AEC).

13. Normal goat serum: Dilute goat serum in PBS to reach 10 %.

14. Neuraminidases: Add 1 mU of *Vibrio cholera* neuraminidase or *Arthrobacter ureafaciens* neuraminidase to 100 μl of PBS (pH 5.0).

15. Lectins: Dilute MALI and MALII at a concentration gradient from 64 to 256 μg/ml in PBS (*see* **Note 5**).

16. Hematoxylin.

17. Aquatex mounting medium.

18. Dako or Immunopen.

19. Coverslips (24 × 32 mm).

20. Coplin jar.

21. Humidity chamber.

3   Methods

Carry out all procedures at room temperature unless otherwise specified. Centrifuge 50 ml tubes in a benchtop centrifuge and Eppendorf tubes in microcentrifuge.

3.1   Expression of Recombinant IBV S1 Protein

Amounts are shown for expression of S1 protein in one T175 flask.

1. Day 1: Seed a T175 culture flask with 1×10^7 HEK 293 T cells in a total volume of 25 ml of DMEM + medium. Incubate the cells at 37 °C for 24 h until the cells reach a confluence of 50–60 % (*see* **Note 6**).

2. Day 2: Prepare reaction mix. For one T175 flask first pipette 15 µg of the expression vector pCD5 containing IBV S1 into DMEM and then pipette PEI into the DMEM. The total volume of DNA, PEI and DMEM should be 1.5 ml per flask to be transfected. Incubate the reaction mix for 15 min. Remove 5 ml of the medium from the cells and add the reaction mix into the medium with the T175 flask in an upright position then gently agitate before repositioning the flask horizontally to incubate at 37 °C for 24 h (*see* **Note 2**).

3. Day 3: Replace DMEM with 20 ml of the 293 SFM + and continue incubation at 37 °C.

4. Day 8: Collect the supernatant into 50 ml tube (usually 7 days after transfection) and centrifuge at $300 \times g$ for 10 min. Transfer into new 50 ml tube and centrifuge another 10 min at $800 \times g$. Transfer the supernatant into a new tube. The supernatant can now be stored at –20 °C or directly proceed to Subheading 3.2, **step 1**.

3.2   Purification of Recombinant IBV S1 Protein

1. Add *Strep*-Tactin sepharose 50 % suspension (*see* **Note 7**) to the supernatant and incubate overnight at 4 °C on a tube roller.

2. The next day centrifuge at $800 \times g$ for 10 min and carefully remove the supernatant without disturbing the bead pellet. Add 500 µl of PBS onto the beads, stir gently with a pipette tip and transfer the beads into a 2 ml Eppendorf tube (*see* **Note 8**).

3. Wash the beads three times using PBS (bead pellet: PBS is 1:1).

4. Centrifuge at $1,800 \times g$ for 10 min for each wash.

5. After the final washing step remove PBS, add elution buffer (*see* **Note 9**) and incubate for 5 min, vortexing every 1–2 min.

6. Centrifuge at $1,800 \times g$ for 10 min and collect the supernatant.

7. To remove remaining beads in the supernatant centrifuge another 10 min at $1,800 \times g$ and transfer the supernatant into a new Eppendorf tube.

8. Determine the protein concentration (*see* **Note 10**).

3.3 Protein Histochemistry (See Figs. 2 and 3)

3.3.1 Deparaffinization and Rehydration of Tissue Sections

1. Prepare glass dishes with xylene, 100 % ethanol, 96 % ethanol, 70 % ethanol, and distilled water in duplicates.

2. Arrange the glass slides in a staining rack and immerse slides in xylene to distilled water (xylene, xylene, 100 % ethanol, 100 % ethanol, 96 % ethanol, 96 % ethanol, 70 % ethanol, 70 % ethanol, distilled water, distilled water).

3. Keep the slides in each dish of xylene for 5 min and in each dish of alcohol and distilled water for 3 min. End with immersing in distilled water.

3.3.2 Antigen Retrieval

1. Place the staining rack with the slides in a heat-resistant jar or a container and add citrate buffer until the fluid level is at least 2 cm above the slides. Close the container with a lid.

2. Boil the sections in citrate buffer for 10 min at 900 kW in a microwave (*see* **Note 11**).

3. Leave the slides in the citrate buffer and allow to cool down for 15–20 min.

4. Transfer the slides into a Coplin jar filled with PBS and keep on a platform rocker for 5 min. Repeat the PBS step twice.

3.3.3 Inactivate Endogenous Peroxidase and Blocking Nonspecific Staining

1. Remove PBS from the last washing step and add 1 % hydrogen peroxide until the sections are properly covered. Close the jar with a lid and incubate for 30 min.

2. Discard hydrogen peroxide add PBS–Tween 0.1 % and rinse the slides for 5 min on a platform rocker. Repeat the PBS–Tween 0.1 % wash step twice.

3. Dry the back of the slides and around the sections using a tissue, and draw lines around the tissues with a Dako or an Immunopen.

4. Place the slides in a humidity chamber and apply sufficient amounts of 10 % normal goat serum to cover the tissues (usually 50–200 µl depending on the size of the section).

5. Close the humidity chamber and incubate for 30 min.

3.3.4 Application of Spike Proteins

1. Premix spike protein to a final concentration of 0.1 mg/ml and *Strep*-Tactin HRP 1:200 in PBS (*see* **Note 12**) in an Eppendorf tube and incubate for 30 min on ice.

2. Drain goat serum from the sections and apply sufficient amounts (usually 50–200 µl depending on the size of the section) of spike protein–*Strep*-Tactin HRP complex to tissues.

3. Incubate the sections with the protein–*Strep*-Tactin HRP complex overnight at 4 °C.

3.3.5 Visualizing and Counterstaining

1. Drain the protein–antibody complex and place the slides in a Coplin jar filled with PBS.

2. Rinse the slides in PBS three times each for 5 min as previously described.

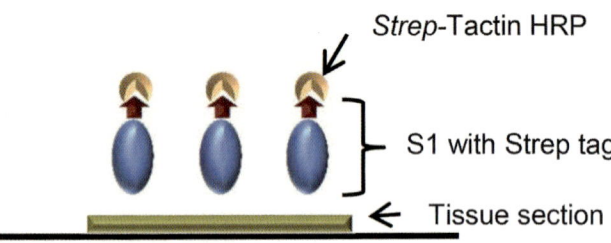

Fig. 2 Schematic representation of protein histochemistry. S1 protein was pre-complexed with *Strep*-Tactin HRP before applying onto tissue section

3. Dry the back of the slides and around the tissues, place the slides in a humidity chamber, and apply AEC dropwise (*see* **Note 13**).

4. Close the chamber and incubate for 15 min.

5. Dip the sections into a Coplin jar with water and place the glass slides in a staining rack.

6. Rinse the slides in tap water for 5 min and immerse in hematoxylin for 40–60 s.

7. Keep the slides in running water for 10 min.

8. Finally place a coverslip to cover the tissues using Aquatex (Fig. 2).

3.4 Protein Histochemistry on Tissues Pretreated with Neuraminidase

1. After treating the slides with hydrogen peroxide (Subheading 3.3.3) place the slides in a humidity chamber and circle the tissue regions with Dako or Immunopen.

2. Dilute 1 mU of neuraminidase (*see* **Note 14**) in 100 μl of PBS (pH 5.0) and apply to tissues within the circle.

3. Close the humidity chamber and keep overnight at 37 °C in an incubator.

4. The next day rinse the slides in PBS–Tween 0.1 % three times each for 5 min, incubate with 10 % goat serum for 30 min and continue with Subheading 3.3.4.

3.5 Protein Histochemistry for Tissues Blocked with Lectins

1. After treating with hydrogen peroxide (Subheading 3.3.3), apply lectins to tissues circled with Dako or Immunopen.

2. Incubate the slides overnight at 4 °C in a humidity chamber.

3. Next day rinse the slides in PBS–Tween 0.1 % three times each for 5 min and continue with Subheading 3.3.4 (Fig. 3).

4 Notes

1. The sequences coding for spike were codon-optimized for expression in mammalian cells, resulting in approximately five times higher production of proteins than using non-optimized viral sequences.

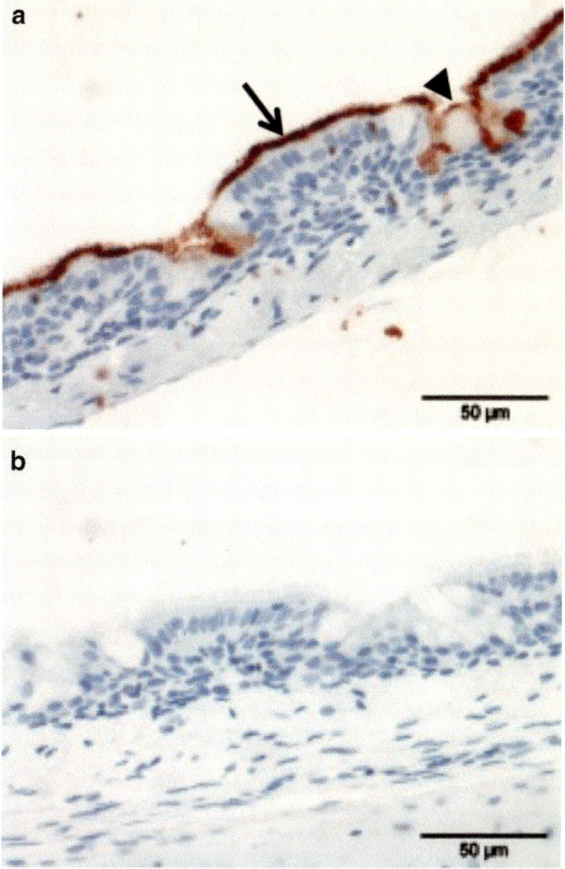

Fig. 3 Protein histochemistry for IBV M41-S1. IBV M41-S1 was applied onto (**a**) untreated chicken trachea and (**b**) chicken trachea treated with neuraminidase. Positive staining (*red*) in cilia and goblet cells is indicated with an *arrow* and *arrowhead*, respectively

2. Transfection of HEK 293T cells with pCD5 expression vector has been described previously [10, 11].

3. Dissolving of PEI in distilled water might take up to 1 or 2 days. The solution should be continuously stirred at 50–60 °C and when it is completely dissolved, filter-sterilize, aliquot, and store at –20 °C. The efficiency of PEI for transfecting HEK 293T cells with DNA is tested by using PEI ratios from 1:5 to 1:20. The number of transfected cells is counted using a fluorescence microscope under 10× magnification. The best ratio to use for subsequent transfection is the ratio that gives the highest percentage (usually 40 %) of transfected cells with lowest toxicity or cell death.

4. Tissues that easily detach during antigen retrieval, including for example trachea, can be mounted onto KP plus slides to reduce the tissue damage.

5. Concentration gradient ensures reaching the optimum amount of lectins required for complete blocking of the binding of recombinant proteins.

6. By seeding 1×10^7 cells per T175 flask we were able to reach 50–60 % confluence after 24 h post seeding. When compared to <50 % or >60 % cell confluence, transfection at 50–60 % confluence results in a significantly higher transfection efficiency and thereby higher amounts of recombinant proteins.

7. Proteins in the supernatant (using 5–10 µl) are analyzed using SDS PAGE followed by western blotting to determine whether the protein is properly produced. In particular we check for any degradation, low or no expression and correct molecular weight (IBV S1 protein is highly glycosylated and migrates around 110 kDa). Upon high amounts of protein in the culture supernatant (usually appearing as thick bands of ≥5 mm in the film) we add 250 µl of 50 % Strep-Tactin sepharose suspension for each 10 ml of supernatant. However, compared to column-based purification minor fraction of proteins were lost with the supernatant after purification with the beads. If necessary, column based purification can be done according to the manufacturer's instructions.

8. Since the beads tend to stick on to the surface of the tube, it is important not to disturb the sediment after centrifugation and while transferring to a 2 ml Eppendorf tube. If necessary, to recover more beads from the surface of the tube add PBS for another one or two times, but limit the total volume to no more than 1.8 ml to prevent spilling of the beads while closing the Eppendorf tube.

9. For every 250 µl of 50 % *Strep*-Tactin sepharose suspension we use 125 µl of elution buffer. Whenever we obtained low protein yields (<4 mg/ml) the proteins were concentrated using Vivaspin according to the manufacturer's instructions.

10. We use ≥2 µl of purified proteins to measure the concentrations in Qubit fluorimeter. We also approximated the protein concentrations compared to a BSA standard after GelCode Blue/Coomassie staining of a SDS PAGE gel.

11. Performing antigen retrieval in the microwave can destroy some tissues (for example tracheal epithelium and cartilage). In such instances transfer the glass slides into a polypropylene Coplin jar filled with citrate buffer, cover with a lid, and keep in a water bath preheated to 80 °C for 45 min.

12. Since *Strep*-Tactin HRP is optimized only for western blotting different lots may complex to a different extent with spike proteins. Therefore, every lot number has to be tested using a prior lot number giving positive signals. Moreover, the amount of *Strep*-Tactin HRP to the total volume (1:200) was opti-

mized for IBV-S1, and has to be optimized accordingly when using a recombinant protein with different molecular weight.

13. Wear gloves when handling AEC. Apply AEC in a fume hood and discard safely the water drained with AEC. For large tissue sections a coverslip can be used to spread AEC drops gently over the tissues, thus minimizing the required amounts of AEC to sufficiently cover tissues.

14. We used both *Vibrio cholera* neuraminidase and *Arthrobacter ureafaciens* neuraminidase. Compared to *Vibrio cholera* neuraminidase, *Arthrobacter ureafaciens* neuraminidase showed more efficient cleaving of sialic acids from tissues embedded in paraffin. It is important to apply sufficient volume of total fluid to prevent drying off the tissues during incubation at 37 °C.

Acknowledgment

We thank Steven van Beurden for critical reading of this chapter.

References

1. Cavanagh D (2007) Coronavirus avian infectious bronchitis virus. Vet Res 38:281–297

2. Winter C, Schwegmann-Wessels C, Cavanagh D et al (2006) Sialic acid is a receptor determinant for infection of cells by avian infectious bronchitis virus. J Gen Virol 87:1209–1216

3. Winter C, Herrler G, Neumann U (2008) Infection of the tracheal epithelium by infectious bronchitis virus is sialic acid dependent. Microb Infect 10:367–373

4. Casais R, Dove B, Cavanagh D et al (2003) Recombinant avian infectious bronchitis virus expressing a heterologous spike gene demonstrates that the spike protein is a determinant of cell tropism. J Virol 77:9084–9089

5. Yamada Y, Liu DX (2009) Proteolytic activation of the spike protein at a novel RRRR/S motif is implicated in furin-dependent entry, syncytium formation, and infectivity of coronavirus infectious bronchitis virus in cultured cells. J Virol 83:8744–8758

6. Cavanagh D, Davis PJ, Darbyshire JH et al (1986) Coronavirus IBV: virus retaining spike glycopolypeptide S2 but not S1 is unable to induce virus-neutralizing or haemagglutination-inhibiting antibody, or induce chicken tracheal protection. J Gen Virol 67: 1435–1442

7. Bosch BJ, van der Zee R, de Haan CA, Rottier PJ (2003) The coronavirus spike protein is a class I virus fusion protein: Structural and functional characterization of the fusion core complex. J Virol 77:8801–8811

8. Wickramasinghe INA, de Vries RP, Gröne A et al (2011) Binding of avian coronavirus spike proteins to host factors reflects virus tropism and pathogenicity. J Virol 85:8903–8912

9. Promkuntod N, van Eijndhoven REW, de Vrieze G et al (2014) Mapping of the receptor-binding domain and amino acids critical for attachment in the spike protein of avian coronavirus infectious bronchitis virus. Virology 448:26–32

10. Bosch B, Bodewes R, de Vries RP et al (2010) Recombinant soluble, multimeric HA and NA exhibit distinctive types of protection against pandemic swine-origin 2009 A(H1N1) influenza virus infection in ferrets. J Virol 84: 10366–10374

11. de Vries RP, de Vries E, Bosch BJ et al (2010) The influenza A virus hemagglutinin glycosylation state affects receptor-binding specificity. Virology 403:17–25

Chapter 15

Identification of Protein Receptors for Coronaviruses by Mass Spectrometry

V. Stalin Raj, Mart M. Lamers, Saskia L. Smits, Jeroen A.A. Demmers, Huihui Mou, Berend-Jan Bosch, and Bart L. Haagmans

Abstract

As obligate intracellular parasites, viruses need to cross the plasma membrane and deliver their genome inside the cell. This step is initiated by the recognition of receptors present on the host cell surface. Receptors can be major determinants of tropism, host range, and pathogenesis. Identifying virus receptors can give clues to these aspects and can lead to the design of intervention strategies. Interfering with receptor recognition is an attractive antiviral therapy, since it occurs before the viral genome has reached the relative safe haven within the cell. This chapter describes the use of an immunoprecipitation approach with Fc-tagged viral spike proteins followed by mass spectrometry to identify and characterize the receptor for the Middle East respiratory syndrome coronavirus. This technique can be adapted to identify other viral receptors.

Key words Receptor, DPP4, Middle East respiratory syndrome coronavirus, Mass spectrometry, Immunoprecipitation

1 Introduction

The first step of the infection cycle of a virus is characterized by the interaction between the viral particle and the cell surface receptor. This interaction is followed by a series of events that lead to the delivery of the viral genome inside the cytoplasm. Viruses can use diverse types of molecules to bind and enter cells. The presence of a receptor is the principal determinant of cell, tissue and organ tropism, host range, and virulence. Therefore, identifying a receptor can give clues on pathogenesis, mode of transmission, zoonotic transmission potential and can lead to the design of targeted intervention strategies.

Helena Jane Maier et al. (eds.), *Coronaviruses: Methods and Protocols*, Methods in Molecular Biology, vol. 1282, DOI 10.1007/978-1-4939-2438-7_15, © Springer Science+Business Media New York 2015

For the last three decades the identification of virus receptors has been a major goal in virology. A group of viruses of which many receptors are known are coronaviruses (CoVs). Coronaviruses infect a wide range of avian and mammalian hosts and they are known for their ability to cross the species barrier [1]. This is exemplified by the 2003 severe acute respiratory syndrome (SARS) pandemic that was caused by the SARS-CoV [2]. In 2012, a novel zoonotic CoV was identified from a patient from Saudi Arabia that presented with a severe pneumonia [3]. This virus belongs to the same genus as SARS-CoV and was named Middle East respiratory syndrome coronavirus (MERS-CoV).

For CoVs, the viral Spike (S) protein primarily determines host and cell tropism. It is a type I membrane glycoprotein that is assembled in trimers in the viral envelope. The S protein can be functionally divided into two distinct subunits, S1 and S2. The S1 subunit binds to a cell surface receptor, whereas S2 facilitates fusion with cellular membranes.

Although virus receptors can be identified using several methods [4–8], we identified the MERS-CoV receptor using Fc-tagged S1 proteins in an immunoprecipitation assay followed by mass spectrometry [9]. This assay is basically similar to the method described by Li et al., for the identification of the SARS-CoV receptor [10]. In this assay, the S1 subunit of MERS-CoV is ligated into a fusion vector to generate an S1-Fc fusion protein, for expression in HEK-293T cells and purification using protein A-sepharose beads. Incubation of the S1-Fc proteins with whole cell lysate of virus-susceptible cells allows the precipitation of the virus receptor with the tagged S1. This complex can then be pulled down from the lysate using protein A-sepharose beads. Subsequently, mass spectrometry is employed to identify candidate protein receptors (Fig. 1). These candidates must be evaluated functionally, which is done using flow-cytometric binding assays, infection blocking experiments using antibodies against the candidate receptor, and finally by attempting to infect non-susceptible cells that have been transfected with the candidate receptor. This method has been successfully employed for the rapid identification of the SARS-CoV and MERS-CoV receptor [9, 10] and is suitable for identification of protein receptors with reasonable affinity. Glycan receptors cannot be identified using the described method; treatment of susceptible cells with glycosidases prior to infection can give an insight into the type of viral receptor. Success of the protein receptor pulldown using the S1-Fc as bait depends on the affinity of S1-receptor interaction. A FACS-based S1-Fc cell-binding assay provides good insight in the strength of this interaction. The FACS-based S1-Fc assay is also instrumental to identify cell lines with high levels of receptor expression that can be used as a source for receptor affinity-isolation. Alternatively, in the absence of a suitable cell line, homogenates of tissue targeted by the virus can also be used for immunoprecipitation of the receptor.

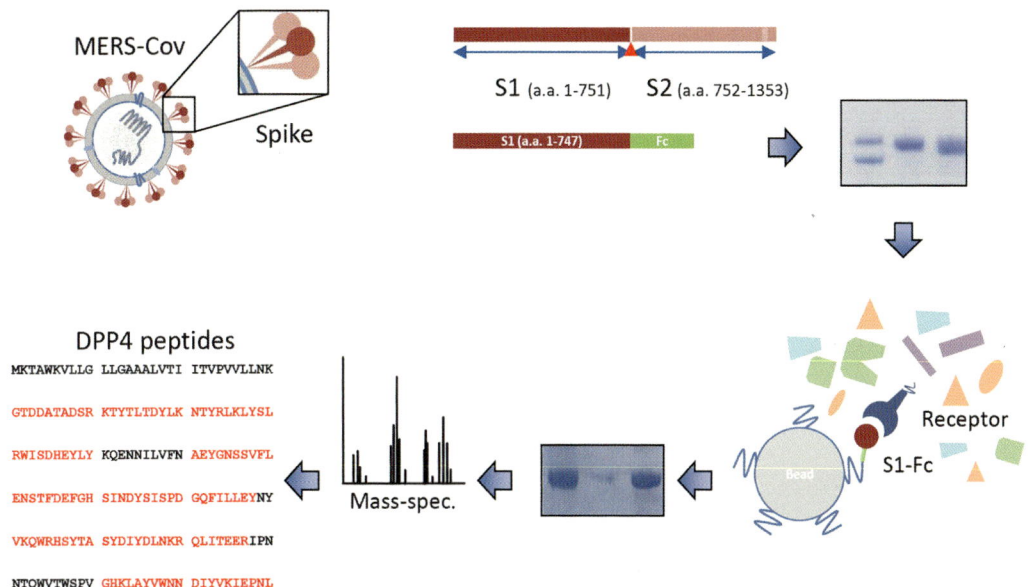

DPP4 peptides

MKTAWKVLLG LLGAAALVTI ITVPVVLLNK

GTDDATADSR KTYTLTDYLK NTYRLKLYSL

RWISDHEYLY KQENNILVFN AEYGNSSVFL

ENSTFDEFGH SINDYSISPD GQFILLEYNY

VKQWRHSYTA SYDIYDLNKR QLITEERIPN

NTQWVTWSPV GHKLAYVWNN DIYVKIEPNL

Mass-spec.

Fig. 1 Schematic drawing of the identification of the MERS-CoV receptor. The S1 subunit of the MERS-CoV S protein is expressed as an IgG Fc-tagged protein in HEK-293T cells and purified using protein A-sepharose beads. Incubating the S1-Fc protein with whole cell lysate of susceptible cells allows the precipitation of the virus protein receptor with the tagged S1. This complex can be pulled down from the lysate using protein A-sepharose beads. Subsequently, mass spectrometry is employed to identify candidate receptors

2   Materials

2.1   RNA Isolation, cDNA Synthesis, PCR, and Cloning

1. The virus used in this protocol is used as an example. A stock of MERS-CoV-EMC isolate was prepared at 10^7 TCID$_{50}$/ml and stored at -70 °C.

2. RNA isolation: viral RNA isolation kit or tissue RNA isolation kit (Qiagen) or equivalent.

3. Reverse transcriptase, e.g., SuperScript II or equivalent.

4. 100 mM dithiothreitol (DTT).

5. RNase inhibitor.

6. 10 mM random primers.

7. Pfu Ultra II Fusion HS DNA polymerase or equivalent.

8. 10 mM dNTPs.

9. 10 mM gene specific forward primer, e.g., MERS-CoV S1 forward primer cgaattcaccATGATACACTCAGTGTTTCTAC: the nucleotides in upper case represent MERS-CoV and those in lower case a suitable restriction endonuclease site for cloning purposes; the example here contains an *Eco*RI site.

10. 10 mM gene specific reverse primer, e.g., MERS-CoV S1 reverse primer cggatccGGTGTGAGAGTACTAGGTGTC:

the nucleotides in upper case represent MERS-CoV and those in lower case a suitable restriction endonuclease site for cloning purposes; the example here contains a *Bam*HI site.

11. PCR cleanup kit.

12. DNA gel extraction kit.

13. Competent *E. coli*, e.g., Top10 competent cells.

14. Cloning vectors: pCAGGS (Addgene), pFUSE-hIgG1-Fc2 (InvivoGen), and pCDNA3.1(+) (Life Technologies) or similar.

15. Restriction enzymes, e.g., *Eco*RI and *Bam*HI.

16. SOC medium: 2 % w/v tryptone, 0.5 % w/v yeast extract, 10 mM NaCl, 2.5 mM KCl, 10 mM MgSO$_4$, 20 mM glucose. Adjust to pH 7.5 using NaOH.

17. T4 DNA Ligase.

18. Maxi prep DNA kit (Qiagen) or equivalent.

19. LB medium: 1 % w/v tryptone, 0.5 % w/v yeast extract, 1 % w/v NaCl. Adjust pH to 7.0 with 5 N NaOH. Sterilize by autoclaving.

20. LB Amp medium: LB medium, 100 μg/ml ampicillin added after sterilization. Store at 4 °C.

21. LB Amp plates: LB medium, 1.5 % agar added prior to sterilization. Sterilize by autoclaving and allow to cool to 45 °C. Add ampicillin to a final concentration of 100 μg/ml. Dispense into 90 mm petri dishes and store at 4 °C.

2.2 Expression of S1-Fc Fusion Proteins

1. HEK-293T cells. HEK-293T growth medium: DMEM, 10 % fetal calf serum (FCS), 100 U/ml penicillin, 100 mg/ml streptomycin, 2 mM glutamine, 1 % nonessential amino acids, and 1 mM sodium pyruvate.

2. T175 cell culture flasks.

3. 293T cell expression medium (as described in ref. 11) 293SF II medium (life Technologies), 1 % GlutaMAX, 0.3 % primatone, 0.2 % glucose, 0.37 % NaHCO$_3$ and 1.5 % DMSO.

4. Serum-free DMEM.

5. 1 mg/ml Polyethylenimine (PEI) stock solution: Add 50 mg PEI to 50 ml of endotoxin-free dH$_2$O, place in a 75 °C water bath and vortex every 10 min until completely dissolved, cool to room temperature, neutralize to pH 7.0, filter-sterilize using a 0.22 μm filter, aliquot and store at –20 °C.

6. Protein-A sepharose CL-4B (GE Healthcare) or equivalent.

7. 1 M Tris–HCl pH 8.0.

8. Protein elution buffer: 0.5 M acetic acid pH 3. Add 29 ml of acetic acid to a beaker, make up to 1 l by adding 971 ml dH$_2$O, adjust to pH 3 using NaOH.

9. 3 M Tris–HCl pH 8.8.

2.3 Immunopreci-
pitaion

1. Huh-7 cells
2. Huh-7 growth medium: RPMI1640, 10 % FCS, 100 U/ml penicillin, and 100 mg/ml streptomycin.
3. 100 mm cell culture dishes.
4. Phosphate-buffered saline (PBS).
5. Protein-A sepharose CL-4B (GE Health care) or equivalent.
6. Protein elution buffer: 0.5 M acetic acid pH 3 as in Subheading 2.2.
7. Lysis buffer: 3.3 mg/ml *n*-decyl-β-D-maltopyranoside (DDM), protease inhibitor cocktail complete Mini (Roche).
8. Rubber policemen or plastic cell scrapers.

2.4 Mass
Spectrometry

1. Tris-Glycine SDS sample buffer 2×: 1.25 ml 1 M Tris–HCl (pH 6.8), 4 ml 10 % (w/v) SDS, 2 ml glycerol, 1 ml 0.1 % (w/v) bromophenol Blue, 1 ml 2 M DTT, make up to 10 ml with distilled water and store at room temperature.
2. 10 % pre-cast Tris-Glycine SDS-PAGE gels (Bio-Rad) or equivalent.
3. Tris-Glycine SDS Running buffer: 250 mM Tris base, 1,920 mM glycine, 1.0 % SDS, pH 8.3.
4. Absolute methanol.
5. Coomassie blue staining solution: 100 mg of Coomassie brilliant Blue R250, 10 ml acetic acid, 50 ml methanol, 40 ml dH_2O.
6. Destaining solution: 10 ml acetic acid, 50 ml methanol, and 40 ml dH_2O.
7. 100 mM ammonium bicarbonate (NH_4HCO_3).
8. Acetonitrile.
9. 50 mM NH_4HCO_3.
10. 20 mM dithiothreitol (DTT).
11. 200 mM Tris(carboxyethyl)phosphine (TCEP).
12. 55 mM iodoacetamide (IAA).
13. Trypsin, mass spectrometry grade.
14. Trypsin stock solution: dissolve 100 μg of trypsin in 1 ml of 1 mM HCl, aliquot into 10 μl samples, store at –80 °C.
15. 0.5 % formic acid in 30 % acetonitrile.
16. Razor blade and tweezers.
17. Filter paper.
18. Mickle gel slicer (Brinkman) or equivalent.
19. Ultrasonic water bath.
20. SpeedVac.

21. EASY-nLC coupled to a Q Exactive mass spectrometer (both Thermo Scientific).

22. ReproSil C18 reversed-phase column (Dr. Maisch GmbH).

2.5 Validation of Receptor Identification

1. Cos-7 cells.

2. Cos-7 cell growth medium: DMEM, 10 % FCS, 100 U/ml penicillin, and 100 mg/ml streptomycin.

3. Trypsin–EDTA: 0.25 % w/v trypsin, 0.02 % w/v EDTA in PBS.

4. Hemocytometer or cell counting chamber.

5. Anti-DPP4 or antibody against other protein of interest.

6. Flow cytometer.

7. 4 % formaldehyde.

8. 10 % normal goat serum or serum corresponding species from which secondary antibody is raised.

9. Anti-SARS nsp4 or antibody against other viral protein.

10. Goat anti-rabbit FITC or other suitable secondary antibody.

11. Fluorescence microscope.

3   Methods

3.1 RNA Isolation

The virus used in this method, MERS-CoV EMC, was described previously [12] and is used as an example.

1. Isolate viral RNA from 140 μl of virus stock at 10^7 TCID$_{50}$/ml using the viral RNA isolation kit, following manufacturer's instructions. The tissue RNA isolation kit was used to isolate RNA from 2×10^7 Huh-7 cells, following manufacturer's instructions (*see* **Note 1**).

3.2 cDNA Synthesis

To convert RNA into cDNA we use SuperScript II reverse transcriptase but other reverse transcriptase enzymes can also be used.

1. For a 20 μl reaction mix, 10 μl RNA, 1 μl 10 mM dNTPs, 1 μl 10 mM random primers, and 1.5 μl dH$_2$O.

2. Incubate at 65 °C for 10 min and then at 4 °C for 2 min (*see* **Note 2**).

3. Place on ice.

4. Make reverse transcriptase (RT) mix (for a 20 μl total reaction volume) as follows: 4 μl 5× SuperScript II reaction buffer, 1 μl 100 mM DTT, 0.5 μl RNase inhibitor (20 U/μl), 1 μl 10 mM random primers (Promega), and 1 μl SuperScript II reverse transcriptase (200 U/μl).

5. Add 6.5 μl of RT mix and incubate as follows: 25 °C for 5 min, 50 °C for 45 min, 70 °C for 20 min.

6. Store at 4 °C.

3.3 PCR Amplification of the S1 Region

We strongly recommend the use of Pfu Ultra II Fusion HS DNA polymerase, although other enzymes may be used. The PCR instructions in this protocol are optimized for the use of this polymerase.

1. Prepare the PCR mix (48 μl reaction volume) as follows: 5.0 μl 10× Pfu Ultra II Fusion HS DNA polymerase buffer, 2.5 μl 10 mM dNTPs, 1.5 μl 10 mM gene specific forward primer, 1.5 μl 10 mM gene specific reverse primer, 1.0 μl Pfu Ultra II Fusion HS DNA polymerase, and 36.5 μl dH$_2$O.

2. Add 2.0 μl of the cDNA reaction mix from Subheading 3.2, **step 6** to the 48 μl PCR mix and mix by pipetting the solution up and down several times.

3. Incubate the PCR cDNA mix as follows: 94 °C for 3 min; then incubate for 39 cycles at: 94 °C for 20 s, 58 °C for 30 s, 72 °C for 2 min, with a final extension at 72 °C for 10 min.

4. Store PCR reaction at 4 °C (*see* **Note 3**).

5. Analyze the PCR products by standard agarose gel electrophoresis.

6. If a single PCR product of the expected size is detected, remove polymerase, dNTPs, and primers using a standard PCR cleanup kit.

7. If multiple products are detected, separate PCR products by gel electrophoresis, remove an agarose slice containing the required product and use a gel extraction kit to isolate the DNA from the agarose slice.

8. Elute the required PCR product from the cleanup or gel purification kit column in dH$_2$O.

9. Analyze the PCR product using an appropriate restriction enzyme or enzymes followed by standard agarose gel electrophoresis to confirm that the PCR product is as expected.

10. Quantify the nucleic acid concentration of the PCR product using NanoDrop 1000 or similar spectrophotometer.

11. Store the PCR product at −20 °C.

3.4 Cloning and Expression of the S1-Fc Fusion Protein

1. The gene fragment encoding the Fc part of human IgG1 is PCR amplified from pFUSE-hIgG1-Fc2 with flanking restriction sites using the forward cgaattcagatctTGAGCCCAAATCTTGTGAC and reverse primer cggatccTCATTTACCCGGAGACAGG. Subsequently, the EcoRI/BamHI digested PCR fragment can be cloned into the EcoRI-BglII digested pCAGGS vector creating the pCAGGS-Fc vector.

2. Digest the S1 PCR product and the pCAGGS-Fc vector by adding the following into separate 1.5 ml tubes: 2 μl of the appropriate 10× restriction buffer, 1 μg of DNA (PCR product or vector), H$_2$O up to 20 μl, 20 U of the appropriate

restriction enzyme (*Eco*RI and *Bam*HI in the example in Subheading 2.1).

3. Mix gently by pipetting and incubate at 37 °C for 1 h.

4. Electrophorese the restriction digests of the vector and S1 PCR product in a agarose gel, identify the products and cut out the gel slices containing the digested products.

5. Purify the DNA products from the agarose slices using a gel extraction kit, elute the DNA into H_2O and quantify the nucleic acid concentrations using a NanoDrop.

6. Ligate the digested S1 PCR product into the pCAGGS-Fc vector by adding the following to a 1.5 ml tube: 2 μl 10× T4 ligation buffer, DNA of digested vector and S1 PCR product in a 1:3 molar ratio, H_2O up to 19 μl, 1 μl T4 DNA ligase, mix gently by pipetting (*see* **Note 4**).

7. Incubate the ligation mixture at 16 °C for 1 h.

8. Transform competent *E. coli* cells by adding 2–5 μl of ligation mixture to 50 μl competent cells.

9. Incubate on ice for 30 min.

10. Heat-shock cells for 30 s at 42 °C in a thermocycler or water bath.

11. Add 250 μl of SOC medium.

12. Incubate at 37 °C for 1 h while shaking.

13. Plate 100 μl on prewarmed LB Amp plates.

14. Incubate at 37 °C overnight.

15. Next day, pick colonies using a sterile toothpick for colony PCR screening and storage.

3.5 Colony PCR

1. Transfer a small amount of a colony into 25 μl of LB medium.

2. Make a PCR mix as follows: 2 μl 10× PCR polymerase buffer, 1 μl 10 mM dNTPs, 0.6 μl 10 mM forward primer, 0.6 μl 10 mM reverse primer, 1 μl of the colony mix, 1 μl PCR polymerase, 13.8 μl dH₂O.

3. Incubate the PCR mix for 30 cycles at 94 °C for 20 s, 58 °C for 30 s, 72 °C for 2 min, with a final extension at 72 °C for 10 min.

4. Analyze the PCR products by standard agarose gel electrophoresis.

5. Inoculate PCR positive clones in 2 ml LB Amp medium and to grow at 37 °C for ~8 h while shaking.

6. From this 2 ml culture, inoculate 500 μl into a 500 ml of LB Amp medium.

7. Allow the bacteria to grow ~8 h at 37 °C while shaking.

8. Next day, extract plasmid DNA from the bacteria using a maxi prep DNA kit, according to manufacturer's instructions.

9. Perform a restriction digest and analyze the products by standard agarose gel electrophoresis to confirm the plasmid DNA is correct.

10. Determine DNA concentration using a NanoDrop or spectrophotometer and prepare a DNA stock of 1 μg/μl.

3.6 Large-Scale Expression and Purification of S1-Fc Fusion Proteins

1. Seed HEK-293T cells in 20T175 flasks in 40 ml of 293T cell growth medium and incubate at 37 °C with 5 % CO_2 for approximately 24 h until 60–70 % confluent.

2. Prepare a working stock of 1 mg/μl PEI. This can be kept at 4 °C.

3. Two hours prior to transfection, remove medium from 293T cells and replace with 30 ml of fresh prewarmed 293T cell growth medium.

4. For each T175 flask, prepare the DNA transfection solution as follows: add 18 μl of 1 μg/μl pCAGGS-MERS-CoV-S1-Fc plasmid DNA (Subheading 3.5, **step 10**) to 3 ml of serum-free DMEM and mix by pipetting.

5. Add 54 μl of 1 mg/μl PEI to the transfection solution and mix.

6. Incubate at room temperature for 30 min.

7. Add the DNA/PEI complex dropwise to the T175 flask and gently swirl to mix.

8. Incubate cells 4–12 h (determine experimentally).

9. Aspirate the medium from the transfected cells and replace with 40 ml of HEK-293T expression medium, incubate at 37 °C with 5 % CO_2 for 6 days.

10. Prepare 50 % (w/v) protein-A sepharose beads: Add 0.25 g of protein-A sepharose CL-4B to a tube, add 10 ml PBS to form a slurry, centrifuge for 2 min at $2,000 \times g$, remove supernatant, and repeat two more times. Pellet the beads for 2 min at $2,000 \times g$ and resuspend in 1.4 ml PBS per tube (50 % w/v), the final volume will be ~2.8 ml.

11. Collect the expression medium from the transfected HEK-293T cells into 50 ml tubes and centrifuge at $2,850 \times g$ for 10 min to remove cell debris.

12. Transfer medium to new 50 ml tubes and centrifuge again at $2,850 \times g$ for 15 min.

13. Transfer cleared medium to new 50 ml tubes and keep it on ice; take a 100 μl aliquot and store at –20 °C.

14. Add 0.5 ml of washed protein-A sepharose beads (50 % w/v) and 800 μl of 1 M Tris–HCl pH 8.0 to each 40 ml supernatant to neutralize the pH and incubate overnight, rotating at 4 °C (*see* **Note 5**).

15. Collect the protein-A sepharose beads by centrifugation at $2,000 \times g$ for 15 min (see **Note 6**).

16. Pool all the protein-A sepharose beads together in a 50 ml tube and wash three times with 10 ml PBS.

17. After the final centrifugation, resuspend the protein-A sepharose beads in 1 ml of 0.5 M acetic acid pH 3 elution buffer and incubate for 1 min at room temperature.

18. Centrifuge the protein-A sepharose beads at $14,000 \times g$ for 10 min and transfer the supernatant to a 1.5 ml tube.

19. Repeat **steps 17** and **18** twice more to elute any remaining S1-Fc protein from the protein-A sepharose beads.

20. To remove any remaining protein-A sepharose beads in the supernatant repeat **step 18** once and transfer supernatant to a fresh tube.

21. To neutralize the pH of the eluted S1-Fc protein, add 200 µl of 3 M Tris–HCl pH 8.8 (final pH 7.5).

22. Quantify the protein concentration using a NanoDrop at 280 nm.

23. To analyze the size of the eluted S1-Fc protein, run 2 µg of the protein in a standard 10 % SDS-PAGE gel.

24. Aliquot the S1-Fc protein and store at –80 °C.

3.7 Immunoprecipitation

3.7.1 Preparation of Cell Lysate

1. Seed 5×10^7 Huh-7 cells in 100 mm dishes with growth medium and incubate at 37 °C for 24 h to allow the cells to become confluent.

2. Wash the adherent cells twice with ice-cold PBS and allow the PBS to drain off.

3. Add 1 ml of DDM lysis buffer onto the cells and gently rock the dish to cover the entire cell sheet.

4. Scrape adherent cells off the dish with either a rubber policeman or a plastic cell scraper and transfer the cell suspension into a fresh centrifuge tube. Gently rock the suspension on either a rocker or an orbital shaker at 4 °C for 15 min to lyse the cells.

5. Centrifuge the lysate at $14,000 \times g$ in a precooled microcentrifuge for 1 min.

6. Immediately transfer the supernatant to a fresh centrifuge tube and discard the pellet.

3.7.2 Preclearing of the Huh-7 cell Lysate

1. Prepare a 50 % (w/v) protein-A sepharose bead slurry as in Subheading 3.6, **step 10**.

2. Add 100 µl of the protein A-sepharose bead slurry to every 1.5 ml of cell lysate and incubate at 4 °C for 10 min on a rocker or orbital shaker (*see* **Note 7**).

3. Remove the beads by centrifugation at $14,000 \times g$ at 4 °C for 1 min and carefully transfer supernatant to a fresh tube.

3.7.3 Immunoprecipitation

1. Add 2.5 µg of the purified S1-Fc fusion protein (Subheading 3.6, **step 24**) to 1.5 ml of the Huh-7 precleared lysate and incubate for 1 h at room temperature on a rocker or an orbital shaker.

2. Use 1.5 ml of the Huh-7 precleared lysate, without the purified S1-Fc fusion protein, as a negative control and incubate as described in **step 1**.

3. Capture any immunocomplexes between the S1-Fc fusion protein and the precleared Huh-7 cell lysate by adding 150 µl of the protein A-sepharose 50 % bead slurry to 1.5 ml of the lysates in Subheading 3.7, **step 2**, gently mix overnight at 4 °C on either a rocker or an orbital shaker.

4. Collect the protein A-sepharose beads by pulse centrifugation (i.e., 5 s in the microcentrifuge at $14,000 \times g$). Discard the supernatant and wash the protein A-sepharose beads twice with DDM lysis buffer and once in PBS alone. Discard the supernatants.

5. Resuspend the protein A-sepharose beads in 200 µl of PBS and store at 4 °C.

3.8 Mass Spectrometry Analysis by Nanoflow LC-MS/MS

3.8.1 SDS-PAGE

1. Prepare 100 mM NH_4HCO_3–acetonitrile wash solution as follows: 1:1 (v:v) of 100 mM NH_4HCO_3 and acetonitrile.

2. Pellet 100 µl of the protein A-sepharose beads (Subheading 3.7.3, **step 5**) by pulse centrifugation and discard the supernatant.

3. Resuspend the protein A-sepharose beads in 30 µl 2× Tris-Glycine SDS sample buffer, mix gently and incubate at 100 °C for 10 min (*see* **Note 8**).

4. Centrifuge at $14,000 \times g$ for 1 min

5. Load 15 µl of supernatant on a 10 % pre-cast Tris-Glycine SDS-PAGE gel, electrophorese the sample for 40 min at 100 V; Alternatively, the supernatant can be transferred to a fresh 1.5 ml tube and stored frozen at −20 °C for later use, frozen supernatants should be reboiled for 5 min directly prior to loading on a gel.

6. Transfer SDS-PAGE gel to a clean cell culture dish or other plastic container and cover with Coomassie blue staining solution. Incubate at room temperature for 45 min while shaking.

7. Destain the gel with destaining solution until bands can clearly be seen and leave the gel in dH_2O in a clean cell culture dish.

8. Cut out the lane of interest using a clean razor blade and tweezers and put the complete lane onto two dH_2O-wetted filter papers (1.5×10 cm).

9. Clean the razor blade of the Mickle gel slicer with methanol and then dH_2O.

10. Place the filter paper with the gel lane on top onto the sled of the Mickle gel slicer and cut the gel lane into 1 mm slices.

11. Depending on the complexity of the protein mixture in the gel lane, transfer two or three adjacent slices to 1.5 ml tubes that contain 600 μl of NH_4HCO_3–acetonitrile wash solution, so that you divide the complete gel lane over 20–30 sample tubes.

3.8.2 Destaining and Washing of Gel Pieces

1. Destain the gel slices by shaking at 4 °C overnight in the 100 mM NH_4HCO_3–acetonitrile wash solution (*see* **Note 9**).

2. Aspirate off the wash solution with a gel loading tip, replace with 0.5 ml of fresh NH_4HCO_3–acetonitrile wash solution, and shake for 1 h at 4 °C.

3. Wash with 200 μl dH_2O once and with 200 μl NH_4HCO_3–acetonitrile wash solution twice for 15 min.

4. Shrink the gel pieces using 200 μl 100 % acetonitrile and flick the tube until the gel pieces turn white.

5. Incubate for 5 min at room temperature.

6. Aspirate off acetonitrile and air-dry the gel slices for 5 min.

3.8.3 Reduction and Alkylation of Proteins (See **Note 10***)*

1. Freshly prepare gel swelling solution as follows: 5 ml 100 mM NH_4HCO_3, 5 ml freshly prepared 20 mM DTT.

2. Freshly prepare alkylation solution as follows: Dissolve 102 mg of Iodoacetamide (IAA) in 5 ml of dH_2O and then add 5 ml of 100 mM NH_4HCO_3.

3. Swell each gel slice in 200 μl gel swelling solution.

4. Incubate for 1 h at 37 °C.

5. Remove the gel swelling solution and add 200 μl alkylation solution to each gel slice.

6. Incubate for 1 h at room temperature in the dark.

7. Wash the gel slices twice with 200 μl of the NH_4HCO_3–acetonitrile wash solution for 15 min.

8. Shrink the gel pieces in 200 μl 100 % acetonitrile.

3.8.4 In-Gel Digestion

1. Just before use, prepare a 10 μg/ml trypsin working solution by diluting the 100 μg/ml trypsin stock solution with 50 mM NH_4HCO_3.

2. Add 10 μl of the 10 μg/ml trypsin working solution to every 1 mm gel slice so that each slice is fully immersed in the trypsin working solution.

3. Incubate the gel slices for 30 min at room temperature.

4. Check that the gel pieces are still fully covered by the trypsin solution, if not, add some more trypsin working solution.

5. Incubate overnight at 37 °C; shaking not necessary.

*3.8.5 Extraction
of Peptides from the Gel*

1. Centrifuge the gel slices for 10 s and remove trypsin.

2. Add 50 μl of the 0.5 % formic acid solution in 30 % acetonitrile peptide extraction solution, sonicate in an ultrasonic bath for 2 min at room temperature, then incubate for 30 min at room temperature.

3. Transfer the supernatant to a clean 1.5 ml tube or into a 96-well plate.

4. Repeat **steps 3** and **4** twice and combine the supernatants.

5. Dry the combined supernatants in a SpeedVac.

3.8.6 Mass Spectrometry

1. Dissolve the peptides in 30 μl of 0.1 % formic acid.

2. Analyze the samples by nanoflow LC-MS/MS on an EASY-nLC coupled to a Q Exactive mass spectrometer, operating in positive mode and equipped with a nanospray source.

3. Trap peptide mixtures on a ReproSil C18 reversed-phase column (column dimensions 1.5 cm × 100 μm, packed in-house) at a flow rate of 8 μl/min^{-1}.

4. Separate peptides on a ReproSil C18 reversed-phase column (column dimensions 15 cm × 50 μm, packed in-house) using a linear acetonitrile gradient from 0 to 80 % (A = 0.1 % formic acid; B = 80 % (v/v) acetonitrile, 0.1 % formic acid) in 70 or 120 min and at a constant flow rate of 200 nl/min^{-1}.

5. Spray the column eluent directly into the electrospray ionization (ESI) source of the mass spectrometer.

6. Acquire mass spectra in continuum mode; fragment the peptides in data-dependent mode by higher-energy collisional dissociation (HCD).

7. For data analysis, create peak lists automatically from raw data files using a software suite such as Mascot Distiller software (Matrix Science), Proteome Discoverer (Thermo), or MaxQuant. Use a database search engine such as Mascot or Andromeda (MaxQuant) for searching peak lists against a Uniprot fasta database that contains all human protein sequences. For control purposes, merge the human protein database with a fasta database containing all for example MERS virus protein sequences. Perform the database search analysis on either an in-house server or a multi-core desktop PC.

8. Human Dipeptidyl peptidase 4 (DPP4 or CD26) tryptic peptides are expected to be detected specifically in MERS-CoV-S1-Fc IP samples, in relation to the examples given.

3.9 Cloning and Expression of Human DPP4 or Appropriate Receptor

3.9.1 RNA Isolation and PCR Amplification of Human DPP4

1. Isolate total RNA from Huh-7 cells and make cDNA as described in Subheadings 3.1 and 3.2.

2. Amplify complete human *DPP4* using Pfu Ultra II fusion HS DNA polymerase using the PCR protocol described in Subheading 3.3 using specific primers for human *DPP4* or the gene sequence of the protein of interest.

3.9.2 Cloning and Expression of Human DPP4

1. Clone the complete *DPP4* gene or gene of interest into the pcDNA 3.1 (+) expression vector into the *Bam*H1 and *Not*I site (pcDNA-hDPP4 plasmid).

2. Check the construct by sequence analysis.
   Prepare cells not susceptible to virus infection (e.g., Cos-7 cells) and transfect them with the DPP4-expression plasmid. Transfect pcDNA-hDPP4 using PEI (plasmid PEI ratio 1:3)

3. At 24 h post transfection the cells can be tested for the cell surface expression of DPP4 using FACS analysis and susceptibility of infection by infecting those cells with MERS-CoV-EMC.

3.9.3 Identification Receptor Expressing Cells by Flow Cytometry Analysis

1. After 24 h of transfection (Subheading 3.9.2, **step 2**), wash the cells once with PBS and add 1 ml of Trypsin EDTA.

2. Incubate at 37 °C for approximately 5 min (until complete disassociation of the cells).

3. Add 1 ml of PBS and mix it by pipetting up and down.

4. Transfer the cell suspension in to a new tube and add additional 5 ml of PBS.

5. Centrifuge at $400 \times g$ for 5 min.

6. Resuspend the cells in 2 ml PBS and count the cells using the counting chamber.

7. Place 5×10^5 cells in a 96 well "v" bottom plate and add S1-Fc or 5 μg/ml antibody against protein under investigation such as goat anti-DPP4 polyclonal antibodies in this example, or without any protein as a control, in volume of 50 μl.

8. Incubate on ice for 30 min.

9. Wash the cells three times with PBS containing 0.5 % BSA.

10. Add 50 μl of FITC-labelled goat anti-human IgG or FITC-labelled rabbit anti goat serum (5 μg/ml) or any other labelled secondary antibody depending on the origin of the antibody in **step 7**.

11. Incubate on ice for 30 min and wash the cells three times with PBS.

12. Resuspend the cells in 190 μl of PBS.

13. Analyze the cells for any fluorescence by flow cytometry (Fig. 2).

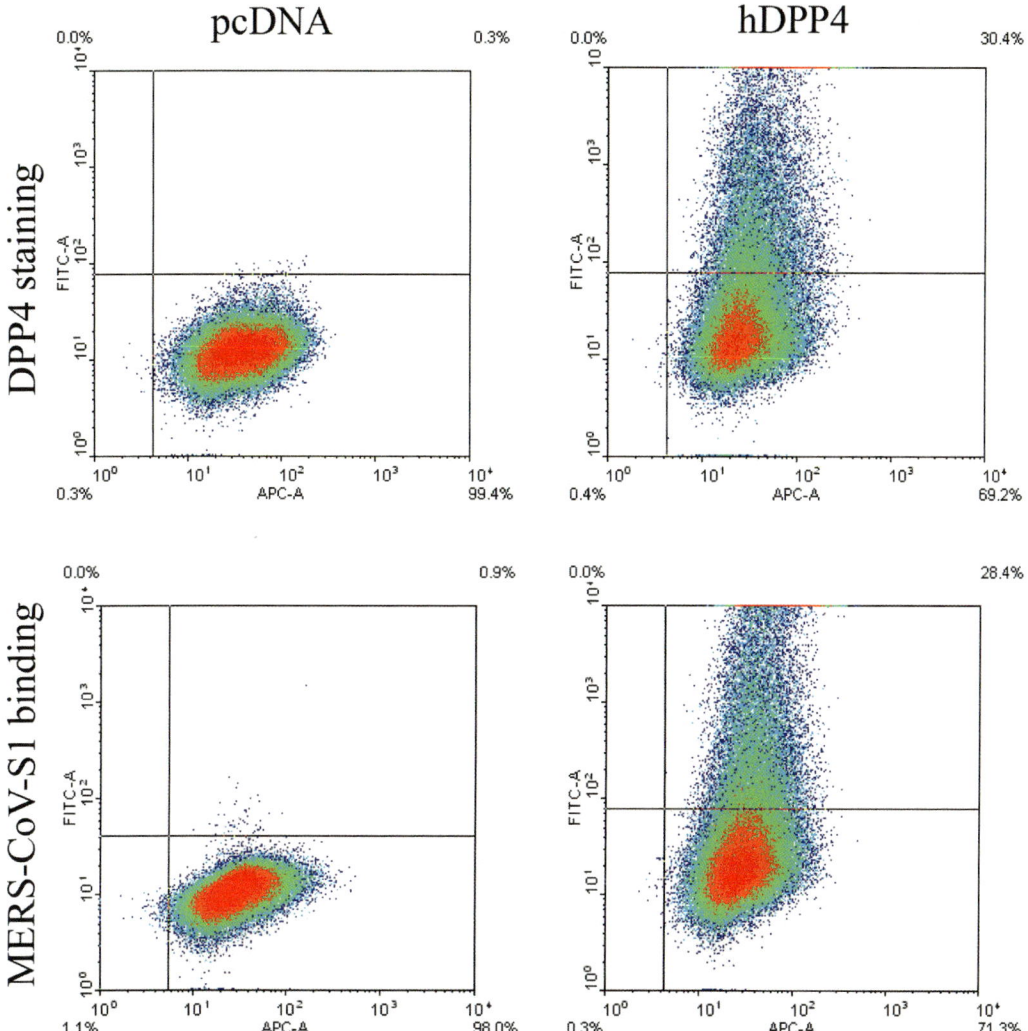

Fig. 2 Human DPP4 expression and MERS-CoV-S1 binding to cells transfected with human DPP4 plasmid or empty plasmid as analyzed by FACS analysis. DPP4 staining (*upper panel*) and MERS-CoV-S1 binding (*lower panel*) are shown

3.9.4 Virus Infection Assay

1. Transfect pcDNA-DPP4 plasmid or plasmid containing the gene under investigation in Cos-7 and empty plasmid as control.

2. After 24 h of transfection, wash the cells with Cos-7 growth medium and incubate the cells with virus under investigation, e.g., MERS-CoV-EMC for 1 h.

3. Wash the cells two times with Cos-7 growth medium containing 1 % FCS to remove any unbound virus and after final wash add 3 ml of fresh medium.

4. Incubate the cells at 37 °C with 5 % CO_2 for 24 h.

5. Fix the cells with 4 % formaldehyde solution for 10 min.

6. Wash the cells three times with PBS.

7. Add 500 μl of 70 % ethanol and keep the plate at 4 °C until immunofluorescent staining.

8. Wash the cells three times with PBS.

9. Add 200 μl of 10 % normal goat serum or serum corresponding to the species from which the secondary antibody in **step 14** is derived.

10. Incubate the cells at 37 °C for 30 min.

11. Remove the 10 % normal goat serum and add 200 μl of any antibody to a specific virus protein, for example rabbit-anti-SARS-CoV nsp4 (5 μg/ml) is cross-reactive for MERS-CoV-EMC.

12. Incubate the cells at 37 °C for 1 h.

13. Wash the cells three times with PBS.

14. Add 200 μl of goat anti-rabbit serum conjugated with FITC (5 μg/ml).

15. Incubate the cells at 37 °C for 1 h.

16. Wash the cells three times with PBS and analyze using a fluorescent microscope (Fig. 3).

a   pcDNA+MERS-CoV   **b**   hDPP4+MERS-CoV

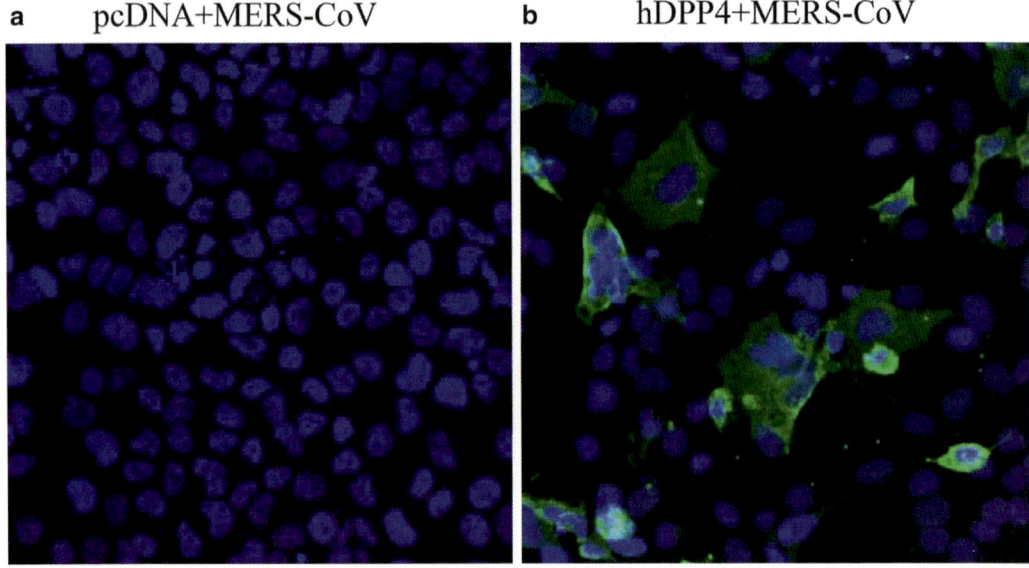

Fig. 3 Characterization of human DPP4 as a functional receptor for MERS-CoV. Human DPP4 plasmid or a control plasmid was transfected in non-susceptible Cos-7 cells and after 24 h the cells were infected with MERS-CoV-EMC. Subsequently, the cells were fixed after 24 h of infection and stained for MERS-CoV. (**a**) Empty plasmid transfected cells. (**b**) Human DPP4 plasmid transfected cells

4   Notes

1. When using another RNA isolation kit, please refer to the recommendations of the respective manufacturer.

2. The amount of RNA that can be used for this reaction volume is 100 pg to 1 µg RNA. If more RNA is used, (e.g., 2 µg), add the appropriate amount of reagents needed.

3. When using another polymerase than Pfu, please refer to the recommendations of the respective manufacturer.

4. Add the ligase last.

5. It is recommended to cut the tip of the pipette-tip off when working with sepharose beads to avoid disruption of the beads.

6. Before aspiration of the medium, take a 100 µl aliquot and store at –20 °C.

7. Preclearing the lysate will reduce nonspecific binding of proteins to the agarose or sepharose when it is used later on in the assay.

8. Be sure that all equipment that you use for running gels (trays, boxes, dishes, tips, etc.) is clean and try to keep equipment that you use for mass spec gels apart from other electrophoresis equipment in your lab. The more the keratin contaminants, the less the protein identifications in the end.

9. A few hours is sufficient.

10. These steps can be skipped if you have alkylated the proteins before running the gel.

Acknowledgement

This work was supported by a grant from the Dutch Scientific Research (NWO; no. 40-00812-98-13066) granted to BJB and BLH. SLS is partly employed by Viroclinics Biosciences.

References

1. Parrish CR, Holmes EC, Morens DM et al (2008) Cross-species virus transmission and the emergence of new epidemic diseases. Microbiol Mol Biol Rev 72:457–470

2. Drosten C, Günther S, Preiser W et al (2003) Identification of a novel coronavirus in patients with severe acute respiratory syndrome. N Engl J Med 348:1967–1976

3. Zaki AM, van Boheemen S, Bestebroer TM et al (2012) Isolation of a novel coronavirus from a man with pneumonia in Saudi Arabia. N Engl J Med 367:1814–1820

4. Dalgleish AG, Beverly PCL, Clapham PR et al (1984) The CD4 (T4) antigen is an essential component of the receptor for the AIDS retrovirus. Nature 312:763–767

5. Greve JM, Davis G, Meyer AM, Forte CP et al (1989) The major human rhinovirus receptor is ICAM-1. Cell 56:839–847

6. Bergelson JM, Shepley MP, Chan BMC et al (1992) Identification of the integrin VLA-2 as a receptor for echovirus 1. Science 255:1718–1720

7. Mendelsohn CL, Wimmer E, Racaniello VR (1989) Cellular receptor for poliovirus: molecular cloning, nucleotide sequence, and expression of a new member of the immunoglobulin superfamily. Cell 56:855–865

8. Yeager CL, Ashmun RA, Williams RK et al (1992) Human aminopeptidase N is a receptor for human coronavirus 229E. Nature 357:420–422

9. Raj VS, Mou H, Smits SL et al (2013) Dipeptidyl peptidase 4 is a functional receptor for the emerging human coronavirus-EMC. Nature 495:251–254

10. Li W, Moore MJ, Vasilieva N et al (2003) Angiotensin-converting enzyme 2 is a functional receptor for the SARS coronavirus. Nature 426:450–454

11. Zeng Q, Langereis MA, van Vliet AL et al (2008) Structure of coronavirus hemagglutinin-esterase offers insight into corona and influenza virus evolution. Proc Natl Acad Sci U S A 105:9065–9069

12. van Boheemen S, de Graaf M, Lauber C et al (2012) Genomic characterization of a newly discovered coronavirus associated with acute respiratory distress syndrome in humans. MBio 3, e00473-12

Chapter 16

Single Particle Tracking Assay to Study Coronavirus Membrane Fusion

Deirdre A. Costello and Susan Daniel

Abstract

Single particle tracking (SPT) of individual virion fusion with host cell membranes using total internal reflection microscopy (TIRFM) is a powerful technique for quantitatively characterizing virus–host interactions. One significant limitation of this assay to its wider use across many types of enveloped viruses, such as coronavirus, has been incorporating non-lipid receptors (proteins) into the supported lipid bilayers (SLBs) used to monitor membrane fusion. Here, we describe a method for incorporating a proteinaceous viral receptor, feline aminopeptidase N (fAPN), into SLBs using cell blebbing of mammalian cells expressing fAPN in the plasma membrane. This receptor binds feline coronavirus (FECV 1683). We describe how to carry out single particle tracking of FECV fusion in this SLB platform to obtain fusion kinetics.

Key words Cell bleb, Single particle virus fusion, Single particle tracking, Supported lipid bilayers, Microfluidics, Fusion kinetics

1 Introduction

A versatile approach for quantitatively studying virus–host cell interactions and viral entry kinetics is single particle imaging using total internal reflection fluorescence microscopy (TIRFM) [1–3] combined with microfluidic handling and fluid supported lipid bilayers (SLBs). There are a number of advantages of using this platform. First, the TIRFM imaging approach allows one to obtain unique single particle data that can be used to discriminate individuals within populations and to identify intermediate mechanistic steps of the entry process that are often masked by ensemble approaches. Second, microfluidic handling enables unique control over the temporal sequence of fusion triggers such as viral binding, exposure to proteases, and pH drop. Third, the supported lipid bilayer that coats the walls of microfluidic channels and acts as a host membrane mimic [4, 5] preserves lipid mobility in the bilayer while its planar geometry removes many experimental complications imposed by live cells. These features facilitate the study of

Helena Jane Maier et al. (eds.), *Coronaviruses: Methods and Protocols*, Methods in Molecular Biology, vol. 1282, DOI 10.1007/978-1-4939-2438-7_16, © Springer Science+Business Media New York 2015

virus–cell interactions and the membrane fusion processes required for viral infection. This convenient platform enables quantitative data collection used for statistical analysis of stochastic virus fusion events and the measurement of membrane fusion kinetics.

One of the biggest drawbacks of the SLB platform for single virion studies of virus entry is the limited range of viruses that are compatible with it. This limitation is imposed by the complexity of the receptor that can be incorporated into the SLB. As such, these platforms have been limited to the study of a few viruses, such as influenza virus [2, 3, 6], Sindbis virus [3], and vesicular stomatitis virus (VSV) [7], that are all known to interact with specific lipids that are easily incorporated into SLBs. However, the receptors for many enveloped viruses, including those in the *Coronaviridae*, are membrane proteins. In this chapter, we describe a method to incorporate membrane proteins into supported planar bilayers for the study of coronavirus fusion using single particle tracking by TIRFM. Here, we focus on one of the best-characterized receptors used by many coronaviruses in the alphacoronavirus genus [8–10], aminopeptidase N (APN).

To provide some perspective, we summarize standard procedures for creating proteoliposomes that can be used to form supported bilayers. Proteins are typically incorporated into vesicles using detergent to solubilize the membrane protein of interest, which is then reconstituted into a vesicle called a proteoliposome [11]. When membrane proteins are solubilized, they are extracted from their native lipid environment, which can expose the hydrophobic transmembrane domains to an aqueous environment. To minimize these energetically unfavorable interactions, the proteins may refold and lead to the incorporation of misfolded proteins into proteoliposomes. The reconstitution process can also lead to randomly oriented proteins in the bilayers. These non-native changes have major implications for pathogenesis: protein conformation in the membrane and its glycosylation are critical to controlling the host–pathogen interaction. To overcome these limitations, we have developed a method of embedding functional, enzymatically active membrane proteins in supported bilayers by using chemically induced cell blebbing [12–15] to create proteoliposomes composed of plasma membrane constituents [16, 17].

Chemical induction of cell blebs results in the productions of proteoliposomes that have never been subjected to detergent solubilization and are ideal for use in biomimetic systems to study virus–host interactions. Cells are first transfected with receptor proteins specific for coronaviruses (or any desired protein), grown to confluency, and then chemically induced to form blebs. To form the planar bilayers, the cell blebs are first adsorbed to a glass surface, and then incubated with liposomes devoid of proteins, but closely matching the lipid composition of the host cell. The rupturing of the liposomes in spaces in between adsorbed blebs induces the rupture of the cell blebs on the surface to form a single planar bilayer [16]. We showed previously [17] that APN in the SLB made from blebs is enzymatically active, oriented properly, and competent to bind CoV prior to membrane fusion. Note that the blebbing procedure applies to many cell types and we have successfully expressed other proteins and incorporated them into SLBs using this approach, including DPP4, the receptor for MERS-CoV.

2 Materials

2.1 Cell Culture, Cell Blebbing, and Proteoliposome Preparation

1. Giant plasma membrane vesicle (GPMV) buffer: 150 mM NaCl, 10 mM HEPE, 2 mM calcium chloride. Adjust pH to 7.4 with hydrochloric acid (HCl).

2. Blebbing buffer: GPMV buffer, 2 mM dithiothreitol (DTT), 25 mM formaldehyde.

3. Phosphate Buffered Saline (PBS): 137 mM NaCl, 2.7 mM KCl, 10 mM $Na_2HPO_4 \cdot 2H_2O$, 1.8 mM KH_2PO_4. Adjust to desired pH with HCl.

4. Biotechnology grade chloroform and methanol.

5. 1-oleoyl-2-palmitoyl-*sn*-glycero-3-phosphocholine (POPC).

6. 1-oleoyl-2-palmitoyl-sn-glycero-3-phosphoethanolamine (POPE).

7. 1,2-dioleoyl-sn-glycero-3-phosphocholine (DOPC).

8. Cholesterol

9. Sphingomyelin.

10. Oregon Green DHPE.

11. Baby hamster kidney (BHK-21) cells.

12. pCAGGS-fAPN (Feline aminopeptidase N plasmid, generous donation from Prof. Kathryn Holmes of University of Colorado).

13. Sonicating water bath.

14. Zetasizer NanoZ (Malvern).

2.2 Microfluidic Setup, Virus Labeling, and SPT Assay

1. Glass coverslips (25 mm × 25 mm; No. 1.5).

2. Hydrogen peroxide (50 % solution).

3. Sulfuric acid.

4. Polydimethylsiloxane (PDMS).

5. Scotch tape.

6. Tygon Microbore tubing (outer diameter: 0.06″, inner diameter: 0.02″).

7. Tube connector (outer diameter: 0.025″, inner diameter: 0.013″, 0.300″ long).

8. 1 ml hypodermic syringes with flat ends.

9. 23 Gauge luer stubs.

10. Syringe pump.

11. Feline Enteric Corona Virus (FECV) 1683.

12. L-1-tosylamido-2-phenylethyl chloromethyl ketone (TPCK) Trypsin.

13. Lipophilic fluorophores such as octadecyl rhodamine B chloride (R18), 1′-dioctadecyl-3,3,3′,3′-tetramethylindocarbocyanine perchlorate (DiI), or octadecyl-rhodamine (Rh110C18) (*see* **Note 1**).

14. Sulforhodamine B (SRB) powder.

15. Lipex extruder, 10 ml (Northern Lipids Inc).

16. 50 and 100 nm polycarbonate filters.

17. Ceramic boat.

18. Plasma cleaner.

19. Inverted Zeiss Axio Observer.Z1 with an α Plan-Apochromat 100× oil objective with a numerical aperture (NA) of 1.46. This microscope is equipped with a Laser TIRF 3 slider (Carl Zeiss, Inc.) and two-channel dual-view imaging system (DV2, Photometrics).

20. Semrock 74 HE GFP/mRFP filter cube.

21. Electron multiplying CCD camera (Hamamatsu ImagEM C9100-13, Bridgewater, NJ).

22. Index-matching liquid (Carl Zeiss, Inc.).

23. Image processing software such as Axiovision and Image J.

24. Data analysis software such as Matlab.

3   Methods

3.1   Cell Culture and Cell Blebbing

1. Cell seeding: Thaw BHK-21 mammalian cells and grow to confluency. Passage and seed cells at a density of 1.5 cells/ml in a 10 cm dish for 24 h at 37 °C and 5 % CO_2 (*see* **Note 2**).

2. Transfection: Following standard transfection protocols, transfect cells with 6 μg of PCAGGS-fAPN and incubate for at least 18 h at 37 °C and 5 % CO_2.

3. Blebbing: Wash cells with 6 ml of GPMV buffer (per 10 cm dish). Add 4 ml of blebbing buffer to each dish. Place dishes rocking gently at 37 °C for 1 h or at room temperature for 2 h (*see* **Note 3**).

4. Place dishes on an inverted microscope and using bright field, look for blebs floating in solution at 100× magnification. Blebs appear as small floating dark spheres.

5. To ensure maximum bleb yield, tap the sides of the dishes gently to help any undetached blebs release into the supernatant. Gently collect the supernatant in a test tube. Place test tube on ice for 15 min to allow any detached cells to the settle to the bottom. Remove all but last 500 μl from the tube and add to a new tube.

6. Dialyze the blebs against two 1-l volumes of GPMV buffer for 24 h to remove DTT and HCHO.

3.2 BHK Fluid (BHKF) Liposome Preparation

1. To form liposomes, mix the appropriate amounts of each POPC–POPE–sphingomyelin–cholesterol–Oregon Green DHPE in the ratio 37.3:34.2:5.7:22.8:0.1. This ratio is chosen to match as close as possible to the lipid content of the BHK cell. Dissolve all components except sphingomyelin in biotechnology grade chloroform before mixing in a scintillation vial. Dissolve sphingomyelin in a 4:1 mixture of chloroform–methanol.

2. Remove the bulk solvent from the vial under a stream of high purity nitrogen gas and then place in a desiccator under vacuum overnight to ensure complete evaporation of all solvent. Protect the lipids and fluorophores from the degradation by room lighting, by wrapping the dessicator in aluminum foil or a dark cloth.

3. Add phosphate-buffered saline at pH 7.4 to the dried lipid film and gently resuspend in a sonicating bath for twenty minutes on the lowest setting. The final lipid concentration should be approximately 2 mg/ml.

4. Extrude the liposomes five times through a 100 nm polycarbonate filter and then twice through a 50 nm polycarbonate filter. The average liposome diameter is typically 100 nm as determined by dynamic light scattering (Zetasizer Nano Z, Malvern).

3.3 Virus Labeling

1. Internal virus labeling: Working in a biosafety hood thaw a vial of coronavirus on ice. Add 10 μl of 20 mM SRB dye to 20 μl of coronavirus. Allow the virus/SRB mixture to incubate for 16–20 h (*see* **Note 4**).

2. Dilute the SRB-labeled coronavirus with 250 μl of GPMV buffer in a microcentrifuge tube. Vortex gently for 30 s to mix. Add 2 μl of TPCK-trypsin (2.5 μg/ml) to the virus/buffer mixture and vortex again for 30 s. Place the tube in a water bath or heat block at 37 °C for 15 min.

3. Viral membrane labeling: Remove trypsin treated virus from the water bath. Add 3 μl of lipophilic fluorophore, e.g., R110C18, to virus–buffer mix and place in a sonicating bath for 1.5 h. Filter out unincorporated dye by centrifuging using a G-25 spin column for 2 min at a speed of $3 \times 1,000 \times g$ (*see* **Note 5**).

3.4 Microfluidic Device Setup

1. Place glass coverslips in slots in a ceramic boat and then put the boat into a glass beaker. Working in a chemical safety hood and wearing appropriate personal protection equipment, measure 43 ml of hydrogen peroxide (50 wt%) in a 1 l-graduated cylinder and pour into the beaker, completely covering the slides. In the same graduated cylinder measure 100 ml of sulfuric acid and then add to the beaker. This mixture is typically called "piranha solution," and the reaction proceeds vigorously for the first few minutes.

2. Allow the reaction to proceed for 10 min. Then very carefully add about 50 ml of deionized water. There will be some bubbling from the beaker as the reaction is quenched.

3. Once the vigorous bubbling has ceased, using a Teflon mitt, pick up the beaker and very carefully decant some of the liquid from the beaker into a chemical waste bottle specifically for piranha waste. Only decant enough liquid so that the glass slides are always covered in liquid.

4. Repeat this rinsing step until approximately 500 ml of deionized water has been used and then transfer the beaker to a sink and rinse with a constant stream of deionized water for ~15 min.

5. Microfluidic devices are formed using polydimethysiloxane (PDMS) in a molding process. In a clean plastic container weigh out the elastomer and cross-linker in a 10:1 ratio (i.e., to make 22 g of PDMS use 20 g of elastomer and 2 g of cross-linker). Using a spatula, mix thoroughly, but be careful not to scratch any plastic from the container into the mixture.

6. After thorough mixing, the sample will be very aerated and the bubbles must be removed by degassing prior to complete cross-linking. Place the plastic container in a clean dessicator and place under vacuum. The PDMS mixture will rapidly climb the sidewalls of the container as the gas escapes. Periodically shutting the dessicator off from the vacuum and allowing slow pressure equalization will prevent the mixture from rising above the container sides. When all the larger gas bubbles are removed, leave the container under vacuum until the mixture is completely degassed.

7. Pour the PDMS mixture gently over the silicon wafer containing the microfluidic channel patterns (*see* **Note 6**). The layer of PDMS poured over the silicon wafer should be no more than 0.5 in. thick.

8. Bake for 3 h at 80 °C.

9. After sufficient cooling has taken place, cut the devices out of the mold and punch holes in the inlet and outlet of the channels using a 23 gauge luer stub.

10. Wash the device with water and then ethanol, especially the inlet and outlet ports to remove any obstructions created during the hole-punching process.

11. Dry the device completely and clean with scotch tape to remove any dust/particulates.

12. Place a dry piranha cleaned slide and PDMS microfluidic device into the chamber of the plasma cleaner. Use oxygen plasma on the highest setting at 600 μm pressure for ~25 s.

13. Following treatment, equalize the pressure in the chamber, then once opened very quickly place the device, channel side down, on top of the glass slide, as close to the center of the glass slide as possible (i.e., far away from the glass edges). Using a pair of tweezers, gently push down on the device to ensure complete contact with the glass.

14. Bake the device at 80 °C for 10 min. This process should result in the glass slide being irreversibly bound to the PDMS mold. If it does not, the device was not clean enough and the above steps should be repeated.

15. Allow the device to cool for 5 min after baking.

16. Cut 2 lengths of Tygon tubing for each channel in the device: 1×23 cm (inlet) and 1×30 cm (outlet).

17. Using a pair of pliers or tweezers insert metal tube connectors into one end of the inlet and outlet tube. Insert the metal end of each tube into the inlet and outlet ports in the assembled microfluidic device. Take care not to apply too much force, which may crack the glass coverslip. Do not insert the tube so that it is touching the bottom of the channel so that the liquid may flow freely into the channels.

18. Place the other end of the inlet tube in a vial containing filtered GPMV buffer and attach a syringe/luer stub to the end of the outlet tube.

3.5 Microscopy and SPT Assay

1. Membrane fusion assays are conducted with total internal reflection fluorescence (TIRF) microscopy using an inverted Zeiss Axio Observer.Z1 with an α Plan-Apochromat 100× oil objective with a numerical aperture (NA) of 1.46. This microscope is equipped with a Laser TIRF 3 slider and two-channel dual-view imaging system to split the image for simultaneous imaging of two emission signals on one CCD chip. In this setup, two lasers are used simultaneously to excite different color fluorophores. Here 561 and 488 nm excitation wavelengths from solid-state lasers are used excite red and green fluorophores, respectively. Excitation laser light is band-pass filtered through a Semrock 74 HE GFP/mRFP filter cube, and then combined with a dichroic mirror before being focused on the outer edge of the back aperture of the objective. The fluorescence emission signal is filtered through a 525/31 and 616/57 nm dual band-pass emission filter and then sent to an electron multiplying CCD camera.

2. Couple the glass coverslip of the microfluidic device to the 100×-objective for TIRF imaging using index-matching liquid.

3. Place device on 10× objective of microscope (or another low magnification).

4. Attach syringes to a syringe pump. Flow buffer through the channels at a flow rate of 100 μl/min for 2 min to set the walls of the device and clear any debris. Wait for 5 min, then transfer the inlet tubes into a vial containing BHK21-fAPN blebs.

5. Visually inspect the tubing inlet to ensure no bubbles or plugs of air have formed. If bubbles do form, run the syringe pump in reverse at 50 μl/min until the air has been pushed back into the vial of buffer and then transfer tubing to the vial containing the cell blebs.

6. Flow blebs at dilution of 1:4 (blebs–GPMV) buffer (*see* **Note 7**) into the channel at a flow rate of 30 μl/min for 5 min.

7. Allow the blebs to incubate on the glass for at least 30 min.

8. Rinse the microchannel with GPMV buffer for 2 min at 100 μl/min.

9. Flow a 0.5 mg/ml solution of BHKF liposomes into the channel for 2 min at 100 μl/min.

10. Incubate the liposomes in the channel with the cell blebs for at least 1 h. Repeat the aforementioned rinsing step with GPMV buffer.

11. Inspect the bilayer under 100× magnification. If the bilayer appears patchy, further incubation with BHKF liposomes may be required (*see* **Note 8**).

12. Dilute 250 μl of fluorescently labeled virus in 800 μl of PBS buffer at pH 7.4.

13. Flow the diluted virus solution into the microfluidic channel at 30 μl/min for 5 min. Allow the virus to bind to the membrane for at least 20 min.

14. Rinse excess virus from the channel with buffer at a flow rate of 100 μl/min for 2 min.

15. Locate a region in the channel that has a uniform bilayer and a high density of bound virions in the green channel (*see* **Note 9**). Switch the camera view to dual-view mode and turn on the red laser. In this setting, virions that have taken up both SRB and R110C18 should be visible in the green and red channels. Ensure that the focus is maintained in both channels.

16. Switch off the red and green lasers once an appropriate location has been found to avoid unnecessary photobleaching. Carefully switch the inlet tubing for the channel from neutral PBS buffer into acidic PBS buffer (<pH 5.5).

17. Flow acidic buffer into the channel at 100 μl/min for 2 min and set up recording software to run for 4–5 min. Set the camera exposure to a maximum of 100 ms. A drop in background intensity in the green channel indicates that the acidic buffer has reached the channel. Hemifusion follows and is marked by a sharp increase in fluorescence of the punctate fluorescent virions and then diffusion of the green fluorophores into the supported bilayer, radiating away from the punctate dot. A sharp drop in intensity in the red channel marks pore formation in the same particle.

3.6 Imaging Processing and Data Analysis

1. Import the image sequence into an image processing software such as Image J.

2. Determine the approximate time at which acidification occurs by tracking Oregon Green intensity.

3. Create a substack video from the original file. The time at which the Oregon Green intensity drops becomes the first frame of the video substack (time zero). For experiments where only hemifusion is being monitored, track the intensity of each particle that fused over the course of the entire video. For experiments monitoring both hemifusion and pore formation identify and track the intensity only of virions that undergo both processes.

4. To analyze data and obtain fusion kinetics, import this data into data analysis software such as Matlab to determine the time at which the maximum intensity (hemifusion) or intensity drop (pore formation) occurs for each fusing particle. These data can then be analyzed in various ways to determine kinetics parameters associated with the fusion process. One commonly used method plots the data as cumulative distributions and fits with a gamma distribution [6]. In this approach, the number of steps in the kinetic process can be resolved and the rate constant for the limiting step. For pH sensitive fusion, these values can vary with the pH of buffer used to initiate fusion.

---

4 Notes

1. Hemifusion experiments can be carried out using commercially available lipophilic fluorophores such as octadecyl Rhodamine B chloride (R18) or 1'-Dioctadecyl-3,3,3', 3'-Tetramethylindocarbocyanine Perchlorate (DiI) from Invitrogen. Synthesis of R110C18 is described in Floyd et al. [2] and Costello et al. [6].

2. Cell Culture. Cell blebbing is not restricted to BHK-21 cells; a variety of cell lines can be used.

3. Cell blebbing. There are a number of ways in which the cells can be incubated during blebbing. Agitation is not strictly required and blebs can be produced at 37 °C if the incubation time is increased to 2 h. Another very effective method is to place the plates in a thin layer of water shaker/heating bath at 37 °C and incubate for 1 h.

4. Timing of experiment. Microfluidic devices and BHKF liposomes may be prepared days or weeks in advance of the experiment so long as they are stored appropriately. Piranha cleaned glass coverslips should not be used more than 24 h after cleaning. For dual labeling experiments with the coronavirus should be incubated with SRB at least 16 h before co labeling with a lipophilic fluorophore such as R110C18. Once the virus has been labeled, the fluorescent probes will only stay stably associated with the virus for several hours.

5. Optimization of virus labeling. When a new batch of dye is synthesized or new bath of virus is obtained it can be necessary to re optimize the amount of R110C18 required to achieve sufficient dequenching at a single particle level. Typically to optimize the dye concentration, bulk dequenching events should be carried out in a fluorimeter. Incubate the virus with varying amounts of dye and filter. Add the labeled virus to a cuvette and take a baseline reading at the appropriate wavelength for the dye being used. To check if the virus is quenched, add 100 μl of a 10 % solution of Triton-X detergent. A significant and rapid increase in fluorescence indicates that the virus was quenched before solubilization of the membrane by the detergent. Once the bulk quenching concentration has been obtained the optimal quenching conditions in the microfluidic should be close to that obtained in the fluorimeter. Further optimization may be required as a result of photobleaching.

6. Microfluidic master. The microfluidic master silicon wafer used for these experiments was designed in the Cornell Nanoscale Facility (CNF). Each channel was designed to be 135 μm wide and 70 μm deep.

7. Bleb concentration. The ratio of blebs to GMPV buffer recommended in this chapter can be subject to change. Depending on the yield of blebs this ratio may need to be changed in order to optimize adsorption in the microfluidic device. It should be noted that completely saturating the channel with blebs will result in an immobile bilayer upon incubation with BHKF liposomes, presumably because there is no space left for the liposomes to fuse between adsorbed blebs.

8. BHKF incubation. The concentration of BHKF liposomes for bleb incubation is recommended to be 0.5 mg/ml; however, increasing the concentration may help the bilayer form faster.

Depending on the concentration of blebs adsorbed it may also be necessary to increase the BHKF incubation time from 1 h to 2 or 3 h.

9. Laser power. Photobleaching can greatly impair visualization of bound dual-label virions. The lowest laser power possible should be determined, typically by trial and error, so that photobleaching can be avoided but the intensity of the bound virions is above background noise in both red and green channel. The Zeiss microscope used in our experiments, the laser power is modulated to 20 % of total intensity available from the lasers. Within this modulation, the laser power was typically further reduced to between 5 and 10 % in both the red and green channels. These values may vary depending on the microscope set up and software used.

Acknowledgement

We thank the National Science Foundation (Grant CBET-1149452) and (Grant CBET-1263701) for supporting this work. Some experiments were performed at the Nanobiotechnology Center and Cornell NanoScale Facility, a member of the National Nanotechnology Infrastructure Network, which is supported by the National Science Foundation (Grant ECS-0335765). We also acknowledge and thank our collaborators, Professor Gary R. Whittaker and Dr. Jean K. Millet, of the Cornell College of Veterinary Medicine, for supplying FECV for these studies and their expertise on coronavirus infection.

References

1. Axelrod D, Burghardt TP, Thompson NL (1984) Total internal reflection fluorescence. Ann Rev Biophys Bioeng 13:247–268

2. Floyd DL, Ragain JR, Skehel JJ et al (2008) Single-particle kinetics of influenza virus membrane fusion. Proc Natl Acad Sci USA 105:15382–15387

3. Wessels L, Elting MW, Scimeca D et al (2007) Rapid membrane fusion of individual virus particles with supported lipid bilayers. Biophys J 93:526–538

4. Brian AA, McConnell HM (1984) Allogeneic stimulation of cytotoxic T cells by supported planar membranes. Proc Natl Acad Sci USA 81:6159–6163, Epub 1984/10/01

5. Castellana ET, Cremer PS (2006) Solid supported lipid bilayers: from biophysical studies to sensor design. Surf Sci Rep 61(10):429–444

6. Costello DA, Lee DW, Drewes J et al (2012) Influenza virus-membrane fusion triggered by proton uncaging for single particle studies of fusion kinetics. Anal Chem 84:8480–8489

7. Matos PM, Marin M, Ahn B et al (2013) Anionic lipids are required for vesicular stomatitis virus g protein-mediated single particle fusion with supported lipid bilayers. J Biol Chem 288:12416–12425

8. Tresnan DB, Levis R, Holmes KV (1996) Feline aminopeptidase N serves as a receptor for feline, canine, porcine, and human coronaviruses in serogroup I. J Virol 70:8669–8674

9. Tusell SM, Schittone SA, Holmes KV (2007) Mutational analysis of aminopeptidase n, a receptor for several group 1 coronaviruses, identifies key determinants of viral host range. J Virol 81:1261–1273

10. Yeager CL, Ashmun RA, Williams RK et al (1992) Human aminopeptidase N is a receptor for human coronavirus 229e. Nature 357:420–422

11. Ciancaglini P, Simão A, Bolean M et al (2012) Proteoliposomes in nanobiotechnology. Biophys Rev 4:67–81

12. Scott RE, Perkins RG, Zschunke MA, Hoerl BJ, Maercklein PB (1976) Plasma membrane vesiculation: a new technique for isolation of plasma membranes. Science 194:743–745

13. Holowka D, Baird B (1983) Structural studies on the membrane-bound immunoglobulin E-receptor complex. 1. Characterization of large plasma-membrane vesicles from rat basophilic leukemia-cells and insertion of amphipathic fluorescent-probes. Biochemistry 22(14):3466–3474

14. Baumgart T, Hammond AT, Sengupta P, Hess ST, Holowka DA, Baird BA et al (2007) Large-scale fluid/fluid phase separation of proteins and lipids in giant plasma membrane vesicles. Proc Natl Acad Sci USA 104(9):3165–3170

15. Sezgin E, Kaiser H-J, Baumgart T, Schwille P, Simons K, Levental I (2012) Elucidating membrane structure and protein behavior using giant plasma membrane vesicles. Nat Protoc 7(6):1042–1051

16. Costello DA, Hsia CY, Millet JK et al (2013) Membrane fusion-competent virus-like proteoliposomes and proteinaceous supported bilayers made directly from cell plasma membranes. Langmuir 29:6409–6419

17. Costello DA, Millet JK, Hsia CY et al (2013) Single particle assay of coronavirus membrane fusion with proteinaceous receptor-embedded supported bilayers. Biomaterials 34:7895–7904

Part V

Studying Virus-Host Interactions

Chapter 17

Studying Coronavirus–Host Protein Interactions

Chee-Hing Yang, Hui-Chun Li, Cheng-Huei Hung, and Shih-Yen Lo

Abstract

To understand the molecular mechanisms of viral replication and pathogenesis, it is necessary to establish the virus–host protein interaction networks. The yeast two-hybrid system is a powerful proteomic approach to study protein–protein interactions. After the identification of specific cellular factors interacting with the target viral protein using the yeast two-hybrid screening system, co-immunoprecipitation and confocal microscopy analyses are often used to verify the virus–host protein interactions in cells. Identification of the cellular factors required for viral survival or eliminating virus infected cells could help scientists develop more effective antiviral drugs. Here we summarize a standard protocol used in our lab to study the coronavirus–host protein interactions, including yeast two-hybrid screening, co-immunoprecipitation, and immunofluorescence microscopy analyses.

Key words Protein–protein interactions, Virus–host interactions, Human coronavirus, Yeast two-hybrid (Y2H), Co-immunoprecipitation, Confocal microscopy analysis

1 Introduction

Virus–host interactions have long been studied since the first discovery of Tobacco Mosaic Virus in 1898. However, due to the limitation of experimental tools to investigate the mass interacting networks was not always simple. The investigation became more productive upon the technical development of protein biochemistry, nucleic acid sequencing, and several high-throughput screening systems. Three common methods used for high-throughput protein interaction analysis are yeast two-hybrid (Y2H) system, affinity purification, and protein chip [1]. Y2H was first described in 1989 for identifying and analyzing various protein–protein interactions in the yeast model [2]. The GAL4 based Y2H system relies on the GAL4 transcription activator, which consists of a DNA binding domain (BD) and a transcription activating domain (AD). The yeast strain used in this system contains several nutrient gene mutations and without adding these nutrients into the culturing media, these yeast cells will not grow. Successful introduction of separate bait and prey plasmids into the mutant yeast strain will

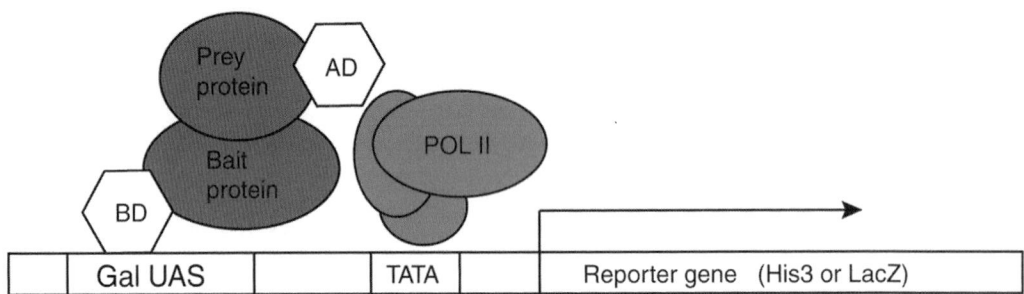

Fig. 1 A schematic view of yeast two-hybrid system. The bait protein fused to the DNA binding domain (BD) of GAL4 protein will bind to the GAL4 UAS. If the GAL4 activating domain (AD)-fused prey protein can interact with the bait protein, then the reporter genes (e.g., LacZ or *His*3) will be activated

provide the lacking nutrients hence the yeast cells will survive. One or more reporter genes will be cloned under the control of the UAS-GAL4 promoter. If the BD-bait fusion protein interacts with the AD-prey fusion protein, the GAL4 activator becomes functional and binds to the GAL UAS promoter region, then activates the reporter gene(s). The expression of one or more reporter genes thus indicates that there are protein–protein interactions between the target protein (bait) and the others (prey). The activation of reporter genes provides a platform to select the yeast clones that have protein–protein interactions. Most reporter genes used in this system are the complements of the mutated nutrient genes, e.g., Histidine, or an enzyme that catalyzes a color change reaction, e.g., LacZ. Figure 1 gives a schematic view of the yeast two-hybrid system. High-throughput screening of cDNA library by Y2H system provides a fast and comprehensive way to identify the possible proteins in the library that could interact with the bait protein. The yeast clones containing the interacting proteins will survive in a nutrient lacking medium due to the activation of the reporter gene.

The identification of virus–host interactions could facilitate the understanding of viral strategy to manipulate cellular functions for its survival or to know how the host controls and eliminates the pathogens. Several studies have used Y2H techniques to study the protein–protein interactions between viruses and host cells [3–6]. A list of the interacting cellular proteins can be established after screening using a viral protein as the bait, providing a framework for further study on the relationship between the virus and the host. However, there are several limitations in this strategy that must be considered. First, due to the modular nature, the fusion proteins are sometimes not folded in a native form; second, the posttranslational modifications of some proteins in mammalian cells are not present in the yeast system, if the interactions between proteins are dependent on these modifications, false-negative results would be observed. In Y2H systems, both the BD-fused and AD-fused proteins must be targeted to the nucleus; therefore, extracellular proteins or organelle targeted proteins may not work in this system. The protocol written in this chapter is used to study protein–protein interactions of cytosolic proteins. Protocols used to study interactions between membrane proteins were depicted previously [7, 8]. We have used the protocol to identify many cellular factors interacting with different viral proteins [9–11], including SARS-CoV nucleocapsid protein [12].

To obtain a more reliable result of the protein–protein interactions, further experiments should be carried out to verify these interactions, e.g., co-immunoprecipitation and confocal microscopy analyses. After Y2H screening, co-immunoprecipitation is a biochemical method often used to verify whether the identified prey protein physically interacts with the bait protein in cells. Figure 2 shows a schematic summary of the co-immunoprecipitation assay. Specific antibodies against X protein or tag 1 peptide will first be coupled to Protein A/G beads, if the X protein is immunoprecipitated, then the Y protein which interacts with the X protein should be precipitated along with it. If the paired proteins of interest (X-Y) are indeed interacting with each other in cells, then these two proteins should also be co-localized in cells. Confocal microscopy analysis is used to verify the co-localization of the two interacting proteins. The co-immunoprecipitation and confocal microscopy analysis protocols written in this chapter have been used in our lab to study several protein–protein interactions [9, 11–16].

2 Materials

2.1 Yeast Strain and Vectors

1. *Saccharomyces cerevisiae* YRG-2 strain, Mata *ura3*-52 *his3*-200 *ade2*-101 *lys2*-801 *trp1*-901 *leu2*-3 112 *gal4*-542 *gal80*-538 LYS2::UAS$_{GAL1}$-TATA $_{GAL1}$-HIS3 URA3::UAS$_{GAL4\ 17mers}$(x3)-TATA$_{CYC1}$-lacZ.

2. pBD-GAL4 Cam phagemid vector (Agilent technology).

3. pAD-GAL4-2.1 Amp phagemid vector (Agilent technology).

4. pACT2 (Clontech).

5. For library-based screening, tissues-specific cDNA libraries cloned into prey vector pACT2 (Clontech).

2.2 Media and Stock Solutions

All the media are prepared with ddH$_2$O. All reagents are autoclaved and stored at room temperature.

1. YEPD liquid medium for general yeast growth: 20 g/L peptone, 10 g/L yeast extract, 0.64 g/L L-tryptophan, add glucose to 2 % (50 ml of a sterile 40 % stock solution).

2. YEPD agar for general yeast growth: 20 g/L peptone, 10 g/L yeast extract, 0.64 g/L L-tryptophan, 2 % agar, add glucose to 2 % (50 ml of a sterile 40 % stock solution).

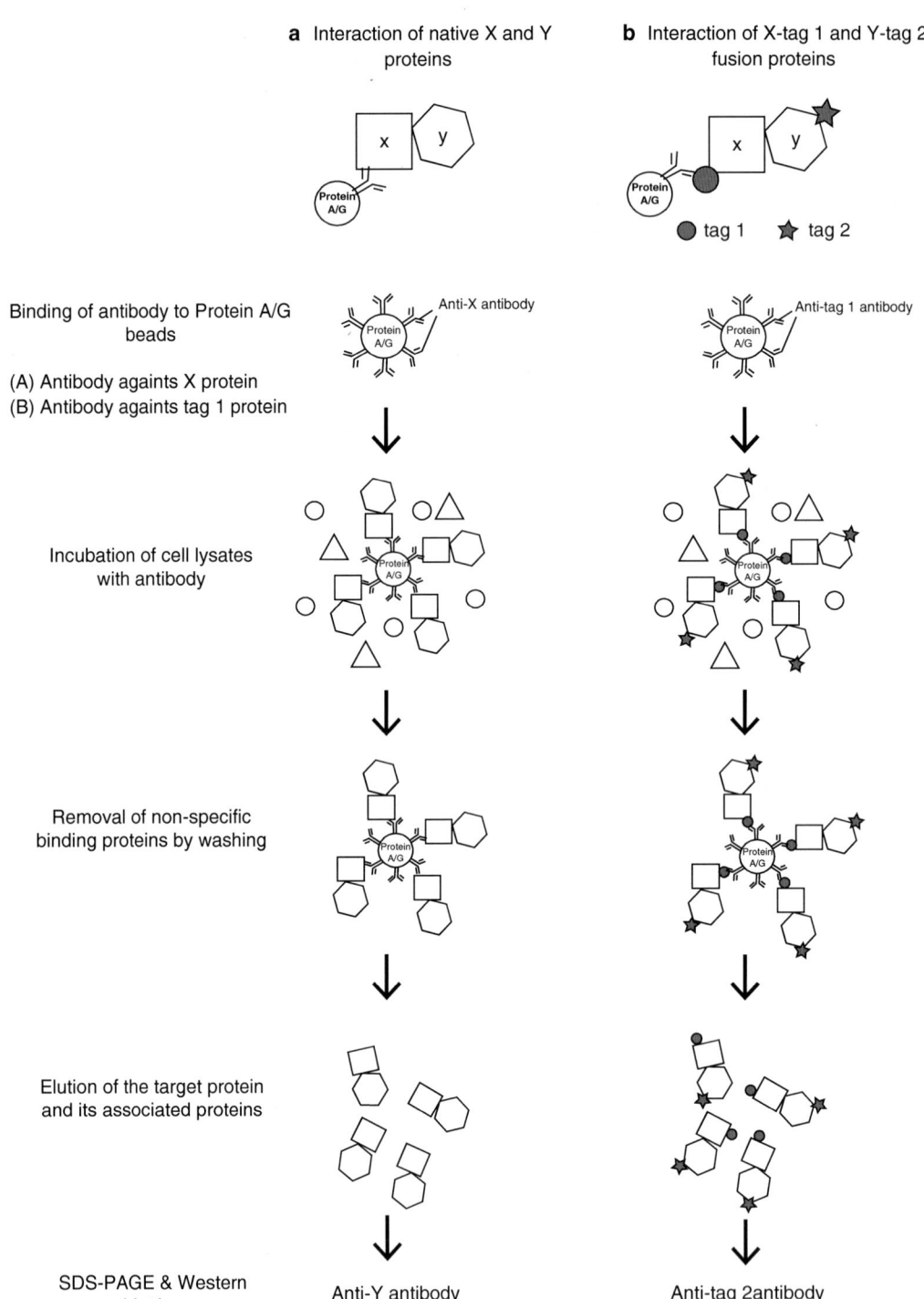

a Interaction of native X and Y proteins

b Interaction of X-tag 1 and Y-tag 2 fusion proteins

● tag 1   ★ tag 2

Binding of antibody to Protein A/G beads

(A) Antibody againts X protein
(B) Antibody againts tag 1 protein

Anti-X antibody

Anti-tag 1 antibody

Incubation of cell lysates with antibody

Removal of non-specific binding proteins by washing

Elution of the target protein and its associated proteins

SDS-PAGE & Western blotting

Anti-Y antibody

Anti-tag 2antibody

Fig. 2 A schematic summary of co-immunoprecipitation assay. (**a**) 16–24 h after viral infection, lysates derived from virus-infected cells will be immunoprecipitated by the antibody against viral protein X. If the cellular protein Y interacts with protein X, it will be co-immunoprecipitated by the anti-X antibody in the presence but not absence of protein X. (**b**) 48 h after transfection of two plasmids encoding the viral X-tag 1 and the cellular Y-tag 2 fusion proteins separately, lysates derived from transfected cells will be immunoprecipitated by the antibody against tag 1. If the cellular protein Y interacts with viral protein X, Y-tag 2 fusion protein will be co-immunoprecipitated by the anti-tag 1 antibody in the presence but not absence of the X-tag 1 fusion protein

Table 1
Preparation of different selection media and agar plates

	−W+G	−W−L+G	−W−L−H+G	−W−L−H+G+3AT	−W−H+G
Uracil	☑	☑	☑	☑	☑
Leucine	☑				☑
Lysine	☑	☑	☑	☑	☑
Histidine	☑	☑			
Adenine	☑	☑	☑	☑	☑
3AT				☑	
Glucose	☑	☑	☑	☑	☑

3. YNP- selection medium (minimal medium necessary for the selection of nutritional mutants): 3 g/L yeast nitrogen base without amino acids and ammonium sulfate, 10 g/L ammonium sulfate, and 0.043 g/L inositol to a volume of 1 L. Prepare different selection media or agar plates by adding different amino acids or 3AT as listed in Table 1. The final concentrations of amino acids added to the selection medium are 0.4 mM uracil, 3.34 mM leucine, 2 mM lysine, 0.26 mM histidine, 0.6 mM adenine, and 50 or 100 mM 3AT.

4. 40 % glucose.

5. 200 mM inositol.

6. 100 mM lysine.

7. 100 mM leucine.

8. 20 mM uracil.

9. 40 mM tryptophan.

10. 40 mM histidine.

11. 40 mM adenine.

12. 1 M 3AT: 3-amino-1,2,4-triazole used as a competitive inhibitor of the His3-gene product.

2.3 Yeast Transformation

1. 2 mg/ml salmon-sperm carrier DNA denatured by boiling for 10 min and placed on ice before transformation.

2. 50 % PEG 3350. Autoclave to sterilize.

3. 10× LiOAc: 1 M lithium acetate, adjust to pH 7.5 with HCl and filter-sterilize. 10× TE buffer: 0.1 M Tris–HCl (pH 7.5), 10 mM EDTA and autoclave.

4. 1× TE/LiOAc: To prepare 10 ml of solution, add 1 ml of 10× LiOAc, 1 ml of 10× TE, and 8 ml of ddH$_2$O dilute to the final concentration of 1×.

5. 1× PEG/LiOAc: To prepare 10 ml of solution, add 1 ml of 10× LiOAc, 1 ml of 10× TE, and 8 ml of 50 % PEG (final concentration, 40 %) dilute to the final concentration of 1×.

2.4 Protein Extraction

1. 2 M lithium acetate (LiAc).

2. 0.5 M NaOH.

3. 5× Laemmli sample buffer: 60 mM Tris–HCl pH 6.8, 10 % glycerol, 2 % SDS, 5 % β-mercaptoethanol, 0.01 % bromophenol blue.

2.5 X-gal Filter Lift Assay

1. Z buffer: 60 mM Na_2HPO_4, 40 mM $NaH_2PO_4·H_2O$, 10 mM KCl, 1 mM $MgSO_4·7H_2O$, adjust to pH 7 with HCl, autoclave and store at room temperature.

2. X-GAL stock solution: 20 mg/ml 5-bromo-4-chloro-3-indolyl-b-D-galactopyranoside in *N,N*-dimethylformamide, stored in dark at −20 °C.

3. X-GAL working solution: Dilute the X-GAL stock solution to 0.2 mg/ml with Z buffer.

4. Sterile filter paper.

5. Liquid nitrogen in suitable container.

6. Hybond-N nylon membrane.

2.6 Plasmid DNA Extraction

1. Solution I: 25 mM Tris–HCl (pH 8), 10 mM EDTA.

2. Solution II: 1 % SDS and 0.2 N NaOH.

3. Solution III: 5 M potassium acetate, adjust to pH 5.5 with Glacial acetic acid.

4. Isopropanol.

5. 75 % ethanol.

6. Competent *Escherichia coli*.

7. LB Amp agar: 10 g/L tryptone, 5 g/L yeast extract, and 10 g/L sodium chloride. Autoclave and allow solution to cool to 55 °C, add ampicillin to final concentration of 100 μg/ml.

2.7 Cell Culture and Plasmid Transfection

1. Vero E6 cells.

2. RPMI 1640 with 10 % FBS, 2 mM L-glutamine, 100 units/ml penicillin, and 0.1 mg/ml streptomycin.

3. Trypsin–EDTA (0.25 %), phenol-red.

4. 1× PBS: Dilute 10× PBS pH 7.4 to 1× with ddH_2O.

5. Transfection reagent: 1 mg/ml PEI (Polyethylenimine 25 kDa linear from Polysciences).

2.8 Co-immunoprecipitation (Co-IP)

1. Modified RIPA buffer: 10 mM Tris–HCl (pH 7.4), 10 mM EDTA, 1 % NP-40, 0.1 % SDS, 150 mM NaCl, and 0.5 % sodium deoxychloride in ddH_2O.

2. Protein A magnetic sepharose.

3. Co-IP binding buffer (1× TBS): 50 mM Tris–HCl (pH 7.5), 150 mM NaCl.

4. Co-IP elution buffer: 0.1 M glycine-HCl (pH 2.5)

5. 5× Laemmli sample buffer: 60 mM Tris–HCl pH 6.8, 10 % glycerol, 2 % SDS, 5 % β-mercaptoethanol, 0.01 % bromophenol blue.

6. Antibody recognizing protein X or tag 1 (Fig. 2)

7. 20× TBS: 500 mM Tris, 3 mM NaCl, adjust to pH 7.0 with HCl, add ddH$_2$O to a total volume of 1 L.

8. Magnetic rack.

2.9 Immunofluo-rescence Analysis

1. Fixing solution: acetone–methanol (1:1).

2. Blocking solution: 1 % skimmed milk, 0.02 % saponin, 0.05 % NaN$_3$ in 1× TBS.

3. Phosphate buffered saline (PBS).

4. Primary antibodies recognizing proteins X and Y or tags 1 and 2 (Fig. 2)

5. Secondary antibodies recognizing Fc region of primary antibodies, conjugated with fluorescent dyes.

6. 4,6′-diamidino-2-phenylindole (DAPI): 1,000× DAPI stock: dissolve 0.2 mg DAPI in 1 ml ddH$_2$O. Use 1× DAPI for staining (dilute by ddH$_2$O).

7. Mounting solution: 50 % glycerol in ddH$_2$O.

8. Glass coverslips.

9. Glass slides.

10. Nail polish or glue.

11. Confocal microscope.

3 Methods

3.1 Growth of Yeast Cells

1. Inoculate YEPD or selection medium with yeast in a test tube (*see* **Note 1**).

2. Incubate at 150–180 rpm at 30 °C for 16–18 h.

3.2 Transformation of BD-Bait Plasmids into Yeast

1. Inoculate the yeast cells from a frozen stock on a YEPD plate and incubate at 30 °C until colonies appear after 3 days. The plates of yeast colonies can be stored at 4 °C for 1 month.

2. Pick one or a few colonies from the plate to 3 ml YEPD liquid medium and incubate them with rotation overnight (16–18 h) at 30 °C (*see* **Note 2**).

3. Centrifuge at 300×*g* for 5 min. Discard the supernatant and resuspend the cell pellet in 1 ml H$_2$O by vortexing.

4. Centrifuge at $300 \times g$ for 5 min and remove the supernatant and resuspend the cell pellet in 1 ml H_2O by vortexing.

5. Centrifuge at $300 \times g$ for 5 min. Discard the supernatant and resuspend the cell pellet with 0.2 ml 1× TE/LiOAc.

6. Add 8 μl plasmid DNA expressing the BD-prey fusion gene and 10 μl carrier DNA to the resuspended cells and mix (*see* **Note 3**).

7. Add 600 μl 1× PEG/LiOAc to the mixture.

8. Mix the contents on a rotator for 30 min at room temperature.

9. Add 10 % DMSO (*see* **Note 4**).

10. Heat-shock the cells at 42 °C for 2 min followed by incubation on ice for 5 min.

11. Centrifuge at $15,000 \times g$ for 30 s. Discard the supernatant and resuspend the cells in 200 μl 1× TE buffer.

12. Plate 100 μl onto a −W+G selection plate and 100 μl onto a −W−L+G selection plate.

13. Incubate the plates at 30 °C for 4–5 days.

3.3 Detection of the Expression of the BD-Bait Fusion Protein

1. Incubate the yeast cells with BD-bait fusion protein in −W+G medium at 30 °C until OD_{600}:1.

2. Centrifuge at $15,000 \times g$ for 1 min. Discard the supernatant and resuspend in 0.2 ml 2 M LiAc.

3. Incubate for 5 min on ice.

4. Centrifuge at $15,000 \times g$ for 1 min. Discard the supernatant and resuspend in 0.2 ml 0.5 M NaOH.

5. Incubate for 5 min on ice.

6. Centrifuge at $15,000 \times g$ for 1 min. Discard the supernatant and resuspend in 100 μl 5× Laemmli sample buffer.

7. Boil the sample at 95 °C for 5 min and centrifuge at $15,000 \times g$ for 2 min to remove the cell debris.

8. Transfer the supernatant into a new tube and store the sample at −80 °C (*see* **Note 5**).

9. Analyze expression of BD-bait fusion protein by standard SDS-PAGE and Western blot using an anti-Gal-BD antibody.

3.4 Self-activation Test for the Bait Protein

Before the library screen, it is important to confirm that the bait protein does not activate the reporter genes without the presence of any prey protein (*see* **Note 6**).

1. Pick a colony of yeast cells with BD-bait fusion protein and transfer the cells to a selection plate with −W−H+G. −W−H+G plates do not contain tryptophan and histidine for the screening of self-activation of bait protein.

2. Incubate the plates at 30 °C for 4–5 days.

3. If BD-bait fusion proteins do not have self-activation activities, the yeast cells will not grow on –W–H+G selection plates. On the other hand, yeast cells will grow if BD-bait fusion proteins have the self-activation activities. Transfer these yeast cells grown on –W–H+G selection plates to a –W–H+G+3AT plate to confirm the BD-bait fusion proteins do have self-activation activities.

4. BD-bait fusion proteins with self-activation activities are not recommended for further library screening. Deletion mapping analysis should be conducted to remove the domain with the self-activation activity.

3.5 Library Screening with Yeast Transformation

1. Pick one or a few yeast colonies transformed with a plasmid expressing the BD-bait fusion protein. Transfer to 3 ml –W+G medium and incubate the cells with rotation overnight (16–18 h) at 30 °C.

2. Centrifuge at $300 \times g$ for 5 min. Discard the supernatant and resuspend the cell pellet in 1 ml H_2O by vortexing.

3. Centrifuge at $300 \times g$ for 5 min and remove the supernatant and resuspend the cell pellet in 1 ml H_2O by vortexing.

4. Centrifuge at $300 \times g$ for 5 min. Discard the supernatant and resuspend the cell pellet with 200 µl 1× TE/LiOAc.

5. Add 8 µl plasmid DNA expressing the BD-prey fusion gene and 10 µl carrier DNA to the resuspended cells and mix.

6. Add 600 µl 1× PEG/LiOAc to the mixture.

7. Mix the contents on a rotator for 30 min at room temperature.

8. Heat-shock the cells at 42 °C for 2 min followed by incubation on ice for 5 min.

9. Centrifuge at $15,000 \times g$ for 30 s. Discard the supernatant and resuspend the cells in 200 µl 1× TE buffer.

10. Plate 100 µl onto a –W–L+G selection plate and 100 µl onto a –W–L–H+G selection plate.

11. Incubate the plates at 30 °C for 4–5 days.

3.6 Validation of Positive Two-Hybrid Clones Using 3AT-Selection Plates

1. Prepare the –W–L–H+G based plates by adjusting the 3AT concentration to 50 mM or 100 mM as selection plates.

2. Transfer the presumably positive yeast clones to plates of –W–L–H+G50 mM 3AT or –W–L–H+G+100 mM 3AT.

3. Incubate the cells at 30 °C, observe the growth of yeast colonies after 3 days (Fig. 3).

4. Only the positive clones will produce histidine and allow the yeast cells to grow on –W–L–H+G selection plates. 3AT acts as a competitive inhibitor of the product of His3 gene,

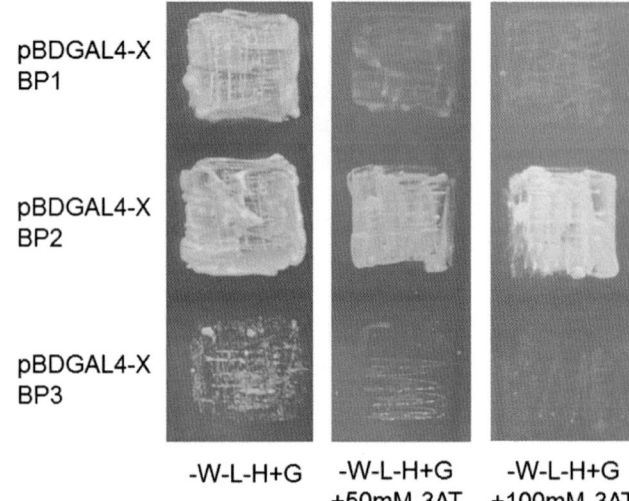

Fig. 3 Yeast growth on selection plates with different concentrations of 3AT. Three different colonies picked from the −W−L−H+G plate while screening for pBD-GAL4-X binding proteins (BP) using cDNA library were then seeded onto selection plates with different concentrations of 3AT (50 or 100 mM). On the plate with 50 mM 3AT, yeast with BP1 and BP2 but not with BP3 survived. Therefore, BP1 and BP2 showed stronger interaction with protein X than that of BP3. In the plate with 100 mM 3AT, BP2 showed stronger interaction with protein X than that of BP1. BP3 was identified as a false positive by this selection system

Imidazoleglycerol-phosphate dehydratase, which is an enzyme catalyzing the production of histidine. Higher expression of histidine in yeast cells will allow the cells survive in the media that contain higher concentration of 3AT (*see* **Note 7**).

3.7 Validation of Positive Two-Hybrid Clones Using X-gal Filter Lift Assay

1. Place a sterile nylon membrane on top of yeast colonies.

2. Remove the nylon membrane and place it on a container with the colony side up.

3. Put the nylon membrane with yeast cells into liquid nitrogen for 15 s and then allow it to thaw at room temperature

4. Repeat the freezing-thawing step three times.

5. Wet the filter paper with 4 ml X-gal working solution and place the nylon membrane on top of the wetted filter paper (prevent bubble in between).

6. Incubate the filter paper at 37 °C for 2.5 h, until colonies turn blue, but no more than 5 h.

3.8 Rescue the Plasmid Encoding AD-Prey Fusion Protein

1. Pick one yeast colony into 3 ml −W−L−H+G medium, incubate the cells with rotation overnight (16–18 h) at 30 °C.

2. Centrifuge at 15,000×*g* for 1 min and then remove the supernatant.

3. Resuspend the pellet in 200 μl solution I.

4. Add 100 μl Solution II and mix gently by inverting the tube ten times

5. Incubate at room temperature for 3 min.

6. Add 150 μl solution III and mix gently by inverting the tube ten times.

7. Centrifuge at $15,000 \times g$ for 10 min and transfer the supernatant into a new microtube

8. Add 400 μl isopropanol to precipitate DNA and incubate at –20 °C for 30 min.

9. Centrifuge at $15,000 \times g$ for 10 min, remove the supernatant, and add 1 ml 75 % ethanol.

10. Centrifuge for 10 min at $15,000 \times g$ and discard the supernatant.

11. Air-dry the pellet and add 50 μl pre-warmed ddH_2O to 60 °C to dissolve the plasmid DNA.

12. Transform the plasmid DNA into competent *E. coli*, according to manufacturer's instructions, and plate the bacteria on LB-Amp plates.

13. Using standard techniques, isolate plasmid DNA and sequence over region of insert to identify interacting protein.

3.9 Co-immunopre-cipitation Assay

1.

(A) Infect the cells with viruses (M.O.I. >3) and incubate for 16–24 h.

    i. Plate 7×10^5 cells in 60 mm culture dish with 3 ml culture medium. Cell density should reach 60–80 % confluent before viral infection (usually, it takes 16–24 h).

    ii. Remove culture medium and wash three times with 1× PBS.

    iii. Add 3 ml serum-free medium containing virus to the plates. More than 2.1×10^6 pfu (plaque forming unit) of virus (M.O.I. >3) should be used to infect the cells. Mix gently by rocking the plates and incubate at 37 °C incubator.

    iv. After 2 h, add serum to the medium to a final concentration of 10 %.

    v. Incubate cells for 16–24 h for further analysis.

(B) Transfect protein expression plasmids into the cells and incubate for 2 days.

    i. Plate 7×10^5 cells in 60 mm culture dish with 3 ml culture medium. Cell density should reach 60–80 % confluent before DNA transfection (usually, it takes 16–24 h).

ii. Dilute 4 μg plasmid with 300 μl serum-free medium in an eppendorf.

iii. Add 10 μl of 1 mg/ml PEI. Vortex to mix and incubate at room temperature for 15 min.

iv. Remove medium from cells and wash three times with 1× PBS.

v. Add 3 ml serum-free medium and premixed DNA-PEI (**step ii**) to the cells. Mix gently by rocking and incubate at 37 °C.

vi. After 2 h, add serum to the medium to a final concentration of 10 %.

vii. Incubate cells for 30–48 h for further analysis.

2. Cells (from A or B), after the removal of culture medium, were washed with 1× PBS three times at room temperature. Then, add 50 μl modified RIPA buffer to the cells and incubate on ice for 5 min.

3. Scrape cells into buffer and centrifuge at 15,000 ×*g* for 10 min.

4. Remove supernatant to a new tube.

5. Add 80 μl of Protein A magnetic beads into a 1.5 ml microtube, place the microtube into the magnetic rack and remove the storage buffer.

6. Add 0.5 ml Co-IP binding buffer to equilibrate the magnetic beads, resuspend and remove the buffer with the magnetic rack.

7. After the equilibration, add 0.6 ml of Co-IP binding buffer to resuspend the beads.

8. Add suitable amount of antibody diluted following the manufacturer's instructions (anti-X or anti-tag 1 in Fig. 2) to the magnetic beads and mix them with end-over-end rotation for 1 h at room temperature.

9. Place the microtube to the magnetic rack and remove the supernatant.

10. Wash the magnetic beads three times with 1 ml Co-IP binding buffer. After adding the buffer, fully mix the buffer with beads by inverting ten times at room temperature, using the magnetic rack to allow removal of buffer.

11. Add the protein samples from the **step 4** to the beads and dilute with Co-IP binding buffer to a total volume of 0.6 ml (*see* **Note 9**).

12. Incubate the protein samples with magnetic beads by end-over-end rotation overnight at 4 °C.

13. Wash the magnetic beads three times with 1 ml modified RIPA buffer. After adding the buffer, fully mix the buffer with beads

by inverting ten times at room temperature, then use the magnetic rack to allow removal of buffer.

14. Remove the supernatant using the magnetic rack and elute the proteins from the magnetic beads by using 30 μl Co-IP elution buffer.

15. Add 6 μl of 5× Laemmli sample buffer to the eluted sample protein.

16. Boil the sample at 95 °C for 5 min and centrifuge at $15,000 \times g$ for 2 min.

17. Collect the samples and perform SDS-PAGE followed by Western blotting using standard procedures. Western blotting should be performed using antibodies specific for protein Y or tag 2 shown in Fig. 2 (*see* **Notes 8** and **10**).

3.10 Immunofluorescence Analysis

1.

  (A) Infect the cells with viruses (M.O.I.=0.01) and incubate for 16–24 h.

    i. Plate 5×10^5 cells into 35 mm tissue culture plates containing glass coverslips. Cell density should reach 60–80 % confluent before viral infection (usually, it takes 16–24 h).

    ii. Remove cell culture medium and wash cells three times with 1× PBS.

    iii. Add 3 ml serum-free medium containing virus to the plates. About 5×10^3 pfu of viruses (M.O.I.=0.01) should be used to infect the cells. Mixed gently by rocking and incubate at 37 °C incubator.

    iv. After 2 h, add serum to the medium to a final concentration of 10 %.

    v. Incubate at 37 °C for 16–24 h for further analysis.

  (B) Transfect protein expression plasmids into the cells and incubate the cells for 2 days.

    i. Plate 5×10^5 cells into 35 mm tissue culture plates containing glass coverslips. Cell density should reach 60–80 % confluent before DNA transfection (usually, it takes 16–24 h).

    ii. Dilute 3 μg plasmid in 200 μl serum-free medium in an eppendorf.

    iii. Add 7 μl of 1 mg/ml PEI, vortex, and incubate at room temperature for 15 min.

    iv. Remove medium from cells and was three times with 1× PBS.

    v. Add 3 ml serum-free medium and premixed DNA-PEI (**step ii**) to the plates. Mix gently by rocking and incubate at 37 °C.

    vi. After 2 h, add serum to the medium to a final concentration of 10 %.

    vii. Incubate cells at 37 °C for 30–48 h for further analysis.

2. Aspirate the culture medium and gently rinse the cells twice in ice-cold 1× PBS.

3. Fix the cells by incubating them in the fixing solution for 10 min at −20 °C.

4. Remove the fixing solution and add 1 ml blocking solution to the coverslips. Incubate for 15 min.

5. Remove the blocking solution.

6. Dilute the primary antibodies (anti-X and anti-Y or anti-tag 1 and anti-tag 2 shown in Fig. 2) in the blocking solution following the manufacturers' instructions. The final volume should be sufficient to cover each coverslip (100 µl in a 24-well plate) (*see* **Note 11**).

7. Cover the plates with Parafilm and incubate the cells with antibodies at 37 °C for 30 min.

8. Wash the coverslips three times with 1 ml 1× PBS at room temperature.

9. Dilute the secondary antibodies conjugated with different fluorescent dyes in blocking solution, following the manufacturers' instructions. Secondary antibodies should recognize the Fc region of the primary antibodies. The final volume should be sufficient to cover each coverslip (100 µl in a 24-well plate).

10. Cover the plates with Parafilm and incubate the cells with antibodies at 37 °C for 30 min.

11. Wash the coverslips three times 1 ml 1× PBS at room temperature.

12. Dilute DAPI in blocking solution, following the manufacturer's instructions and add to the coverslips to stain the nucleus of the cells. The final volume should be sufficient to cover each coverslip (100 µl in a 24-well plate).

13. Cover the plates with Parafilm and incubate the cells with DAPI solution at room temperature for 15 min.

14. Prepare a microscope slide for each coverslip. Add a drop of mounting solution to each slide.

15. Pick up each coverslip with forceps and place it on the mounting solution with the cell-side facing down.

16. Apply nail polish or glue along the edges of the coverslips to seal them to the slides.

17. Observe under a confocal microscope.

4   Notes

1. *Saccharomyces cerevisiae* is a single-cell eukaryote frequently used in scientific research. The laboratory yeast strains usually carry several nutrient gene mutations; therefore, complementation of exogenous amino acids could provide the nutrients for yeast cells to survive. Two most common media used in the Y2H system are the full medium (YEPD) which contains all the nutrients for yeast, and the selection medium which contains only several selective amino acids added to the minimal medium.

2. Yeast cells used for transformation should be in log phase. Transformation efficiency in this phase is better than that in stationary phase.

3. Positive and negative controls should be included in all tests.

4. Addition of 10 % DMSO will increase the transformation efficiency.

5. Whenever Western blotting analysis was conducted to verify protein expression, adding protease inhibitors in the samples could prevent the degradation of protein samples.

6. Self-activation should be tested before using the bait proteins for library screening to avoid false-positive clones. It is not necessary to test the self-activation of prey proteins.

7. In the GAL4 based two-hybrid system, LacZ and *His*3 genes are usually used as reporter systems to verify the protein–protein interactions. In our lab, 3-Amino-1,2,4-triazole (3AT) is added to the selection plate as a competitive inhibitor of the *His*3-gene product. Results of yeast growth in plates with different concentrations of 3AT are shown in Fig. 3. X-Gal filter lift assay is also serves as a double confirmation test to avoid false-positive clones [10].

8. If Y protein interacts physically with X protein, Y protein will be immunoprecipitated by anti-X antibody in the presence but not absence of X protein. To avoid the cross-reaction of anti-X, a negative control should be included, i.e., Y protein in the cell lysate without X protein would not be precipitated by the anti-X antibody.

9. When co-IP is performed (Subheading 3.9), add 9/10 of the protein samples for the immunoprecipitation assay (the 12th step). 1/10 of the protein samples serve as an input control for Western blotting analysis.

10. It is annoying to have the heavy and light chains of immunoglobulins in the background of Western blotting analysis following immunoprecipitation, especially when the size of the target protein is close to that of the heavy chain or light chain. Cross-linking of the antibody to protein A/G beads using a cross-linker, e.g., disuccinmidyl suberate, should prevent the co-elution

of heavy and light chains of immunoglobulins. Alternatively, a commercial immunoprecipitation kit, like EasyBlot (Genetex), provides a secondary antibody which specifically reacts with the native, non-reduced form of IgG. This will decrease the interference of heavy chain and light chain of IgG.

11. If the primary antibodies against two proteins (or peptide tags) are derived from the same animal species, it is necessary to conjugate a fluorescent dye directly to one primary antibody. Incubate the other primary antibody without a conjugated fluorescent dye first, followed by the secondary antibody against this primary antibody. Then, add the primary antibody with a conjugated fluorescent dye to the sample for incubation.

References

1. Cho S, Park SG, Lee DH et al (2004) Protein-protein interaction networks: from interactions to networks. J Biochem Mol Biol 37:45–52

2. Fields S, Song O (1989) A novel genetic system to detect protein-protein interactions. Nature 340:245–246

3. McCraith S, Holtzman T, Moss B et al (2000) Genome-wide analysis of vaccinia virus protein-protein interactions. Proc Natl Acad Sci U S A 97:4879–4884

4. de Chassey B, Navratil V, Tafforeau L et al (2008) Hepatitis C virus infection protein network. Mol Syst Biol 4:230

5. Calderwood MA, Venkatesan K, Xing L et al (2007) Epstein-Barr virus and virus human protein interaction maps. Proc Natl Acad Sci U S A 104:7606–7611

6. Konig R, Zhou Y, Elleder D et al (2008) Global analysis of host-pathogen interactions that regulate early-stage HIV-1 replication. Cell 135:49–60

7. Snider J, Kittanakom S, Damjanovic D et al (2010) Detecting interactions with membrane proteins using a membrane two-hybrid assay in yeast. Nat Protoc 5:1281–1293

8. Lentze N, Auerbach D (2008) Membrane-based yeast two-hybrid system to detect protein interactions. Curr Protoc Protein Sci. Chapter 19, Unit 19.17 (editorial board, John E Coligan et al)

9. Fang CP, Li ZC, Yang CH et al (2013) Hepatitis C virus non-structural protein 3 interacts with cytosolic 5′(3′)-deoxyribonucleotidase and partially inhibits its activity. PLoS One 8:e68736

10. Yang CH, Li HC, Jiang JG et al (2010) Enterovirus type 71 2A protease functions as a transcriptional activator in yeast. J Biomed Sci 17:65

11. Ma HC, Lin TW, Li H et al (2008) Hepatitis C virus ARFP/F protein interacts with cellular MM-1 protein and enhances the gene transactivation activity of c-Myc. J Biomed Sci 15:417–425

12. Wei WY, Li HC, Chen CY et al (2012) SARS-CoV nucleocapsid protein interacts with cellular pyruvate kinase protein and inhibits its activity. Arch Virol 157:635–645

13. Chen SC, Lo SY, Ma HC et al (2009) Expression and membrane integration of SARS-CoV E protein and its interaction with M protein. Virus Genes 38:365–371

14. Hsieh YC, Li HC, Chen SC et al (2008) Interactions between M protein and other structural proteins of severe, acute respiratory syndrome-associated coronavirus. J Biomed Sci 15:707–717

15. Ma HC, Ku YY, Hsieh YC et al (2007) Characterization of the cleavage of signal peptide at the C-terminus of hepatitis C virus core protein by signal peptide peptidase. J Biomed Sci 14:31–41

16. Ma HC, Ke CH, Hsieh TY et al (2002) The first hydrophobic domain of the hepatitis C virus E1 protein is important for interaction with the capsid protein. J Gen Virol 83:3085–3092

Chapter 18

A Field-Proven Yeast Two-Hybrid Protocol Used to Identify Coronavirus–Host Protein–Protein Interactions

Pierre-Olivier Vidalain, Yves Jacob, Marne C. Hagemeijer, Louis M. Jones, Grégory Neveu, Jean-Pierre Roussarie, Peter J.M. Rottier, Frédéric Tangy, and Cornelis A.M. de Haan

Abstract

Over the last 2 decades, yeast two-hybrid became an invaluable technique to decipher protein–protein interaction networks. In the field of virology, it has proven instrumental to identify virus–host interactions that are involved in viral embezzlement of cellular functions and inhibition of immune mechanisms. Here, we present a yeast two-hybrid protocol that has been used in our laboratory since 2006 to search for cellular partners of more than 300 viral proteins. Our aim was to develop a robust and straightforward pipeline, which minimizes false-positive interactions with a decent coverage of target cDNA libraries, and only requires a minimum of equipment. We also discuss reasons that motivated our technical choices and compromises that had to be made. This protocol has been used to screen most non-structural proteins of murine hepatitis virus (MHV), a member of *betacoronavirus* genus, against a mouse brain cDNA library. Typical results were obtained and are presented in this report.

Key words Murine hepatitis virus, Host–pathogen interactions, Yeast two-hybrid, Interactomics, Proteomics

1 Introduction

The yeast two-hybrid system (Y2H) was first developed in 1989 by Fields and Song [1], and quickly became a popular technology to detect protein–protein interactions [2]. Although multiple flavors of this system have been developed, it is in essence a complementation assay based on the reconstitution of a functional transcription factor mediated by protein interaction. Indeed, many transcription factors such as Gal4 exhibit a modular organization with a DNA binding domain (DB) that can be separated from the transactivation domain (AD) [3, 4]. In the Y2H system, a first protein of interest is fused to DB and a second one is fused to AD. When co-expressed in yeast, these two hybrid proteins called "bait" and

Helena Jane Maier et al. (eds.), *Coronaviruses: Methods and Protocols*, Methods in Molecular Biology, vol. 1282, DOI 10.1007/978-1-4939-2438-7_18, © Springer Science+Business Media New York 2015

"prey", respectively, reconstitute a functional transcription factor if they physically interact. This activates a set of reporter genes, including selectable markers. For example, interaction-dependent reporter gene *HIS3* enables yeast growth on a synthetic culture medium depleted of histidine (–His). This provides a system to co-express a bait of interest with a whole set of prey proteins, and then select for HIS3 positive yeast colonies that express interacting protein pairs.

Collections of prey proteins can be expressed in yeast from cDNA libraries cloned in frame with the AD sequence (AD-cDNA library). However, such libraries often have a complexity in the range of five million independent clones for complex organisms like human, which adds to the strong enrichment bias for house-keeping proteins like actin. This implies that when performing an Y2H screen, several times more yeast transformants must be obtained to cover all possible bait–prey combinations and probe underrepresented preys. This can be difficult to achieve by standard yeast transformation and requires large amounts of the AD-cDNA plasmid preparation. This limitation is partially overcome by taking advantage of haploid Mat-a and Mat-α yeast to mate and thus form diploid cells [5]. Haploid yeast of opposite mating type are pre-transformed with the DB plasmid and the AD-cDNA library, and then mated to obtain a large number of diploids co-expressing bait and prey proteins. In this experimental setting, up to 50 million diploids that represent ten times the original complexity of the AD-cDNA library can be easily generated. Besides the technical benefit of yeast mating, it is now possible to generate normalized libraries by pooling thousands of prey plasmids originating from large collections of full-length ORFs like the human ORFeome [6–8]. Since prey plasmids are represented at equimolar concentrations in such libraries, their full coverage by yeast transformation or yeast mating is by far easier to reach. However, only full-length prey proteins are expressed from such normalized libraries and because isolated protein domains often better interact in the Y2H system, this can be a source of undetected interactions (false negatives; see below). This is in contrast with AD-cDNA libraries that usually encode full-length proteins but also protein fragments generated by random priming and premature arrest of reverse-transcription when building the library from cellular mRNAs.

Although multiple protein–protein interaction assays have been developed in the last decade [9], the Y2H system is often preferred because it does not require protein purification steps, which can be technically challenging, and is amenable to high-throughput settings [10–12]. As a matter of fact, Y2H system takes advantage of yeast genetic power to dissect complex problems and perform high-throughput genetic screens by opposition to biochemical screens. Most importantly, this assay essentially provides information on binary protein–protein interactions in contrast to protein complex analysis by mass spectrometry, which does not distinguish direct from indirect partners. Nevertheless, technical limitations of the Y2H system must be considered when performing a screen. Contrary to common thought, Y2H screens generate high-quality datasets with relatively low rates of false-positive interactions when properly performed. In particular, this implies to properly evaluate for each bait construct the level of self-transactivation in yeast, and titrate down this activity at an appropriate stringency with a selective medium containing a competitive inhibitor of HIS3 gene product when performing the screen. Another important point is the elimination of satellite prey plasmids that often contaminate positive yeast colonies, and this is achieved by their serial passage on selective medium [13]. Finally, it is established that filtering out Y2H interactions supported by only one or two positive yeast colonies is essential to enrich datasets for high-quality interactions [7]. When following these recommendations, about 80 % of the interactions identified by Y2H properly retest in another experimental systems [7, 14, 15]. Although this validates Y2H data at a biophysical level, this does not imply that identified interactions are biologically relevant and participate to a specific biological process. This should be kept in mind since functional validation can be a daunting task, and represents the true bottleneck for such interaction-mapping approaches.

As a matter of fact, undetected interactions or false negatives are more problematic. The sensitivity of this assay has been estimated to 20–30 %, meaning that Y2H detects at best a quarter of the interactions from a positive control set [16]. Misfolding, mislocalization, poor expression levels, or the lack of appropriate post-translational modifications of bait and prey proteins that are both tagged and forced to enter the yeast nucleus can explain the high false-negative rate of this assay. In addition, and even if the two hybrid proteins properly interact, steric constraints often prevent the formation of a functional transcription factor to drive reporter gene expression. To some extent, this can be circumvented by using isolated protein domains, different Y2H systems, and by swapping DB and AD tags to both extremities of bait and prey proteins [17, 18]. Beyond sensitivity of the assay, the usual incompleteness of prey libraries that are often missing several binding partners of the bait, and the multiplicity of isoforms, are the main source of false negatives. This is why screening several AD-cDNA libraries from different tissues can be advantageous to cover as much as possible the complexity of the proteome [19, 20]. Finally and as already discussed above, it can be difficult to fully probe the complexity of some libraries despite the use of yeast mating protocols and the production of millions of diploid yeast [14].

In the last decade, Y2H has been extensively used to map virus–host interactions. The main objective for different research groups was to address the lack of information in literature, and obtain proteome-scale pictures of virus infection networks [21, 22]. So far, SARS-coronavirus (SARS-CoV) is the only member of Coronaviridae family for which interactions with host factors were investigated systematically by Y2H [23]. This led to identify 132 SARS–host interactions in the high-confidence dataset (which was only partially disclosed), and an extensive mining of literature also retrieved an additional list of 27 SARS-CoV–host interactions. In particular, this report identified nsp1 interaction with several members of immunophilin and calcipressin families, which led to demonstrate SARS sensitivity to cyclosporin A. More recently, another Y2H screen performed with the C-terminal domain of the spike glycoprotein (S) identified Ezrin as a binding partner and a restriction factor of SARS-CoV [24].

Since 2005, a technological platform is up and running in our laboratory at Institut Pasteur, and dedicated to virus–host interaction mapping for a large panel of RNA viruses. Our Y2H protocol is a combination of tools and techniques greatly inspired by previous reports from Yves Jacob [25] and Marc Vidal's group [10]. Here, we detail our Y2H protocol (Fig. 1), and present results obtained with both structural and non-structural proteins of Mouse Hepatitis Virus (MHV, strain A59). MHV is a *Coronaviridae* from *betacoronavirus* genus—like SARS-CoV—that encodes for 8 structural proteins and 16 non-structural proteins (nsps), which are poorly characterized at the functional level. In this screen we focused on those viral proteins that are known or suspected to be involved in RNA replication and transcription; nsp1–16 (with the exception of the very small nsp11) and the nucleocapsid protein N. The very large nsp3 was divided into three parts of approximately similar size (nsp3a, 3b, and 3c). In total, 15 MHV full-length proteins or isolated domains (including nsp3a and 3b) were used as bait to screen a mouse brain cDNA library. Because nsp3c, nsp6, and nsp12 segments could not be cloned in the Y2H vector, corresponding screens were not pursued. In total, 1,410 positive yeast colonies were recovered, and potential interactors identified by PCR analysis and sequencing. High-quality sequences were obtained for 1,096 positive yeast colonies, and retrieved interactions were filtered using statistical criteria to generate a high-quality dataset. In total, 39 novel interactions were identified with no precedent in literature (Table 1). Interestingly, nsp2 was found to interact with three MARK proteins (Microtubule Affinity-Regulating Kinase), suggesting a role in microtubule assembly, and this echoes Pfefferle et al. report that showed ORF9b and nsp13 of SARS-CoV binding with MARK2 and 3 [23]. In addition, nsp7 was found to bind the small glutamine-rich tetratricopeptide repeat-containing protein alpha (SGTA), like ORF7a from SARS-CoV [26].

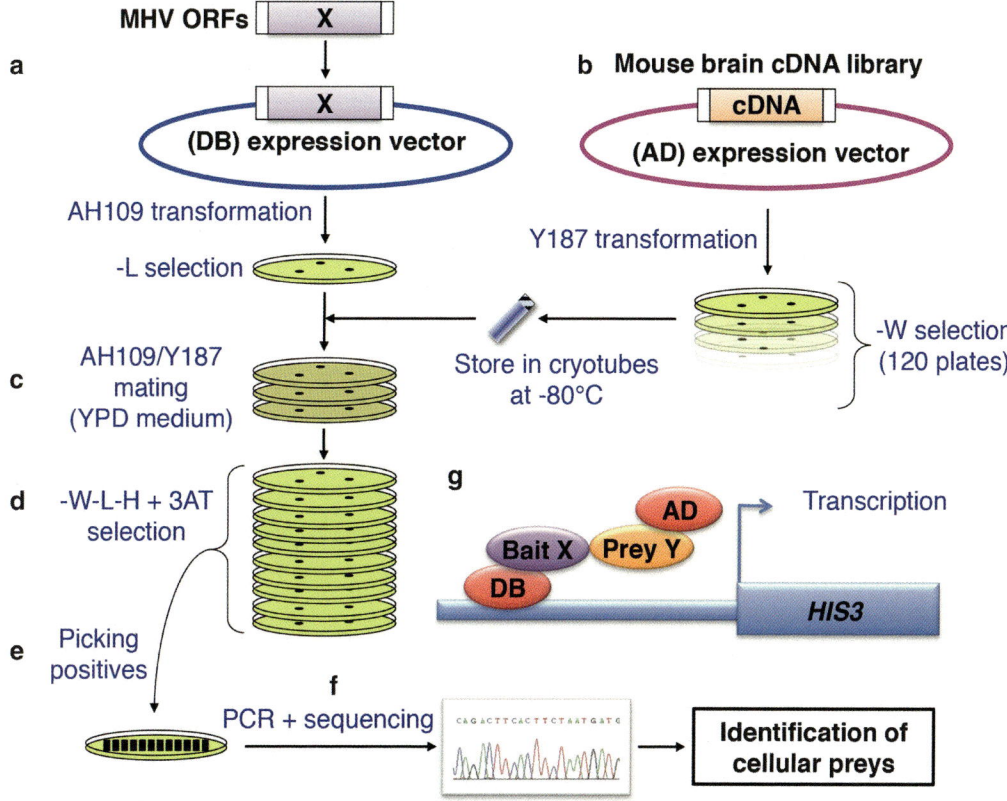

Fig. 1 Summary of the yeast two-hybrid screening pipeline. Cloning and transformation of the viral bait construct in AH109 yeast strain is shown in the *upper left* (A). Cloning and transformation of the prey AD-cDNA library in Y187 yeast is shown in the *upper right* (B). *Lower part* of the figure is showing successively yeast mating (C), growth on selective medium (D), picking of positives (E), amplification and sequencing of interacting preys (F). A schematic of *HIS3* reporter gene transactivation by DB-X bait interaction with AD-Y prey is also presented (G)

2 Materials

2.1 Yeast Strains, Plasmids, and Media

1. Yeast strain AH109 (*MATa, trp1-901, leu2-3, 112, ura3-52, his3-200, gal4Δ, gal80Δ, LYS2::GAL1$_{UAS}$-GAL1$_{TATA}$-HIS3, GAL2$_{UAS}$-GAL2$_{TATA}$-ADE2, URA3::MEL1$_{UAS}$-MEL1$_{TATA}$-lacZ*) (Clontech).

2. Yeast strain Y187 (*MATα, ura3-52, his3-200, ade2-101, trp1-901, leu2-3, 112, gal4Δ, met-, gal80Δ, URA3::GAL1$_{UAS}$-GAL1$_{TATA}$-lacZ*) (Clontech). The two haploid strains AH109 and Y187 are of opposite mating types, which enables library screens by mating.

3. pDEST32 (Life Technologies) or pPC97-GW (provided by Dr. Vidal) yeast two-hybrid vectors, containing sequence corresponding to viral protein of interest in frame with Gal4DB.

4. Mouse brain cDNA library cloned into yeast two-hybrid vector pPC86 (Life Technologies), or similar.

Table 1
Matrix of MHV–host protein–protein interactions identified by Y2H

Ensembl gene IDs	Gene name	N	nsp1	nsp2	nsp3a	nsp3b	nsp4	nsp5	nsp7	nsp8	nsp9	nsp10	nsp13	nsp14	nsp15	nsp16
29392	Rilpl1															63
64181	Rab3ip															33
46204	Pnma2															22
5417	Mprip														14	17
20776	Fbf1														20	10
32267	Usp28															3
71650	Ganab													20		
3573	Homer3												77			
19856	Fam184a												6			
38170	Pde4dip												4			
34602	Mon2												3			
42156	Dzip1												3			
42766	Trim46												3			
28249	Sdcbp											175				
39197	Adk										12					
68329	Htra2										6					
20375	Rufy1										4					
25855	Prkar1b									43						
4937	Sgta								150							
32405	Pias1								23							
976	Heatr6								3							
37259	Dzank1							35								
46818	Ddit4l							18								
52539	Magi3							17								
34354	Mtmr3							6								
4931	Apba3							5								
55128	Cgrrf1							5								
29312	Klhl8							4								
8348	Ubc					5										
31788	Kifc3					5										
23353	Centg3					4										
64351	mt-Co1					3										
28367	Txn1				28											
24969	Mark2			29												
26620	Mark1			26												
7411	Mark3			16												
37443	Ccdc21			6												
Number of Interactions supported by <3 positive yeast colonies		10	0	2	52	31	2	6	1	3	0	2	2	18	5	19

First and second columns correspond, respectively, to Ensembl gene IDs and canonical gene names for interacting cellular proteins. Columns 3–16 provide, for indicated MHV proteins, numbers of positive yeast colonies obtained for each cellular protein. Last row corresponds to numbers of interactions supported by less than three positive yeast colonies, which were filtered out as explained in Subheading 3.8, **step 6**

5. Nonselective medium (YPD) agar plates: 1 % yeast extract w/v, 2 % Bacto-Peptone w/v, 100 mg/l adenine hemisulfate, 2 % glucose w/v, 2 % agar w/v (*see* **Note 1**). Store at 4 °C.

6. Liquid YPD medium: 1 % yeast extract w/v, 2 % Bacto-Peptone w/v, 100 mg/l adenine hemisulfate, 2 % glucose w/v (*see* **Note 1**). Store at room temperature.

7. Amino acid powder: 10 g L-alanine, 10 g L-arginine, 10 g L-aspartic acid, 10 g L-asparagine, 10 g L-cysteine, 10 g L-glutamic acid, 10 g L-glutamine, 10 g L-glycine, 10 g L-isoleucine, 10 g L-lysine, 10 g L-methionine, 10 g L-phenylalanine, 10 g L-proline, 10 g L-serine, 10 g L-threonine, 10 g L-tyrosine, 10 g L-valine, 10 g of adenine hemisulfate. Mix and grind carefully in a mortar. Store at room temperature.

8. 100 mM L-leucine drop-out solution sterilized through 0.22 μm filter. Store at room temperature.

9. 40 mM L-tryptophan drop-out solution sterilized through 0.22 μm filter. Store at 4 °C in the dark.

10. 20 mM uracil drop-out solution sterilized through 0.22 μm filter. Store at room temperature.

11. 100 mM L-histidine drop-out solution sterilized through 0.22 μm filter. Store at room temperature.

12. 3-aminotriazole powder (3-AT).

13. Petri dishes.

2.2 Yeast Transformation Reagents

1. TE/LiAc solution: 10 mM Tris–HCl (pH 8.0), 1 mM EDTA, 100 mM lithium acetate (*see* **Note 2**).

2. TE/LiAc/PEG solution: 10 mM Tris–HCl (pH 8.0), 1 mM EDTA, 100 mM lithium acetate, 35.2 % polyethyleneglycol (PEG 3350) (*see* **Note 2**).

3. 10 mg/ml salmon sperm DNA denatured by boiling in water for 10 min and chilled on ice.

4. Glycerol.

5. Cryotubes.

6. Glass beads, autoclaved prior to use.

2.3 PCR Amplification of AD-cDNA from Positive Yeast Colonies

1. Zymolyase 20T.

2. La Taq PCR kit (Takara) or similar (*see* **Note 3**).

3. pPC86-For (5′-GACGGACCAAACTGCGTATA-3′) and pPC86-Rev (5′-ACCAAACCTCTGGCGAAGAA-3′) primers.

4. 96-well E-gel (Life Technologies) or 1 % agarose gel.

3   Methods

For convenience, yeast cultures are manipulated in a biosafety level-2 cabinet to avoid contaminations. However, yeast cultures can be manipulated on a regular bench, even without a flame, depending on air quality in laboratory spaces.

3.1   Production of Selective Medium Agar Plates

1. Make up base medium as follows: 0.29 % amino acid powder w/v, 0.38 % yeast nitrogen base (without amino acid and ammonium sulfate) w/v, 1.12 % ammonium sulfate w/v. Adjust pH to 5.9 with NaOH.

2. Make a stock of 4 % agar.

3. Autoclave both solutions (*see* **Note 1**).

4. Supplement base medium with 40 % glucose solution to a final concentration of 4 %.

5. Mix the base medium and agar at a 1:1 ratio.

6. Add 8 ml/L of the appropriate amino acid drop-out solutions and 3-AT whenever required (*see* **Note 4**). For example, uracil, histidine, and tryptophan should be added to obtain some synthetic medium lacking leucine (–L medium).

7. Pour in 15 cm petri plates and dry for 3–4 days. Store at 4 °C.

3.2   Establishment of cDNA Library into Yeast Culture

A mouse brain cDNA library cloned in yeast two-hybrid vector pPC86 is first established into yeast strain Y187 (*see* **Note 5**).

1. Inoculate 500 ml of nonselective YPD medium at 0.007 optical density (OD) at 600 nm with a fresh yeast culture.

2. Grow overnight in a shaker at 30 °C.

3. Determine the OD at 600 nm. Harvest culture when OD is in a 0.4–0.5 range.

4. Centrifuge cells in 10×50 ml tubes at $750 \times g$ for 5 min. Discard supernatant and resuspend each yeast pellet in 40 ml water.

5. Centrifuge cells at $750 \times g$ for 5 min. Discard supernatant, and resuspend each yeast pellet in 40 ml TE/LiAc.

6. Centrifuge cells at $750 \times g$ for 5 min. Discard supernatant and pool the ten yeast pellets in 5 ml TE/LiAc.

7. Add 500 µl heat-denatured carrier DNA and 150 µg cDNA library to competent yeast, and mix carefully.

8. Dispense 250 µl of this preparation in twenty 2 ml tubes, and then add 1.6 ml TE/LiAc/PEG.

9. Mix by gently vortexing the tube and incubate for 45 min at 30 °C.

10. Heat-shock for 20 min at 42 °C.

11. Centrifuge at $750 \times g$ for 5 min. Discard supernatant and fill tube with water without resuspending the pellet.

12. Centrifuge at $750 \times g$ for 5 min. Discard supernatant and resuspend the pellet in 1 ml water. Pool the 25 transformation reactions in a single 50 ml tube.

13. Spread 200 µl of yeast using glass beads onto one hundred twenty-five 15 cm petri dishes with –W agar.

14. Make 1/100 and 1/1,000 dilutions in water and plate on two 15-cm petri dishes with –W agar to determine the transformation efficiency.

15. Grow cells for 3 days at 30 °C. Calculate the total number of yeast transformants (*see* **Note 6**).

16. Add 5 ml YPD medium to each plate and scrape cells into medium using for example a Pasteur pipette bent using a flame. Pool into a 2-l flask.

17. Add glycerol to obtain a 20 % (w/v) solution, and determine the final OD. Calculate the volume V of yeast suspension required to perform one two-hybrid screen considering that $V = 60/\text{OD}$.

18. Aliquot in cryotubes with the volume required for one yeast two-hybrid screen. Store at –80 °C.

3.3 Yeast Transformation with Bait Constructs Expressing Viral Proteins Fused to Ga4-DB

1. In a 100 ml flask, inoculate 50 ml of nonselective YPD medium with a patch of fresh AH109 yeast cells scooped from a YPD plate stored at room temperature (few days old at most).

2. Grow overnight in a shaker at 30 °C.

3. Determine the OD at 600 nm. Take the appropriate volume of yeast culture considering that 1 ml at 5 OD is sufficient to perform ten transformations.

4. Centrifuge cells at $750 \times g$ for 5 min. Discard supernatant and resuspend yeast in >500 µl water per transformation.

5. Centrifuge cells at $750 \times g$ for 5 min. Discard supernatant, and resuspend yeast in >100 µl TE/LiAc solution per transformation.

6. Centrifuge cells at $750 \times g$ for 5 min. Discard supernatant and resuspend the pellet in 20 µl of TE/LiAc per transformation.

7. Add 2 µl heat-denatured carrier DNA and 50–250 ng Gal4-DB plasmid.

8. Add 120 µl TE/LiAc/PEG and mix by gently vortexing the tube.

9. Incubate for 45 min at 30 °C.

10. Heat-shock for 15 min at 42 °C.

11. Centrifuge at $750 \times g$ for 5 min. Discard supernatant, and resuspend the pellet in 20 µl water.

12. Spot 10 μl on a petri dish with –L agar.

13. Grow cells for 3 days at 30 °C.

3.4 Testing the Transactivation of HIS3 Reporter Gene by Gal4-DB Bait Constructs

Before performing a screen, determine basal transactivation of *HIS3* reporter gene for each viral bait protein fused to Gal-DB (*see* **Note 7**).

1. With a loop, take a small patch of transformed yeast from Subheading 3.3, **step 13**.

2. Dilute in 1 ml water, and spot 10 μl on petri dishes with –L and –L–H agar. After evaporation of water, the remaining yeast layer should be almost transparent and barely visible (if not, increase yeast dilution and repeat).

3. Incubate cells for 5 days at 30 °C.

4. Read plates. If no growth is observed on –L–H agar when compared to –L plate, then perform the screen on –L–W–H plates.

5. If yeast growth is observed on –L–H plates, repeat the same experiment by plating yeast on –L–H agar supplemented with increasing concentrations of 3-AT (*see* **Note 7**).

6. Determine a minimal 3-AT concentration sufficient to block yeast growth. Then, prepare –L–W–H plates containing the ad hoc concentration of 3-AT to perform the screen.

3.5 Yeast Mating and Selection of Positive Yeast Colonies

1. In a 100 ml flask, inoculate 50 ml YPD medium with a patch of fresh AH109 yeast expressing the bait protein of interest (*see* **Note 8**).

2. Grow overnight in a shaker at 30 °C.

3. Determine OD at 600 nm. Expected value should be in a 2–6 range. Calculate the volume V of AH109 yeast culture required to perform the screen considering that $V = 72/OD$ (*see* **Note 9**).

4. Thaw one vial containing the cDNA library in Y187 yeast and transfer into a 50 ml tube containing 10 ml fresh YPD medium.

5. Incubate 10 min in a shaker at 30 °C.

6. Add AH109 yeast with the bait protein to the tube containing Y187 yeast with the cDNA library. Mix by inverting.

7. Centrifuge at $750 \times g$ for 5 min. Discard supernatant and resuspend the pellet in 1.5 ml of YPD medium.

8. Spread 500 μl onto three YPD plates using beads.

9. Incubate for 4.5 h at 30 °C to allow yeast mating (*see* **Note 10**).

10. Add 8 ml of water to each YPD plate and resuspend yeast with a scraper (made, for example, from a Pasteur pipette bent by heating in a flame).

11. Repeat the same procedure to wash the three YPD plates at least once more and pool yeast suspensions into one 50 ml tube.

12. Centrifuge at $750 \times g$ for 5 min. Discard supernatant and resuspend the pellet in 6 ml water.

13. Take 4 μl of yeast suspension to prepare a 1/10,000 dilution in water that will be used to determine mating efficiency.

14. Using beads, spread 500 μl yeast suspension onto 12 –L–W–H plates containing the ad hoc concentration of 3-AT (determined in Subheading 3.4, **step 6**). In addition, spread 500 μl of the 1/10,000 dilution on a –L–W plate.

15. Incubate for 6 days at 30 °C.

16. In order to determine the efficiency of mating, count yeast colonies on the –L–W plate. Then, multiply by $10,000 \times 12$ to obtain the total number of diploids generated during the screen (*see* **Note 11**).

3.6 Identification Interacting Partners

1. Cherry-pick positive yeast colonies from the 12 screening plates with a sterile toothpick or tip, and patch them on fresh –L–W–H plates containing the ad hoc concentration of 3-AT (determined in Subheading 3.4, **step 6**) to maintain selection pressure on *HIS3* reporter gene. It is best to organize positive yeast colonies at the standard 96-well format with the help of a grid paper placed underneath the petri dish.

2. Grow at 30 °C for 3–4 days (*see* **Note 12**).

3. To eliminate contaminations with satellite AD-cDNA plasmids, which can be present in yeast aside plasmids encoding for *bona fide* interactors, purify positive colonies by replication every 3–4 days on fresh selective medium over 3 weeks. This can be quickly achieved with the extremity of tips mounted on a multichannel or using an automated platform, as displayed in Fig. 2a.

3.7 PCR Identification of Positive Colonies

1. Prepare a 2.5 mg/ml solution of zymolyase 20T in water and dispense 50 μl per well in a 96-well PCR plate.

2. For each positive colony, take a patch of yeast and resuspend in the zymolyase solution in one well.

3. Incubate for 5 min at 37 °C and then for 5 min at 95 °C to inactivate the enzyme.

4. Perform a PCR using pPC86-For and pPC86-Rev with LaTaq or similar, according to manufacturer's instructions (*see* **Notes 3** and **13**). An amplification cycle of 94 °C for 1 min then 35 cycles of 98 °C for 10 s and 68 °C for 5 min followed by a single incubation of 72 °C for 10 min should be performed.

5. Analyze PCR products on a 1 % agarose gel, e.g., a 96-well E-Gel (Fig. 2b).

6. Sequence PCR products with pPC86-For primer using standard procedures (*see* **Note 13**).

a

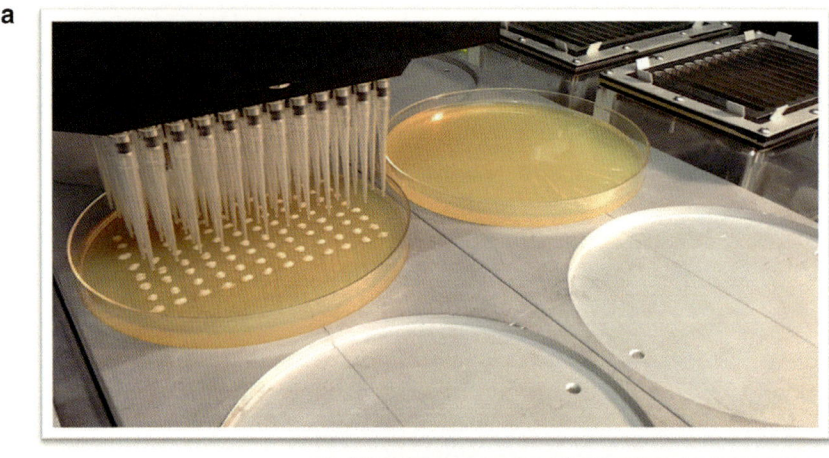

b

Loading well

PCR product

Primer dimer

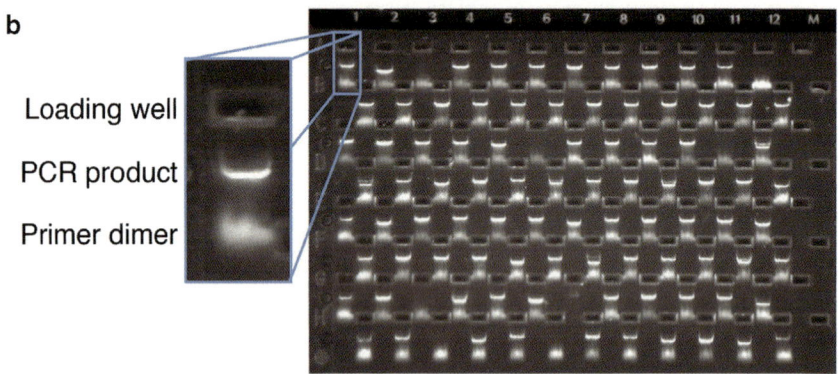

Fig. 2 Selection and replica plating of positive yeast colonies. (**a**) Automated replica plating of positive yeast colonies arrayed in a 96-well format using a TECAN platform. Notice the customized stand for 15 cm petri dishes that fits into the standard 96-well plate holder. Yeast colonies are replica plated from mother to daughter plates by a 96-Multi Channel Arm (MCA) by simple tip-touching without aspiration or dispense. (**b**) PCR products corresponding to AD-cDNA were amplified from positive yeast colonies and analyzed on a 96-well E-gel

3.8 Data Analysis

1. Analyze trace files to assign quality scores and generate accurate sequence files.

2. Trim plasmid adaptor sequences.

3. Use BLAST to probe the mouse mRNA and protein sequence databank at EMBL, and determine which host protein corresponds to each prey sequence.

4. Build an Excel spreadsheet with three columns including, for each positive yeast colony, a sequence ID and the corresponding bait and prey protein names.

5. Use the pivot table function of Excel to build an interaction matrix showing the number of positive yeast colonies for each bait–prey combination.

6. Use Data > filter function in Excel to eliminate interactions supported by less than three positive yeast colonies. This is essential

to remove most technical false positives from the final dataset. Table 1 shows an example of the final results obtained for a screen of MHV proteins.

7. The identified proteins may be subjected to further analysis. For example, use the STRING database (Search Tool for the Retrieval of Interacting Genes/Proteins) to determine, for mouse and other organisms, known interaction data between the proteins listed in Table 1 (Fig. 3). Upload protein IDs on the STRING website (http://string-db.org/), and follow the instructions.

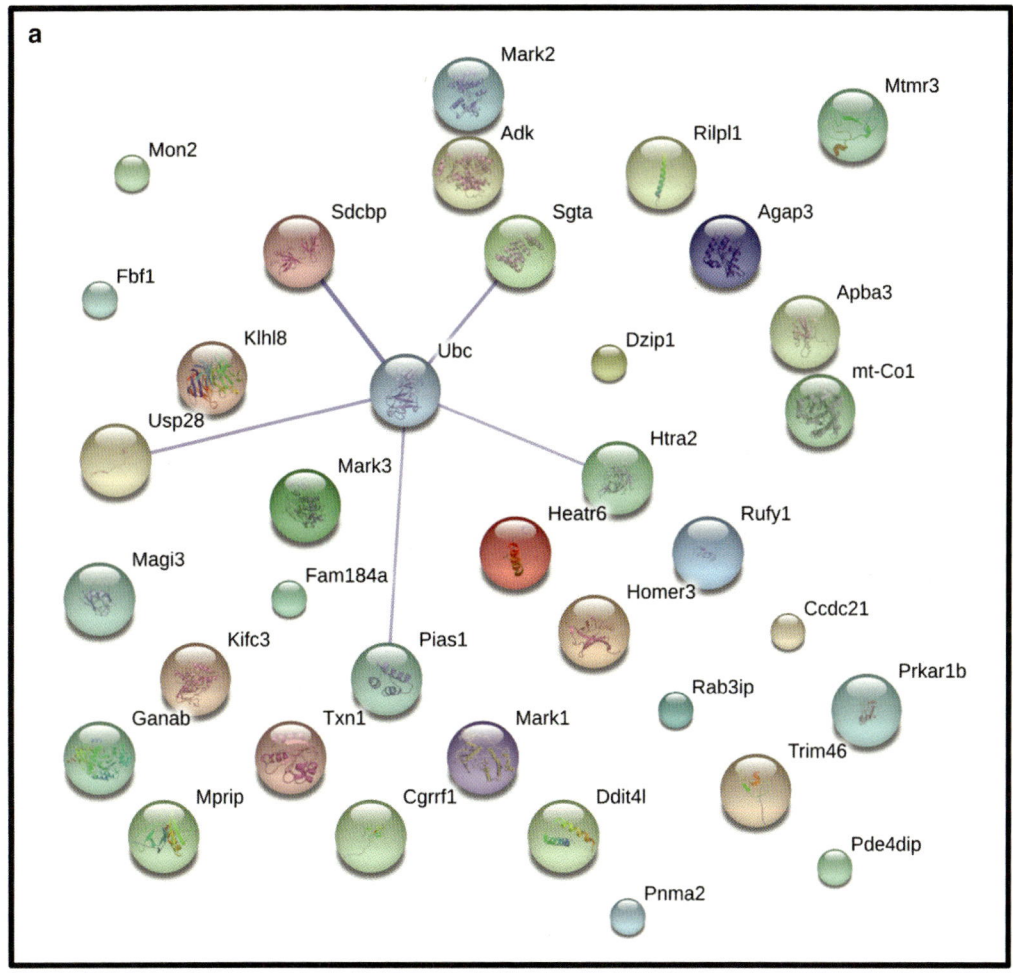

Fig. 3 Confidence view of STRING analyses for all interacting host proteins identified in the different screens and listed in Table 1 [29]. Confidence view of STRING analyses of host proteins that were found to interact with the MHV proteins using either *Mus musculus* (**a**) or *Homo sapiens* (**b**) as input organism

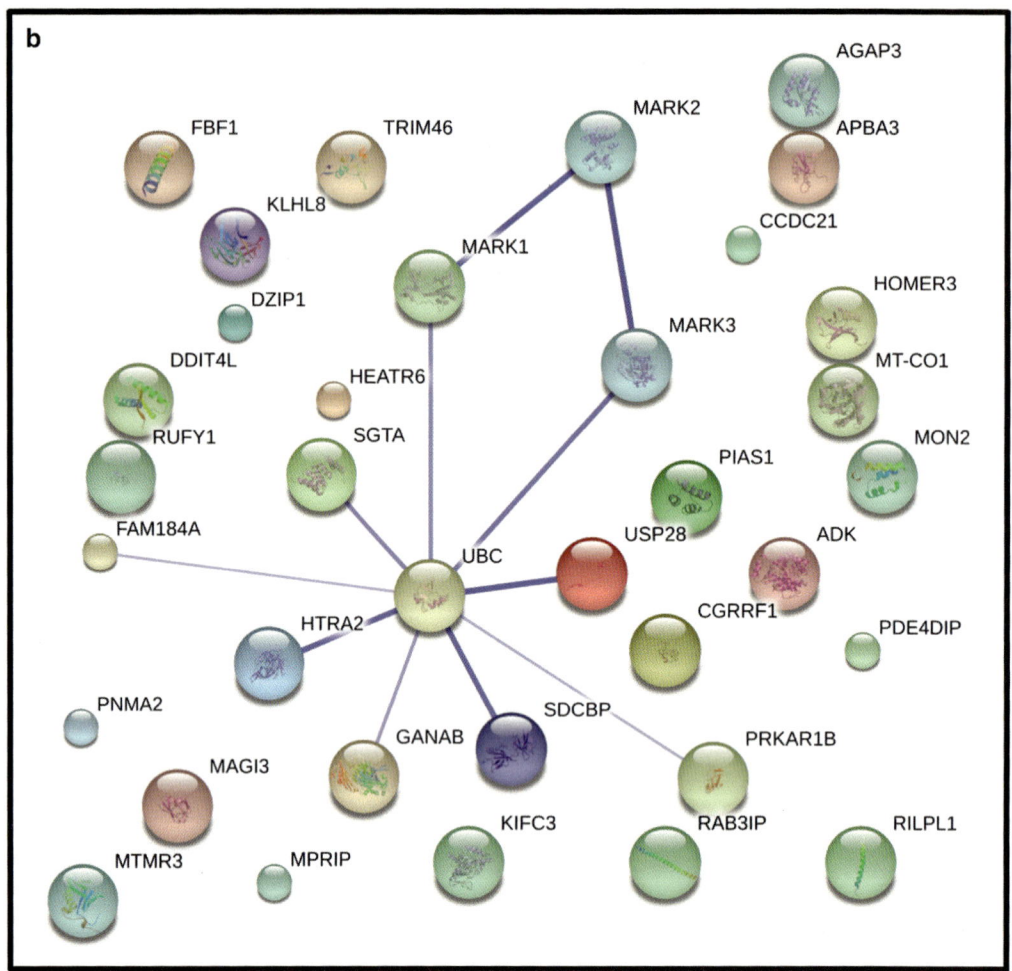

Fig. 3 continued

4 Notes

1. When making agar plates, culture medium and agar should be autoclaved separately to avoid chemical interactions when heating. Glucose solution should not be autoclaved, filter sterilize though a 0.22 μm filter and add to medium after autoclaving.

2. Although Tris–EDTA, lithium acetate, and PEG solutions are stored at room temperature, fresh TE–LiAc and TE–LiAc–PEG solutions should be prepared each time before use.

3. Readers should be aware that over a dozen of PCR enzymes tested, only La Taq from Takara gave us satisfactory results. However, other untested PCR enzymes may also be suitable.

4. Synthetic medium lacking leucine (–L) or tryptophan (–W) are used to select yeast transformants for Gal4-DB and Gal4-AD expression plasmids, respectively. Medium lacking both leucine and tryptophan (–L–W) is used to select yeast transformants for both Gal4-DB and Gal4-AD expression plasmids. Finally, synthetic medium lacking leucine, tryptophan and histidine (–L–W–H) is used to select yeast transformants for both Gal4-DB and Gal4-AD expression plasmids and the activation of the two-hybrid reporter gene *HIS3*. Appropriate amount of 3-AT can be added to increase stringency of the screen (*see* **Note 7**).

5. High transformation efficiency in yeast can be challenging to achieve. The procedure described herein should be tested on a small scale before proceeding to a large-scale transformation.

6. For a good coverage, it is usually accepted that the total number of yeast transformants should correspond to three times the original complexity of the cDNA library. However, this should be considered as a general guideline.

7. A significant fraction of bait proteins transactivate *HIS3* reporter gene in AH109 yeast when expressed in fusion to Gal4-DB. Amino acid stretches with acidic and proline residues in the bait protein are often associated with transactivation [27]. However, transactivation level should be experimentally determined since other poorly defined parameters are also involved. 3-AT, which is a competitive inhibitor of HIS3 enzyme, is used to titrate down yeast growth when the bait protein alone is a transactivator. Concentrations of 5, 10 and 20 mM are usually sufficient, but could be increased up to 200 mM.

8. Bait vectors used herein contain a yeast centromere (CEN) in addition to an autonomously replicating sequence (ARS). Thus, they will be maintained in yeast for several generations when growth is performed on non-selective YPD medium, but yeast could be grown on selective –L medium as well. Situation is different when using plasmids containing a 2μ replication origin. In that case, yeast transformants must be permanently maintained on selective medium or the bait plasmid will be quickly lost.

9. A screen is performed by mixing AH109 yeast transformed with the bait-encoding plasmid with Y187 yeast cells transformed with the AD-cDNA library in a final 1.2 ratio. It has been reporter that a 2.5 ratio could increase mating efficiency as determined by Soellick et al. [28].

10. Although 4.5 h is sufficient to achieve yeast mating, incubation should not last for too long. After few hours, diploid yeast cells start to divide and this artificially increases numbers of diploids

and positive yeast colonies. To increase mating efficiency, yeast can be resuspended and then spread on YCM (1 % yeast extract, 1 % Bacto-Peptone, 2 % dextrose) at pH 4.5 [28].

11. With this protocol, mating efficiency is usually close to 40–80 million diploids, which represent 8–16 times the original complexity of the mouse cDNA library we used. However, we empirically found that saturation is reached when the number of diploids is superior to 40 times the original complexity of the cDNA library. Thus, the screen should be repeated 3–4 times whenever saturation needs to be reached.

12. The number of positive yeast colonies is highly dependent on the experimental design of the screen and in fact, a significant fraction of Y2H screens generate no positives. In such a situation, screens should be performed with isolated domains of the original bait protein, by swapping DB and AD tags to extremities of bait and prey proteins, and using other prey libraries or alternative Y2H systems [17, 18].

13. PCR on yeast can be challenging when using low-copy plasmids such as pPC86. Besides, the amplification success rate critically depends on the length and nucleoside composition of AD-cDNA sequences corresponding to prey proteins. Thus, PCR success rate is highly variable from one screen to another, and together with poor quality and nonspecific PCR amplification products, could significantly decrease the number of exploitable AD-cDNA sequences. In the MHV screen that is presented in this manuscript, high-quality sequences were obtained for 78 % of positive yeast colonies.

Acknowledgment

This work was supported by Institut Pasteur and CNRS (Centre National de la Recherche Scientifique).

References

1. Fields S, Song O (1989) A novel genetic system to detect protein-protein interactions. Nature 340:245–246

2. Uetz P (2012) Editorial for "The Yeast two-hybrid system". Methods 58:315–316

3. Fischer JA, Giniger E, Maniatis T et al (1988) GAL4 activates transcription in Drosophila. Nature 332:853–856

4. Keegan L, Gill G, Ptashne M (1986) Separation of DNA binding from the transcription-activating function of a eukaryotic regulatory protein. Science 231:699–704

5. Fromont-Racine M, Rain JC, Legrain P (2002) Building protein-protein networks by two-hybrid mating strategy. Methods Enzymol 350:513–524

6. Rual JF, Hirozane-Kishikawa T, Hao T et al (2004) Human ORFeome version 1.1: a platform for reverse proteomics. Genome Res 14:2128–2135

7. Li S, Armstrong CM, Bertin N et al (2004) A map of the interactome network of the metazoan C. elegans. Science 303:540–543

8. Maier RH, Maier CJ, Onder K (2011) Construction of improved yeast two-hybrid libraries. Methods Mol Biol 729:71–84

9. Stynen B, Tournu H, Tavernier J et al (2012) Diversity in genetic in vivo methods for

protein-protein interaction studies: from the yeast two-hybrid system to the mammalian split-luciferase system. MMBR 76:331–382

10. Walhout AJ, Vidal M (2001) High-throughput yeast two-hybrid assays for large-scale protein interaction mapping. Methods 24:297–306

11. Mohr K, Koegl M (2012) High-throughput yeast two-hybrid screening of complex cDNA libraries. Methods Mol Biol 812:89–102

12. Roberts GG 3rd, Parrish JR, Mangiola BA et al (2012) High-throughput yeast two-hybrid screening. Methods Mol Biol 812:39–61

13. Vidalain PO, Boxem M, Ge H et al (2004) Increasing specificity in high-throughput yeast two-hybrid experiments. Methods 32:363–370

14. Venkatesan K, Rual JF, Vazquez A et al (2009) An empirical framework for binary interactome mapping. Nat Methods 6:83–90

15. Bourai M, Lucas-Hourani M, Gad HH et al (2012) Mapping of Chikungunya virus interactions with host proteins identified nsP2 as a highly connected viral component. J Virol 86:3121–3134

16. Braun P, Tasan M, Dreze M et al (2009) An experimentally derived confidence score for binary protein-protein interactions. Nat Methods 6:91–97

17. Boxem M, Maliga Z, Klitgord N et al (2008) A protein domain-based interactome network for C. elegans early embryogenesis. Cell 134:534–545

18. Caufield JH, Sakhawalkar N, Uetz P (2012) A comparison and optimization of yeast two-hybrid systems. Methods 58:317–324

19. Kim MS, Pinto SM, Getnet D et al (2014) A draft map of the human proteome. Nature 509:575–581

20. Wilhelm M, Schlegl J, Hahne H et al (2014) Mass-spectrometry-based draft of the human proteome. Nature 509:582–587

21. Vidalain PO, Tangy F (2010) Virus-host protein interactions in RNA viruses. Microbes Infect 12:1134–1143

22. Friedel CC, Haas J (2011) Virus-host interactomes and global models of virus-infected cells. Trends Microbiol 19:501–508

23. Pfefferle S, Schopf J, Kogl M et al (2011) The SARS-coronavirus-host interactome: identification of cyclophilins as target for pan-coronavirus inhibitors. PLoS Pathog 7: e1002331

24. Millet JK, Kien F, Cheung CY et al (2012) Ezrin interacts with the SARS coronavirus Spike protein and restrains infection at the entry stage. PLoS One 7:e49566

25. Muller M, Cassonnet P, Favre M et al (2013) A comparative approach to characterize the landscape of host-pathogen protein-protein interactions. J Vis Exp. doi:10.3791/50404

26. Fielding BC, Gunalan V, Tan TH et al (2006) Severe acute respiratory syndrome coronavirus protein 7a interacts with hSGT. Biochem Biophys Res Commun 343:1201–1208

27. Titz B, Thomas S, Rajagopala SV et al (2006) Transcriptional activators in yeast. Nucleic Acids Res 34:955–967

28. Soellick TR, Uhrig JF (2001) Development of an optimized interaction-mating protocol for large-scale yeast two-hybrid analyses. Genome Biol 2:RESEARCH0052

29. Jensen LJ, Kuhn M, Stark M et al (2009) STRING 8—a global view on proteins and their functional interactions in 630 organisms. Nucleic Acids Res 37:D412–D416

Chapter 19

Investigation of the Functional Roles of Host Cell Proteins Involved in Coronavirus Infection Using Highly Specific and Scalable RNA Interference (RNAi) Approach

Jean Kaoru Millet and Béatrice Nal

Abstract

Since its identification in the 1990s, the RNA interference (RNAi) pathway has proven extremely useful in elucidating the function of proteins in the context of cells and even whole organisms. In particular, this sequence-specific and powerful loss-of-function approach has greatly simplified the study of the role of host cell factors implicated in the life cycle of viruses. Here, we detail the RNAi method we have developed and used to specifically knock down the expression of ezrin, an actin binding protein that was identified by yeast two-hybrid screening to interact with the Severe Acute Respiratory Syndrome Coronavirus (SARS-CoV) spike (S) protein. This method was used to study the role of ezrin, specifically during the entry stage of SARS-CoV infection.

Key words RNA interference (RNAi), Small interfering RNA (siRNA), Ezrin, Severe Acute Respiratory Syndrome Coronavirus (SARS-CoV), Virus–host interactions

1 Introduction

The discovery of RNA interference (RNAi) represents a quantum leap in the fields of molecular and cellular biology [1, 2]. RNAi technologies are powerful tools that are widely used to investigate the biological function of specific proteins either in vitro or in vivo. In particular, RNAi has successfully been used in virology to study the role of specific host proteins in the life cycle and replication of viruses. The introduction into cells of small interfering RNAs (siRNA), 20–25 nucleotide short double-stranded RNAs that are specific to target mRNA sequences and allow for sequence-specific degradation of the mRNA, is a relatively fast, simple and robust method to specifically downregulate protein expression and study their function [3]. In our studies, siRNA has proven very useful to validate the functional relevance of cellular proteins that were identified by yeast two-hybrid screens as binding partners of influenza

Helena Jane Maier et al. (eds.), *Coronaviruses: Methods and Protocols*, Methods in Molecular Biology, vol. 1282,
DOI 10.1007/978-1-4939-2438-7_19, © Springer Science+Business Media New York 2015

and coronavirus structural proteins [4, 5]. We describe herein a method to efficiently knock down protein expression of a cellular actin binding protein, ezrin, and measure the knockdown efficiency. This method was successfully used to investigate the role of ezrin during host cell entry and infection of the Severe Acute Respiratory Syndrome Coronavirus (SARS-CoV) [5]. While the method described here is specific to the downregulation of ezrin expression, it can easily be modified and adapted to study the function of other cellular proteins during viral infection. The methodology described also forms the basis for larger scale experiments such as siRNA library screenings (*see* **Note 1**), which we have successfully established to study host cell factors involved in viral assembly and release [6].

2   Materials

2.1   siRNA Components

1. 1× siRNA buffer: 60 mM KCl, 0.2 mM MgCl$_2$, 6 mM HEPES, pH 7.5 using 2 M KOH.
2. 20 μM ezrin-specific small interfering RNAs in 1× siRNA buffer (Table 1 and *see* **Notes 2–4**).
3. 20 μM negative control non-targeting siRNA in 1× siRNA buffer.
4. DharmaFECT 1 transfection reagent (GE Dharmacon) or similar.

2.2   Cell Culture Components

1. HeLa-F5 cells (*see* **Note 5**).
2. 96-well cell culture-treated plates (*see* **Note 6**).
3. Phosphate buffer saline (PBS).
4. Dulbecco's Modified Eagle Medium (DMEM High Glucose GlutaMax™—Life Technologies) or equivalent.

Table 1
Forward sequences of siRNA used to knock down ezrin

siRNA duplex	Forward sequence (5′–3′)	Nucleotides
1	GCUCAAAGAUAAUGCUAUGUU	21
2	GGCAACAGCUGGAAACAGAUU	21
3	CAAGAAGGCACCUGACUUUUU	21
4	GAUCAGGUGGUAAAGACUAUU	21

The sequences were designed based on the VIL2 or EZR gene sequence (NCBI accession number: NM_003379). The siRNA duplexes were used in transfections as an equimolar pooled mix

5. DMEM-C: DMEM, 10 % fetal bovine serum (FBS), 100 U/ml penicillin, 100 μg/mL streptomycin. Pass solution through a 0.22 μm filtration unit before use and store at +4 °C.

6. DMEM-T: DMEM, 10 % FBS. (*see* **Note 7**).

7. Polystyrene vials.

2.3 Western Blot Components

1. Lysis buffer: RLT buffer (Qiagen). Allows for isolation of proteins as well as nucleic acids.

2. Protein sample loading buffer (LDS sample buffer): 106 mM Tris–HCl, 141 mM Tris Base, 2 % lithium dodecyl sulfate (LDS), 10 % glycerol, 0.51 mM EDTA, 0.22 mM SERVA blue G250, 0.175 mM phenol red.

3. Sample reducing agent (10×): 500 mM dithiothreitol (DTT).

4. Polyacrylamide gel for protein electrophoresis: Novex Bis-Tris 4–12 % gradient gel, ten wells.

5. Gel running buffer (NuPAGE MOPS SDS running buffer): 50 mM MOPS, 50 mM Tris Base, 0.1 % SDS, 1 mM EDTA.

6. Protein ladder: Novex Sharp pre-stained protein standard (Life technologies), or equivalent.

7. Electrophoresis and blotting module: XCell SureLock Mini-Cell (Life technologies), or similar.

8. Transfer buffer (NuPAGE transfer buffer): 25 mM bicine, 25 mM Bis-Tris (free base), 1 mM EDTA, 10 % ethanol.

9. Filter paper.

10. Polyvinylidene fluoride (PVDF) membrane.

11. Blotting sponge pads.

12. Tris-buffer saline (10× TBS): 200 mM Tris base, 1.5 M NaCl, pH to 7.6 with 12 N HCl.

13. TBST: TBS, 0.1 % Tween 20 (*see* **Note 8**).

14. Blocking solution: 5 % milk in TBST.

15. Rabbit polyclonal anti-ezrin (generous gift from Dr. Monique Arpin, Institut Curie, France).

16. Mouse monoclonal anti-glyceraldehyde-3-phosphate dehydrogenase (GAPDH).

17. Horse radish peroxidase (HRP) conjugated goat anti-rabbit.

18. HRP conjugated rabbit anti-mouse IgG.

19. Enhanced chemiluminescence (ECL) compounds.

20. Gel Doc system capable of reading chemiluminescent signals, e.g., Bio-Rad Gel Doc XR system. Alternatively, membranes can be exposed on X-ray films. Exposure times may vary from a few seconds to several minutes.

21. ImageJ or similar software for band quantification.

3   Methods

The following procedures should be performed in a Class II biosafety cabinet, unless otherwise noted. The siRNA transfection method below describes the procedure for transfecting a specific set of siRNAs (ezrin-targeting or non-targeting siRNAs). As siR-NAs are fragile, they should be kept on ice as much as possible. Perform each siRNA treatment condition in triplicates. In our experiments, because cells that undergo siRNA treatment will subsequently be virally infected, care should be taken at all steps to ensure the cells being treated are in the best condition and viability assessed as much as possible (*see* **Note 9**).

3.1   siRNA Transfection (for Each Type of siRNA)

1. Seed 3.6×10^3 cells per well of a 96-well plate. Incubate at 37 °C for 16–18 h.

2. Dilute stock siRNA solution 1/10, to 2 μM, with 1× siRNA buffer.

3. Mix 2 μM siRNA solution with serum-free DMEM to obtain a 1 μM solution using a 1:1 ratio. For mock siRNA condition, replace siRNA solution by DMEM.

4. Perform gentle up–down pipetting and incubate tube at room temperature for 5 min.

5. Dilute to 1/50 DharmaFECT 1 transfection reagent with DMEM in a polystyrene vial.

6. Perform gentle up–down pipetting and incubate tube at room temperature for 5 min.

7. Mix siRNA-DMEM solution with transfection reagent-DMEM solution using a 1:1 ratio, by adding the transfection reagent-DMEM solution to the siRNA DMEM solution.

8. Incubate at room temperature for 20 min.

9. Dilute siRNA-transfection reagent mix with DMEM-T using a 1/5 dilution.

10. Aspirate culture medium from wells.

11. Add 100 μl of transfection mix per well.

12. Incubate at 37 °C cell culture incubator for 48 h.

13. Check for cytotoxicity or cell morphological changes routinely by observing transfected cells under microscope.

14. Optional: Repeat siRNA transfection to increase knockdown efficiency and incubate for another 48 h (*see* **Note 10**). Check for cytotoxicity or cell morphological changes under microscope (Fig. 1).

Non-targeting
siRNA

Ezrin
siRNA

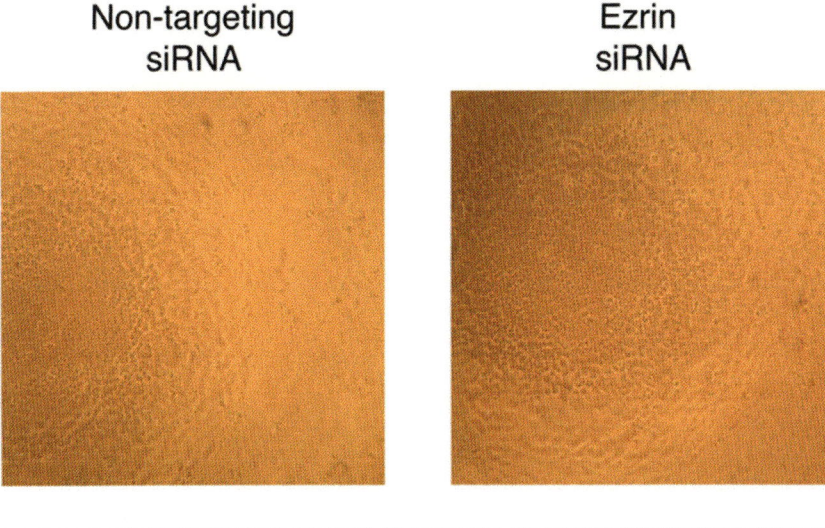

HeLa-F5

Fig. 1 Cell morphology and density after siRNA transfections. 3.6×10^3 HeLa-F5 cells were seeded in wells of a 96-well plate. The cells were then transfected with either non-targeting or ezrin-targeting siRNAs and incubated at 37 °C cell culture incubator for 48 h. A second round of siRNA transfection was then performed and cells were incubated at 37 °C cell culture incubator for 48 h. The cells were then observed under an inverted microscope using at 10× objective and pictures of representative fields were taken

3.2 Measurement of siRNA Knockdown Efficiency

A crucial step in any siRNA transfection experiment is to assess the level of knockdown of expression induced by such treatment. The section below describes how to evaluate siRNA knockdown by measuring the lowering of ezrin protein expression using a Western blot approach (*see* **Note 11**).

1. Lyse cells by adding 100 μl lysis buffer per well (*see* **Note 12**), and incubate plate at room temperature for 15 min. Perform gentle up down pipetting to ensure lysis is complete.

2. Add 40 μl 4× Loading buffer, 16 μl 10×DTT, and 4 μl H_2O to each sample.

3. Boil samples at 95 °C for 5 min.

4. Cool samples down on ice and perform a quick centrifugation to bring down condensate. At this point, samples can be run through a gel or stored at –20 °C for later analysis.

5. Place the pre-cast Bis-Tris polyacrylamide gel in electrophoresis module (*see* **Note 13**).

6. Fill inner and outer compartments of electrophoresis module with the 1× running buffer.

7. Load 10 μl of molecular weight standard ladder and load 10 μl of samples in each lane.

8. Run electrophoresis using following settings: 200 V with constant voltage for 45–60 min.

9. Cut out Whatman paper and PVDF membrane to the gel dimensions.

10. Dehydrate PVDF membrane in 100 % ethanol for 1 min.

11. Rehydrate membrane in H_2O for 5 min, and then incubate it in 1× transfer buffer for 5 min.

12. Soak blotting paper and sponge pads in 1× transfer buffer for at least 5 min.

13. Remove gel from plastic encasing and immerse gel in 1× transfer buffer.

14. Prepare transfer stack by layering (from bottom to top) two blotting sponge pads, three blotting paper cutouts, gel, PVDF membrane. Gently roll out bubbles with roller, e.g., a small pipette. Continue stack by adding three blotting paper cutouts and two sponge pads.

15. Place the stack in transfer module and perform transfer inside electrophoresis module filled with 1× transfer buffer using the following settings: 170 mA constant current for 1–2 h.

16. Remove membrane from stack and place in TBST solution.

17. Block membrane in blocking solution for 1 h at room temperature.

18. Prepare primary antibody solutions by diluting them in blocking solution: rabbit polyclonal anti-ezrin: 1/1,000; goat anti-GAPDH 1/10,000.

19. Cut a straight line on the membrane at 50 kDa marker. Incubate top part with anti-ezrin antibody solution and the bottom part with anti-GAPDH antibody solution for 1 h each.

20. Wash membranes three times 10 min in TBST.

21. Prepare secondary antibodies by diluting them in blocking solution: HRP goat anti-rabbit IgG 1/1,000; HRP rabbit anti-goat IgG 1/1,000.

22. Incubate membranes in the corresponding secondary antibody solutions for 45 min.

23. Wash membranes three times 10 min in TBST.

24. Mix ECL solutions using a 1:1 ratio and add 1–2 mL of mixed solution to membrane surface.

25. Incubate for 1 min, remove excess moisture, and perform band detection using gel doc or film and developer (Fig. 2).

26. Perform band quantification analysis using Gel Analysis module of ImageJ or similar software. For each lane, the GAPDH

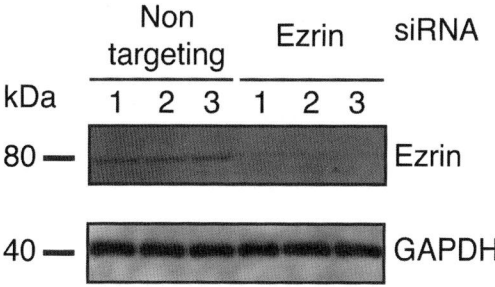

Fig. 2 Assessment of ezrin protein expression knockdown induced by siRNA transfections. 3.6×10^3 HeLa-F5 cells were seeded in wells of a 96-well plate. The cells were transfected twice with non-targeting or ezrin-targeting siRNAs. For each condition, cells from one well were lysed and analyzed for ezrin and GAPDH housekeeping protein content by Western blot. The Western blot shown here displays three independent replicates for either non-targeting or ezrin targeting siRNAs

(housekeeping protein) band serves as a loading control. Normalize ezrin band relative intensity (I_{ezrin}) to GAPDH band relative intensity (I_{GAPDH}):

$$\frac{I_{ezrin}}{I_{GAPDH}}$$

27. The percentage of ezrin protein expression knockdown by siRNA treatment, compared to non-targeting control is calculated by the following equation (KD, expressed in %, Table 2) :

$$KD = \left[1 - \frac{\left(\dfrac{I_{ezrin}}{I_{GAPDH}} \right)_{ezrin\ siRNA\ lane}}{\left(\dfrac{I_{ezrin}}{I_{GAPDH}} \right)_{non-targeting\,siRNA\ lane}} \right] \times 100$$

4   Notes

1. The method described here focuses on knocking down expression of a single cellular gene, ezrin, to uncover its functional role in SARS-CoV infection. This methodology forms the basis for larger-scale functional analyses as it can easily be scaled up to perform siRNA-based functional screen studies to identify host genes involved in the replicative cycle of viruses. Such siRNA-based screening approach has been successfully developed in our lab using a siRNA library screen of 122 cellular

Table 2
Knockdown analysis after siRNA transfection

		Band intensity	Knockdown (%)	Average knockdown (%)	Standard deviation (%)
Ezrin siRNA 1	I_{ezrin}	254			
	I_{GAPDH}	26,582	95.2		
Non targeting siRNA 1	I_{ezrin}	4,602			
	I_{GAPDH}	23,119			
Ezrin siRNA 2	I_{ezrin}	265			
	I_{GAPDH}	22,011	93.7	95.0	1.2
Non targeting siRNA 2	I_{ezrin}	4,433			
	I_{GAPDH}	23,098			
Ezrin siRNA 3	I_{ezrin}	196			
	I_{GAPDH}	18,857	96.0		
Non targeting siRNA 3	I_{ezrin}	5,042			
	I_{GAPDH}	19,165			

HeLa-F5 cells were transfected twice with non-targeting or ezrin-targeting siRNAs and protein content was assessed by Western blot. Western blot band intensities were analyzed with ImageJ and the knockdown efficiency (KD) was calculated

    genes involved in membrane trafficking to reveal host factors that are functionally implicated in dengue virus assembly and release [6].

2. siRNAs should be handled with special care. RNA molecules are prone to degradation by RNases. Gloves and RNAse-free pipette tips should be used. Tubes containing siRNAs should be kept on ice as much as possible. Store siRNAs at –20 °C and aliquot stock solutions to avoid multiple freeze–thaw cycles.

3. The ezrin siRNAs used here are in the form of an equimolar mix of four siRNA duplexes. This allows for robust knockdown of expression because it increases the odds of binding to target mRNA sequence and silencing to ensue. It is important to note however that individual siRNAs can also be used to silence specific mRNA expression. This alternative approach has the advantage to minimize potential off-target effects of pooled siRNA mixes. We have validated the use of individual siRNA treatment for ezrin silencing, and successfully used an individual siRNA duplex to silence annexin A6 to investigate its functional role in influenza virus infection [4]. Another consideration when performing siRNA studies is the potential for functional redundancies found in families of closely related proteins. In another dengue virus study from our lab, we have found that, while individually silencing the closely related small GTPases Arf4 or Arf5 had minimal effect on dengue virus secretion from cells, combined silencing of these two GTPases allowed for marked decrease in secretion [7].

4. siRNAs are used at a final concentration of 100 nM for transfection of cells. This concentration was determined to be the best compromise between siRNA knockdown efficiency and cell viability by prior optimization experiments that tested increasing concentrations of siRNAs. As optimal siRNA concentration varies depending on cell type used and target gene, we recommend performing such optimization during the setting up of any siRNA assay.

5. The choice of cell lines for conducting siRNA transfections is an important step during the setup and optimization of the assay. HeLa-F5 cells were chosen in our experiments because they robustly express ACE-2, the SARS-CoV receptor, and have been shown previously to be susceptible to SARS-CoV S-mediated viral entry. Furthermore, we have conducted preliminary siRNA transfection on a panel of cell lines, which included HeLa-F5, and found that those cells could be efficiently transfected, with ezrin protein expression levels significantly reduced after siRNA treatment.

6. In the experiment described herein, siRNA transfections and subsequent assays were performed in 96-well cell culture-treated plates. Depending on the experiment planned using the siRNA-treated cells, the format can be adapted to larger culture plates. In that case, the number of cells and volumes of reagents used will have to be proportionally scaled up.

7. Avoiding addition of penicillin/streptomycin to the transfection medium (DMEM-T) is important because lipid-based transfection reagents, such as DharmaFECT 1 increases the permeability of the cell plasma membrane. If present in the transfection medium, there is a greater risk for cellular uptake of the antibiotics with potentially higher cytotoxicity and lower transfection efficiency.

8. Tween 20 is a viscous solution. To ensure that the correct volume of Tween 20 is added to the TBS buffer, cut a 1,000 µL pipette tip, gently aspirate Tween 20 and add to TBS buffer. Pipette up–down gently and eject tip in buffer. Add stir bar and let solution stir with pipette tip for ~30 min, or until Tween 20 has completely dissolved.

9. The siRNA transfection procedure involves many steps of aspiration of supernatants and addition of solutions on cells. To avoid detaching cells, care should be taken to avoid letting the cells be without medium for more than a few minutes. Also, when adding new medium, solutions should not be pipetted directly on cells, but on the walls of the wells. Media should be pre-warmed at 37 °C as much as possible, as cold solutions can easily detach cells. We performed routine cell viability assays to determine the cytotoxic effects of siRNA treatments, using Trypan blue exclusion assay after treatments.

10. The repeat of the siRNA transfection step 48 h after the first one depends on several parameters including the efficiency of knockdown after 48 h, the turnover of the targeted mRNA, or the half-life of the protein product. We have observed through a series of tests that the optimal conditions for the knockdown of ezrin was to perform two successive siRNA transfections, 48 h apart.

11. If antibodies for the protein of interest are unavailable, an alternative would be to perform a quantitative RT-PCR assay, using specific primers, to measure levels of corresponding messenger RNAs.

12. After HeLa-F5 cells have undergone two successive siRNA transfections, the siRNA efficiency control plate is used to assess the quality of knockdown. Analysis of ezrin or GAPDH (housekeeping protein) protein content from one well of a 96 well plate is sufficient for detection by Western blot.

13. An alternative to the pre-cast Bis-Tris polyacrylamide gel electrophoresis (PAGE) system is the use of gels prepared in the laboratory using gel casters. This allows customizing gels to the most appropriate percentage of polyacrylamide for the protein to be analyzed. 8–12 % polyacrylamide separating gel (8, 10 or 12 % Acrylamide–Bis-Acrylamide, 400 mM Tris–HCl pH 8.8, 0.1 % SDS, 0.04 % ammonium persulfate (APS), 0.07 % tetramethylethylenediamine (TEMED) added last, ddH$_2$O) and 4 % polyacrylamide stacking gel (4 % Acrylamide–Bis-Acrylamide, 100 mM Tris–HCl pH 6.8, 0.1 % SDS, 0.06 % APS, 0.1 % TEMED added last, ddH$_2$O) are commonly used. Laboratory-made running, loading, and transfer buffers can also be prepared in the laboratory following standard Western blotting procedures.

References

1. Napoli C, Lemieux C, Jorgensen R (1990) Introduction of a chimeric chalcone synthase gene into petunia results in reversible co-suppression of homologous genes in trans. Plant Cell 2:279–289

2. Fire A, Xu S, Montgomery MK et al (1998) Potent and specific genetic interference by double-stranded RNA in Caenorhabditis elegans. Nature 391:806–811

3. Sen GL, Blau HM (2006) A brief history of RNAi: the silence of the genes. FASEB J 20:1293–1299

4. Ma H, Kien F, Manière M et al (2012) Human annexin A6 interacts with influenza A virus protein M2 and negatively modulates infection. J Virol 86:1789–1801

5. Millet JK, Kien F, Cheung C-Y et al (2012) Ezrin interacts with the SARS coronavirus spike protein and restrains infection at the entry stage. PLoS One 7:e49566

6. Wang PG, Kudelko M, Lo J et al (2009) Efficient assembly and secretion of recombinant subviral particles of the four dengue serotypes using native prM and E proteins. PLoS One 4:e8325

7. Kudelko M, Brault JB, Kwok K et al (2012) Class II ADP-ribosylation factors are required for efficient secretion of dengue viruses. J Biol Chem 287:767–777

Chapter 20

Transcriptome Analysis of Feline Infectious Peritonitis Virus Infection

Parvaneh Mehrbod, Mohammad Syamsul Reza Harun, Ahmad Naqib Shuid, and Abdul Rahman Omar

Abstract

Feline infectious peritonitis (FIP) is a lethal systemic disease caused by FIP virus (FIPV). There are no effective vaccines or treatment available, and the virus virulence determinants and pathogenesis are not fully understood. Here, we describe the sequencing of RNA extracted from Crandell Rees Feline Kidney (CRFK) cells infected with FIPV using the Illumina next-generation sequencing approach. Bioinformatics analysis, based on *Felis catus* 2X annotated shotgun reference genome, using CLC bio Genome Workbench is used to map both control and infected cells. Kal's Z test statistical analysis is used to analyze the differentially expressed genes from the infected CRFK cells. In addition, RT-qPCR analysis is used for further transcriptional profiling of selected genes in infected CRFK cells and Peripheral Blood Mononuclear Cells (PBMCs) from healthy and FIP-diagnosed cats.

Key words FIPV, CRFK, PBMCs, Transcriptome, Kal's Z test, RT-qPCR, Gene expression

1 Introduction

The use of a next-generation sequencing approaches in RNA sequencing has facilitated understanding and defining the expression profiles of cellular responses during pathogen infection. This method has been proven to be helpful in explaining the pathogenesis of various viruses [1, 2], including Feline Immunodeficiency Virus (FIV) [3, 4]. Furthermore, the increasing availability of complete genome sequences for a number of model organisms makes host transcriptome analysis a valuable tool for elucidating mechanisms of virus pathogenesis and host responses to virus infection. Feline infectious peritonitis virus (FIPV) is thought to be the causative agent of feline infectious peritonitis (FIP). Understanding the molecular basis of FIPV pathogenesis will provide valuable information to devise effective treatments and formulate vaccines with higher efficacy. Once established, focus can be directed at disrupting the virulent determinants or formulating

Helena Jane Maier et al. (eds.), *Coronaviruses: Methods and Protocols*, Methods in Molecular Biology, vol. 1282, DOI 10.1007/978-1-4939-2438-7_20, © Springer Science+Business Media New York 2015

new vaccine or even designing gene therapy treatment. Facilitating this, the complete 1.9X cat genome, using the Whole Genome Shotgun (WGS) approach, provides valuable information for bio-informatics analysis of feline host responses following pathogen infection. Moreover, the cat genome contigs were aligned, mapped, and annotated to NCBI annotated genome sequences of six index mammalian genomes (human, chimpanzee, mouse, rat, dog and cow) using MegaBLAST [5]. In this chapter we describe a pipeline for transcriptome analysis using FIPV infection of feline cells in culture as an example. Specifically, mRNA harvested from CRFK cells 3 h post infection with FIPV strain 79-1146 were sequenced using Illumina next-generation sequencing technology. The generated data was then analyzed using CLC bio Genomic Workbench, where the genes were compared to *Felis catus* 1.9X annotated shotgun reference genome. Kal's Z-test on expression proportions [6] was used to determine significantly expressed genes. Genes expressed with a False Discovery Rate (FDR) <0.05 and >1.99- and <-1.99-fold change were considered for further analysis.

2   Materials

2.1   RNA Extraction and Next-Generation Sequencing

1. CRFK cell line.
2. Virus strain: FIPV 79–1146.
3. Maintenance medium : Minimal essential medium, 10 % fetal bovine serum, 200 μM nonessential amino acids, 200 units/ml penicillin, 200 μg/ml streptomycin, 0.5 μg/ml antimycotic.
4. Dulbecco's Phosphate Buffered Saline containing no calcium and magnesium (D-PBS).
5. TrypLE™ Express solution or 0.25 % trypsin, 0.02 % EDTA in PBS.
6. RNase AWAY (Life Technologies) or similar.
7. RNeasy mini kit (Qiagen) or similar.
8. UV/Visible spectrophotometer.
9. Illumina Agilent RNA 6000 Nano kit.
10. Agilent 2100 Bio-analyzer.
11. Paired-End Sequencing Preparation Kit (Illumina), or similar.

2.2   Validation of Results by RT-qPCR

1. QiaAmp Viral RNA Mini Kit (Qiagen) or similar.
2. RNeasy kit (Qiagen).
3. RNase AWAY.
4. Quantitative PCR machine (e.g., CFX 96 Real-Time PCR Detection System (Bio-Rad)).
5. RNeasy miniplus kit (Qiagen) or similar.
6. NanoDrop Nanophotometer or spectrophotometer.

7. SensiFAST™ SYBR No-ROX One Step kit (Bioline).

8. Primers specific for genes of interest and reference genes.

9. BD Vacutainer (BD) EDTA-K2 tubes.

10. Ficoll-Paque Plus.

3   Methods

3.1   Transcriptome Analysis of FIPV Infected Cells

To take advantages of this technology, simultaneous analysis of virus–host interactions is investigated in one single experiment where both the transcription of viral genomes and host cell responses are scrutinized. Figure 1 illustrates the work flow for transcriptome analysis of this study.

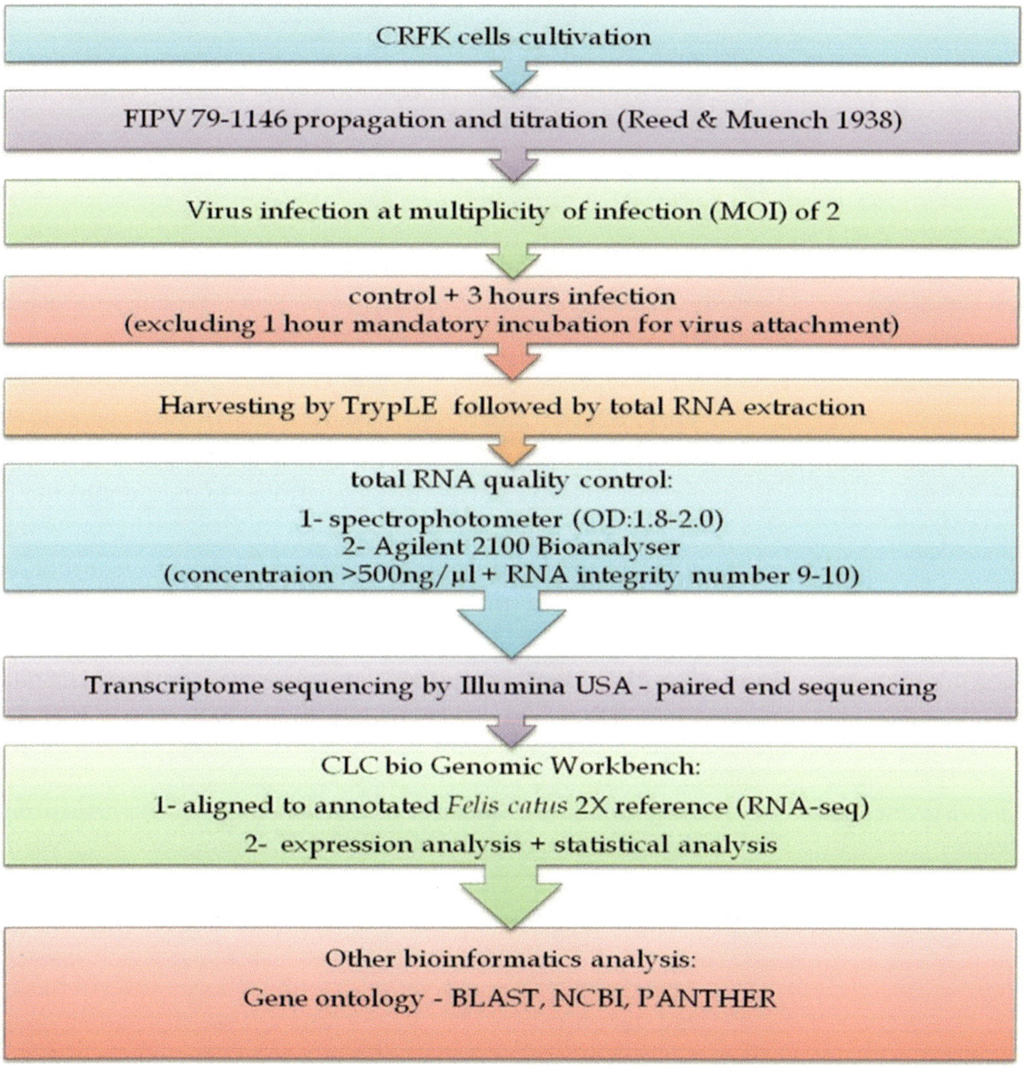

CRFK cells cultivation

FIPV 79-1146 propagation and titration (Reed & Muench 1938)

Virus infection at multiplicity of infection (MOI) of 2

control + 3 hours infection
(excluding 1 hour mandatory incubation for virus attachment)

Harvesting by TrypLE followed by total RNA extraction

total RNA quality control:
1- spectrophotometer (OD:1.8-2.0)
2- Agilent 2100 Bioanalyser
(concentraion >500ng/µl + RNA integrity number 9-10)

Transcriptome sequencing by Illumina USA - paired end sequencing

CLC bio Genomic Workbench:
1- aligned to annotated *Felis catus* 2X reference (RNA-seq)
2- expression analysis + statistical analysis

Other bioinformatics analysis:
Gene ontology - BLAST, NCBI, PANTHER

Fig. 1 Work flow for transcriptome analysis of CRFK cells infected with FIPV 79–1146

3.1.1 Infection of Cells

1. Seed CRFK cells into 75 cm² flasks and incubate at 37 °C with 5 % CO_2 until cells reach 60–70 % confluency.

2. Remove media and wash the cells with D-PBS.

3. Infect the flasks with virus at multiplicity of infection (MOI) 2 in 2 ml, or 2 ml D-PBS as a mock control, and incubate at 37 °C with 5 % CO_2 for 1 h to allow attachment. Perform inoculations in duplicates, one for RNA extraction and the other for CPE visualization.

4. Add 10 ml of maintenance medium with 10 % FBS and incubate the flasks for 3 h.

5. Following 3 h of incubation, remove inoculum, wash the cells with D-PBS.

6. Add 2 ml TrypLE and incubate for 1–2 min until cells detach.

7. Transfer cells to a centrifuge tube and pellet the cells by centrifugation at $120 \times g$ for 5 min and discard supernatant.

8. Add 10 ml D-PBS and repeat centrifugation in order to remove every trace of medium and TrypLE, which could reduce RNA yield.

9. Discard the supernatants and store the cell pellets at –80 °C until RNA purification.

3.1.2 RNA Purification from Infected Cells

The RNeasy kit was used to extract and purify RNA samples in this study (*see* **Note 1**) but other RNA extraction protocols may also be suitable.

1. Spray all micropipettes, gloves, working area, and other things with RNase AWAY to remove any RNase and DNA contamination.

2. Extract RNA using RNeasy kit according to the manufacturer's instruction.

3. Aliquot the eluted RNAs (500 μl) into three different tubes to avoid repeated thawing and freezing of the sample which could affect the quality of the RNA.

4. Use two tubes for quality control analysis with spectrophotometer and Illumina Agilent 2100 bio-analyzer and store the third one at –80 °C for sequencing.

3.1.3 Total RNA Quality Analysis

1. Determine the quality of extracted RNA by measuring the absorbance at 260 and 280 nm in UV/Visible spectrophotometer.

2. Consider the samples with the absorbance ratio value (A260/A280) of 1.8 to 2.0 for further analysis with Illumina Agilent 2100 Bio-analyzer to determine both the RNA quality and quantity.

3. In order to ensure the samples are of highest quality and quantity for transcriptome sequencing, use Agilent RNA 6000

Nano kit together with Agilent 2100 Bio-analyzer to conduct quality and quantity analysis to the extracted total RNA samples (*see* **Note 2**).

4. Load and prime gel–dye mixtures, then load RNA 6000 Nano marker, ladder, and samples in the specified manner.

5. Vortex the chip and insert in the Agilent 2100 Bio-analyzer machine. Analyze the chips based on the method recommended by the manufacturer. Verify whether the run is successful and whether the sample is properly prepared and handled by means of properly pipetted into the wells (*see* **Note 3**).

3.1.4 Preparing Samples for Paired-End Sequencing

Perform the following steps using the reagents provided in the paired end sample preparation kit, according to manufacturer's instructions.

1. Fragment genomic DNA into fragments of less than 800 bp.

2. Perform end repair of DNA fragments to generate 5′-phosphorylated blunt ends.

3. Add an "A" base to the 3′ ends to make 3′-dA overhang.

4. Ligate adapters to the ends of the DNA fragments.

5. Purify ligation products by removing un-ligated adapters.

6. Enrich the Adapter-Modified DNA Fragments by PCR.

7. Obtain the Genomic DNA library.

3.1.5 CLC Bio Genomics Workbench (GWB) Software Analysis

Figure 2 represents the work flow of raw data analysis by CLC bio GWB software and the settings opted for each process. Unless stated otherwise, the settings for the raw data analysis were based from CLC bio manual (*see* **Note 4**).

1. Conduct RNA-seq analysis with the settings stated in the Fig. 2, using trimmed raw sequences. Quality trim limit = 0.01, ambiguity trim maximum value = 2, Adapter trim = Illumina adapter, trim both strands. Map to annotated reference—*Felis catus* 2X, Minimum length fraction and minimum similarity fraction = 0.9, Maximum number of hits/read = 1, Type of organism = eukaryote, Paired settings = default.

2. Reads Per Kilobase of transcript per Million (RPKM) are chosen as the expression value for comparison (*see* **Note 5**).

3. Once finished, conduct quality control to the expression analysis control versus 3 h infection (Cv3) result.

4. Subsequently, conduct statistical analysis to the Cv3 expression analysis based on CLC bio support recommendation.

5. Determine upregulated and downregulated genes from the final Cv3 expression analysis result by selecting the criteria as in Fig. 2. Upregulated = proportions fold change > 1.99 and Downregulated = proportions fold change < −1.99.

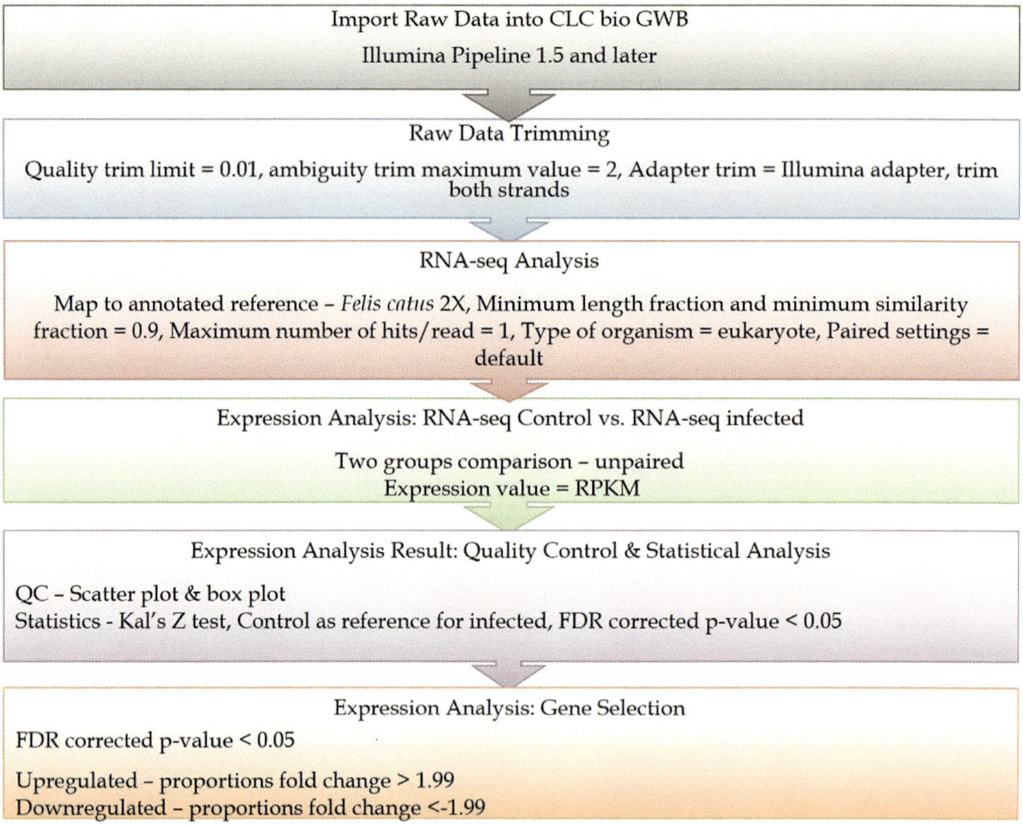

Import Raw Data into CLC bio GWB

Illumina Pipeline 1.5 and later

⬇

Raw Data Trimming

Quality trim limit = 0.01, ambiguity trim maximum value = 2, Adapter trim = Illumina adapter, trim both strands

⬇

RNA-seq Analysis

Map to annotated reference – *Felis catus* 2X, Minimum length fraction and minimum similarity fraction = 0.9, Maximum number of hits/read = 1, Type of organism = eukaryote, Paired settings = default

⬇

Expression Analysis: RNA-seq Control vs. RNA-seq infected

Two groups comparison – unpaired
Expression value = RPKM

⬇

Expression Analysis Result: Quality Control & Statistical Analysis

QC – Scatter plot & box plot
Statistics - Kal's Z test, Control as reference for infected, FDR corrected p-value < 0.05

⬇

Expression Analysis: Gene Selection

FDR corrected p-value < 0.05

Upregulated – proportions fold change > 1.99
Downregulated – proportions fold change <-1.99

Fig. 2 Work flow and settings for CLC bio GWB software analysis

3.1.6   Basic Local Alignment Search Tool (BLAST)

1. Once data has been obtained, import the raw data (~17.3 Gigabyte) into the CLC bio GWB. Once imported, subject the raw data to sequence reads trimming by quality trimming, ambiguity trimming and adapter trimming with the settings as in Fig. 2. The program uses the modified-Mott trimming algorithm for this purpose (see manufacturers instructions) (*see* **Note 6**).

2. BLAST the list of genes that were upregulated and downregulated, using the built-in BLAST program in the CLC bio GWB.

3. Based on the BLAST result, select homologous sequence with the lowest e-value, highest score and lowest percentage of gaps to the query sequence as the gene identity.

4. Briefly, opt blastn: DNA sequence and database program and references mRNA sequences (refseq_rna) or nucleotide collection (nr) database for analysis. In silico analysis which is also a part of bioinformatics analysis is able to analyze the interactions of different genes by integrating data available on bioinformatics databases (*see* **Note 7**).

3.2 Validation of Results by Real-Time RT-qPCR Analysis

3.2.1 Real-Time RT-qPCR Analysis of FIPV Infected CRFK Cells

1. In order to validate the transcriptome results, analyze the expression profiles of genes of interest (for examples A3H, PD-1, and PD-L1 genes) using real-time PCR.

2. Seed CRFK cells into 6-well plates and incubate at 37 °C 5 % CO_2 until 60–70 % confluent.

3. Wash cells once with D-PBS and then infect with FIPV strain 79–1146 at MOI 2 in 1 ml, or with 1 ml D-PBS as a negative control.

4. Incubate cells at 37 °C 5 % CO_2 for desired time.

5. Extract viral RNA from cells using QiaAmp Viral RNA Mini Kit, according to the manufacturer's instructions.

6. Design primers for qPCR specific for the genes of interest, e.g., Table 1.

7. Perform the RT-qPCR reactions using for example SensiFAST™ SYBR No-ROX One Step kit on Real-Time System, with Thermal Cycler. The reaction mixture of 20 μl contains 10 μl 2× SensiFAST SYBR No-ROX One-Step mix, 0.5 μl forward and reverse primers (5 nM for GADPH, PD-L1, and A3H, 3 nM for PD-1, and 10 nM for YWHAZ), 0.2 μl RT, 0.4 μl RiboSafe RNase inhibitor, 2.4 μl H_2O, and 6 μl extracted RNA.

8. The RT-qPCR reaction conditions are as follows; one cycle at 45 °C for 10 min, one cycle at 95 °C for 2 min, and 35 cycles at 95 °C for 5 s; then 57 °C (YWHAZ), 58 °C (PD-L1), 59 °C (GAPDH), 64 °C (A3H), and 65 °C (PD-1) for 20 s; and finally, at 72 °C for 5 s (Table 2). One cycle for the dissociation curve for all reactions is added and melting curve analysis is performed.

Table 1
Sequence of primers for RT-qPCR

Target gene	Accession number	Sequence 5'–3'	References
GAPDH	NM 001009307	F: AGTATGATTCCACCCACGGCA R: GATCTCGCTCCTGGAAGATGGT	[7]
YWHAZ	EF458621	F: ACAAAGACAGCACGCTAATAATGC R: CTTCAGCTTCATCTCCTTGGGTAT	[9]
PD-1	EU295528	F: GAGAACGCCACCTTCGTC R: TGGGCTCTCATAGATCTGCGT	[10]
PD-L1	EU246348	F: CGATCACAGTGTCCAAGGACC R: TCCGCTTATAGTCAGCACCG	[10]
A3H	EF173020	F: ACCCACAATGAATCCACTACAG R: AGGCAGTCTTTGTGAATTAGGG	[11]

F forward primer, *R* reverse primer

Table 2
Amplification program for one step RT-qPCR assay

No.	Step	Temperature (°C)	Time
1	Reverse transcription	48	45 min
2	Initial denaturation	94	02 min
3	Denaturation	94	15 s
4	Annealing	62	15 s
5	Extension	72	15 s
6	Repeat **steps 3–5** for 35 cycles		
7	Final extension	72	05 min

9. Analyze the data generated from the technical triplicate experiment with 2ΔΔCT method, selecting appropriate reference genes, e.g., GAPDH and /or YWHAZ, using Bio-Rad CFX Manager.

3.2.2 Real-Time RT-qPCR Analysis of Peripheral Blood Mononuclear Cells from FIP Diagnosed Cats

Besides FIPV infected cell culture, feline Peripheral Blood Mononuclear Cells (PBMCs) can be used to analyze the transcriptome results, providing valuable in vivo results for comparison.

1. Select five healthy cats and five infected cats. Perform the sampling according to internationally recognized guidelines and recommended by the Animal Care and Use Committee.

2. Draw 2–5 ml of cat blood and store at 4 °C in BD Vacutainer® EDTA-K2 tubes.

3. Isolate PBMCs using the Ficoll-Paque™ Plus method, according to the manufacturer's protocol.

4. Isolate total RNA from PBMCs using an RNeasy mini plus kit, as described by the manufacturer.

5. Measure and assess the RNA quantity and purity using NanoDrop.

6. Store the isolated RNA samples at –80 °C for further analysis, or immediately use for real-time RT-qPCR analysis, as **steps 6–9** in Subheading 3.2.1.

4   Notes

1. The wash step is performed in order to remove traces of serum, calcium, and magnesium that would inhibit the action of dissociation agent.

2. http://www.chem.agilent.com/Library/usermanuals/Public/G2938-90034_RNA6000Nano_KG.pdf.

3. The first feature of a successful total RNA run is that the electropherogram must contain three peaks where one peak represents marker peak while the others two are 18S and 28S ribosomal peaks. Absence of one or both of the ribosomal peaks indicates poor sample preparation or poor sample pipetting technique. The second feature of a successful run is a complete ladder electropherogram. A complete ladder electropherogram must feature one marker peak and six RNA peaks where all seven peaks are well resolved.

4. Adapters used by common high-throughput sequencing vendors such as Illumina and SOLiD were predefined and are available by the software. Removing the adapters will increase the specificity of the raw sequence reducing false match.

5. Perform expression analysis based on the method described by Mortazavi et al. in 2008 [7] and CLC bio manual, CLC bio tutorials and recommendations from CLC bio support services.

6. In short, high quality trim value allowed low quality base or base with low Phred quality score to be included in the sequence. The ambiguity trimming trims the sequence ends based on the presence of ambiguous nucleotides usually denoted as N making the sequence more specific.

7. *Other Bioinformatics analysis.* A gene in eukaryotic organism is commonly regulated by other genes and proteins in its system. The interactions among genes expressed and between gene expressed and other genes can be elucidated by means of computer or in silico analysis. In silico analysis, a part of bioinformatics analysis, able to do this by integrating data with available data on bioinformatics databases. Such integration will allow a researcher to make accurate predictions and designing experiments to test the hypothesis. The main objective of in silico analysis is gene ontology which is defined as the process of elucidating associated pathways, molecular function, biological process, cellular components and protein products of a gene [8].

The bioinformatics database used to analyze gene identity and gene interaction is Protein Analysis Through Evolutionary Relationships Classification System or in short known as PANTHER (http://www.pantherdb.org/). It is a unique resource that classifies genes by their functions, using published scientific experimental evidence and evolutionary relationships to predict function even in the absence of direct experimental evidence and is a part of the Gene Ontology Reference Genome Project (http://www.geneontology.org/GO.refgenome.shtml#curation).

PANTHER provides tools for gene expression analysis for data interpretation (http://www.pantherdb.org/tools/genexAnalysis.jsp). Multiple gene lists will be mapped to PANTHER molecular function, biological process, and cellular component categories as well as to biological pathways. The gene expression data interpretation is conducted by comparing genes in a given list and statistically compares the list to the reference list to look for under and over represented functional categories. The step-by-step method and the statistical test employed are described in detail at http://www.pantherdb.org/tips/tips_binomial.jsp. For the statistical analysis, a cutoff of less than 0.05 is selected as significant p-value.

Acknowledgments

The authors wish to thank laboratory personnel at Virology Lab, Faculty of Veterinary Medicine, UPM and Laboratory of Vaccines and Immunotherapeutics, Institute of Bioscience, UPM. This project was funded by Fundamental Research Project No: 01-11-08-6390FR, Ministry of Higher Education, Malaysia. The funding source has no role in this study.

References

1. Nanda S, Havert MB, Calderón GM et al (2008) Hepatic transcriptome analysis of hepatitis C virus infection in chimpanzees defines unique gene expression patterns associated with viral clearance. PLoS One 3:e3442

2. Assarsson E, Greenbaum JA, Sundström M et al (2008) Kinetic analysis of a complete poxvirus transcriptome an immediate early class of gene. Proc Natl Acad Sci U S A 105: 2140–2145

3. Dowling RJ, Bienzle D (2005) Gene-expression changes induced by Feline immunodeficiency virus infection differ in epithelial cells and lymphocytes. J Gen Virol 86: 2239–2248

4. Ertl R, Birzele F, Hildebrandt T et al (2011) Viral transcriptome analysis of feline immunodeficiency virus infected cells using second generation sequencing technology. Vet Immunol Immunopathol 143:314–324

5. Zhang Z, Schwartz S, Wagner L et al (2000) A greedy algorithm for aligning DNA sequences. J Comput Biol 7:203–214

6. Kal AJ, Van Zonneveld AJ, Benes V et al (1999) Dynamics of gene expression revealed by comparison of serial analysis of gene expression transcript profiles from yeast grown on two different carbon sources. Mol Biol Cell 10:1859–1872

7. Mortazavi A, Williams BA, McCue K et al (2008) Mapping and quantifying mammalian transcriptomes by RNA-Seq. Nat Methods 5:621–628

8. Ashburner M, Ball CA, Blake J et al (2000) Gene ontology: tool for the unification of biology. The Gene Ontology Consortium. Nat Genet 25:25–29

9. Livak K, Schmittgen TD (2001) Analysis of relative gene expression data using real-time quantitative PCR and the 2(–Delta Delta C(T)) method. Methods 25:402–408

10. Li MM, Emerman M (2011) Polymorphism in human APOBEC3H affects a phenotype dominant for subcellular localization and antiviral activity. J Virol 85:8197–8207

11. Harun MSR, Choong OK, Selvarajah GT et al (2013) Transcriptional profiling of feline infectious peritonitis virus infection in CRFK cells and in PBMCs from FIP diagnosed cats. Virol J 10:329

Chapter 21

Quantification of Interferon Signaling in Avian Cells

Joeri Kint and Maria Forlenza

Abstract

Activation of the type I interferon (IFN) response is an essential defense mechanism against invading pathogens such as viruses. This chapter describes two protocols to quantify activation of the chicken IFN response through analysis of gene expression by real-time quantitative PCR and by quantification of bioactive IFN protein using a bioassay.

Key words Interferon, Bioassay, Chicken, Real-time quantitative PCR

1 Introduction

The type I interferon response (IFN response) is an important part of the immune reaction against viruses. Interferon alpha and beta (IFNα and IFNβ) are the prototypical type I interferons and can be produced by most animal cells. Production of IFNα/β is triggered upon stimulation of pattern recognition receptors, such as Toll like receptors (TLRs) or Rig-I like receptors (RLRs). Upon production, IFNα and IFNβ are rapidly secreted to the extracellular compartment, where they can bind to the ubiquitously expressed IFN receptor. Binding of IFN to the receptor activates the JAK/STAT signaling pathway, leading to the formation of the ISGF3 transcription complex consisting of a STAT1, STAT2, and IRF9. In the nucleus, the ISGF3 complex induces transcription of hundreds of IFN-stimulated genes (ISGs) [1]. Many of these genes encode proteins that interfere with the replicative cycle of viruses at various stages (reviewed in ref. 2). The IFN response is a potent antiviral mechanism, and therefore, most viruses have been evolutionarily selected to counteract it and coronaviruses are no exception (reviewed by Zhong et al. [3]).

In this chapter we describe two protocols to quantify activation of the IFN response. We have found these protocols useful to study if and how viruses counteract the IFN response in chicken cells. The first protocol describes how to quantify activation of the

Helena Jane Maier et al. (eds.), *Coronaviruses: Methods and Protocols*, Methods in Molecular Biology, vol. 1282,
DOI 10.1007/978-1-4939-2438-7_21, © Springer Science+Business Media New York 2015

IFN response at the transcriptional level using real-time quantitative PCR (RT-qPCR) on *Ifn* and IFN-stimulated genes. The second protocol describes quantification of bioactive type-I IFN protein (both IFNα and IFNβ) by the use of a reporter cell line. This bioassay can be used to quantify IFN secreted in response to virus infection and, when combined with transcription analysis of *Ifna* and *Ifnβ* these assays can provide an integral picture of activation of the chicken IFN response.

1.1 Quantitation of Transcription of Chicken Ifn-Related Genes

Similar to most mammalian cell lines, activation of the interferon response in most chicken cells is characterized by upregulation of *Ifnβ*. Like the human genome, the chicken genome encodes only one copy of the *Ifnβ* gene, whereas at least ten isoforms of *Ifna* are present [4, 5]. Similar to mammals, production of chicken IFNα is mainly mediated by monocytes; other cells mainly produce IFNβ in response to viral infection [6]. Because avian coronaviruses replicate mainly in epithelial cells, we monitor activation of the type I interferon response by quantification of *Ifnβ*. Similar to mammalian cells, *Ifnβ* is upregulated upon activation of either TLR or Rig-I like receptors (RLRs), but not in response to stimulation with IFN. Concomitant with *Ifnβ*, many ISGs are also upregulated, indicating that the term interferon stimulated genes is somewhat misleading [7]. Studying the expression of ISGs can be useful, therefore we have provided a list of avian-specific primers for use in real-time quantitative PCR (RT-qPCR; Table 1) [8–10]. Protocols for RNA isolation, cDNA synthesis and real-time PCR are plenty and every lab has its own protocols. In this chapter we describe briefly the methods used in our lab. For a detailed overview of RT-qPCR techniques and theoretical background, please refer to Forlenza et al. [11].

1.2 Quantitation of Chicken Type I IFN Protein Using a Bioassay

The chicken interferon bioassay was developed in the laboratory of Prof. P. Staeheli [12]. It is based on a quail cell line (CEC-32) that contains the luciferase gene downstream of a part of the inducible chicken *mx* promoter. Stimulation of these cells with type I interferon readily induces activation of the *mx* promoter and subsequent production of the firefly luciferase enzyme. Firefly luciferase can be easily quantified using commercially available luciferase assay kits. Here we provide a step-by-step protocol for measuring IFN concentrations using this bioassay.

2 Materials

2.1 RNA Isolation, cDNA Synthesis, and RT-qPCR on Avian Cells

1. Cells and virus, as per experiment.

2. RNeasy Mini Kit (QIAgen).

3. RNase-free DNase set (QIAgen).

Table 1
Chicken-specific real-time qPCR primers, including accession numbers of the sequences used to design the primers

gene	sense	sequence (5'-3')	Accession nr.	reference
IFNβ	FW	GCTCTCACCACCACCTTCTC	ENSGALT00000039477	
	RV	GCTTGCTTCTTGTCCTTGCT		
IFNα	FW	ATCCTGCTGCTCACGCTCCTTCT	XM_004937096	8
	RV	GGTGTTGCTGGTGTCCAGGATG		
IRF3	FW	CAGTGCTTCTCCAGCACAAA	NM_205372	
	RV	TGCATGTGGTATTGCTCGAT		
IRF1	FW	CAGGAAGTGGAGGTGGAGAA	ENSGALG00000006785	
	RV	TGGTAGATGTCGTTGGTGCT		
TLR7	FW	TTCTGGCCACAGATGTGACC	NM_001011688	8
	RV	CCTTCAACTTGGCAGTGCAG		
TLR3	FW	TCAGTACATTTGTAACACCCCGCC	NM_001011691	8
	RV	GGCGTCATAATCAAACACTCC		
MDA5	FW	TGGAGCTGGGCATCTTTCAG	GU570144	
	RV	GTTCCCACGACTCTCAATAACAGT		
Mx	FW	TTGTCTGGTGTTGCTCTTCCT	ENSGALT00000025999	
	RV	GCTGTATTTCTGTGTTGCGGTA		
OAS	FW	CACGGCCTCTTCTACGACA	NM_205041	9
	RV	TGGGCCATACGGTGTAGACT		
IL8	FW	TTGGAAGCCACTTCAGTCAGAC	NM_205498	9
	RV	GGAGCAGGAGGAATTACCAGTT		
PKR	FW	CCTCTGCTGGCCTTACTGTCA	NM_204487	10
	RV	AAGAGAGGCAGAAGGAATAATTTGCC		
ADAR	FW	TGTTTGTGATGGCTGTTGAG	AF403114	
	RV	AGATGTGAAGTCCGTGTTG		
ISG12	FW	TAAGGGATGGATGGCGAAG	NM_001002856	
	RV	GCAGTATCTTTATTGTTCTCAC		
MHC-I	FW	CTTCATTGCCTTCGACAAAG	NM_001031338	9
	RV	GCCACTCCACGCAGGT		
IFNAR2	FW	GCTTGTGTTCGTCAGCATT	ENSGALT00000036778	9
	RV	TTCGCAATCTTCCAGTTGT		
Firefly Luciferase	FW	TGTTGGGCGCGTTATTTATC	X65316	
	RV	AGGCTGCGAAATGTTCATACT		

4. Bioanalyser (Agilent Technologies) or agarose gel electrophoresis equipment.

5. Spectrophotometer (NanoDrop or equivalent).

6. DNase I, Amplification Grade.

7. Reverse Transcriptase (Invitrogen SuperScript® III or equivalent).

8. PCR machine (for cDNA synthesis).

9. Nuclease-free water.

10. Luciferase mRNA.

11. Random hexamers.

12. 2× SYBR® Green I mix.

13. Quantitative-PCR machine (Qiagen Rotor-Gene Q or equivalent).

14. Primers (Table 1).

2.2 Quantitation of Chicken Type I IFN Protein Using a Bioassay

1. Culture medium: DMEM, 10 % FCS, 100 U/ml penicillin, 100 μg/ml streptomycin.

2. Stimulation medium: DMEM, 1 % FCS, 100 U/ml penicillin, 100 μg/ml streptomycin.

3. CEC-32 chicken IFN reporter cells in 96-well plates at 70–90 % confluency (provided by P. Staeheli, *see* **Note 1**).

4. Recombinant chicken interferon alpha (chIFNα; Labome).

5. Multichannel pipet (8 × 200 μl).

6. Firefly luciferase assay buffer (Promega Bright-Glo™ or equivalent).

7. Luminometer.

3 Methods

3.1 RNA Isolation, cDNA Synthesis, and RT-qPCR on Avian Cells

1. Perform the experiment in 24-well plates (*see* **Note 2**). Infect or treat cells as desired.

2. When appropriate, cells are lysed by adding 350 μl RLT buffer spiked with 1 ng/sample of luciferase mRNA prior to RNA isolation (*see* **Note 3**).

3. Total RNA is isolated using the RNeasy Mini Kit according to the manufacturer's instructions, including an on-column DNase treatment with RNase-free DNase.

4. Verify RNA integrity on a 1 % agarose gel or using a Bioanalyser.

5. Determine RNA concentration using a spectrophotometer (NanoDrop or equivalent).

6. Prior to cDNA synthesis, perform a second DNase digestion step using DNase I.

7. Synthesis of cDNA is performed on 0.5–1 μg total RNA using Reverse transcriptase and random hexamers according to the manufacturer's instructions. Incubation steps are performed in a regular PCR machine or, alternatively in a water bath.

8. After cDNA synthesis, samples are diluted 1:50 in nuclease-free water before qPCR analysis.

9. Per sample, prepare a master mix containing 7 μl 2× SYBR Green I Mix and 2 μl primer mix (2.1 μM forward and reverse primer).

10. Combine 9 μl master mix and 5 μl diluted cDNA per PCR tube.

11. Real-time quantitative PCR is performed on a qPCR machine, such as Rotor-Gene Q, 35–40 cycles, 60 °C annealing temperature, 20 s extension time.

12. Cycle thresholds and amplification efficiencies are calculated using the software pertaining to the qPCR machine, such as Rotor-Gene 6000.

13. Using Eq. 1, the relative expression ratio of the target gene is calculated using the average reaction efficiency for each primer set and the cycle threshold (C_t) deviation of sample vs. control at time point 0 h (*see* **Note 4**).

$$R_{\text{Ifn}\beta} = \frac{E_{\text{Ifn}\beta}^{(Ct_{\text{Ifn}\beta \, (\text{calibrator})} - Ct_{\text{Ifn}\beta \, (\text{sample})})}}{E_{\text{GAPDH}}^{(Ct_{\text{GAPDH}\beta (\text{calibrator})} - Ct_{\text{GAPDH} \, (\text{sample})})}} \tag{1}$$

With:
R = fold change of the target gene relative to the control
Calibrator = control cells at time point 0 (zero)
E = average amplification efficiency for that set of primers
C_t = cycle threshold

3.2 Quantification of Chicken Type I IFN Protein Using a Bioassay

1. If the samples contain virus, heat-inactivate at 56 °C for 30 min prior to performing the assay. This treatment inactivates coronaviruses but retains bioactivity of type I IFN (*see* **Note 5**).

2. Fill a sterile 96-well plate with 50 μl stimulation medium/well.

3. Add 50 μl chIFNα standard (50 U/ml) or test sample to the first row (vortex before adding).

4. Make serial twofold dilutions in the plate using a multichannel pipet (Fig. 1, *see* **Note 6**).

5. Remove the medium from the CEC-32 cells which have been cultured in the 96-well plate (*see* **Note 7**).

6. Transfer the content of the plate containing the diluted samples and standard to the CEC-32 cells (*see* **Note 8**).

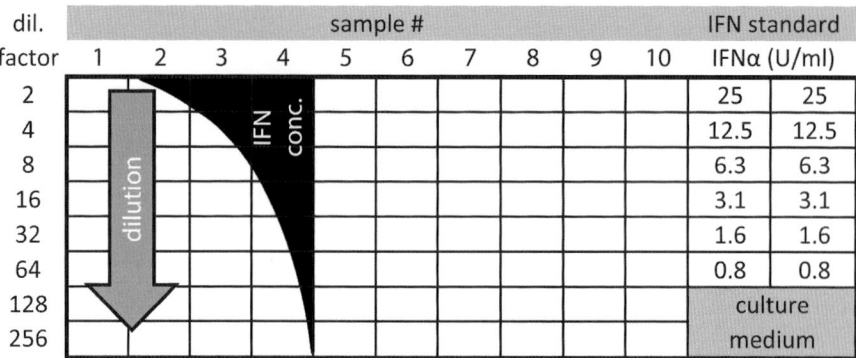

dil.	sample #										IFN standard	
factor	1	2	3	4	5	6	7	8	9	10	IFNα (U/ml)	
2											25	25
4											12.5	12.5
8											6.3	6.3
16											3.1	3.1
32											1.6	1.6
64											0.8	0.8
128											culture	
256											medium	

Fig. 1 Layout of a 96-well plate to accommodate ten samples and an interferon standard

7. Incubate plates at 37 °C and 5 % CO_2 for 6 h.

8. Use a firefly luciferase assay kit to detect luciferase activity, according to manufacturer's instructions.

3.3 Calculation of IFN Concentration from Luminescence Data

To calculate the units of interferon in the original sample, a work-flow is provided in Fig. 2.

1. Transfer the measurements from the luminometer to a spread-sheet program (Microsoft Excel or equivalent).

2. Calculate the average value of the background luminescence and subtract this value from all wells (Fig. 2, point A).

3. Calculate the average of the wells incubated with the diluted interferon standard and plot them in a scatter plot. This graph is the standard curve (B).

4. Make a new graph using only the data points that fall within the linear range of the standard curve, usually 1–12.5 or 1–6 U/ml.

5. Plot a linear trend line through these data points and display the equation on the chart (C).

6. Next, all luminescence values that fall within the linear range of the standard curve are selected (here 2–12 U/ml).

7. Calculate the IFN concentration in each well using the equation from the standard curve (D).

8. Multiply by the dilution factor to obtain the concentration of IFN in the undiluted samples (E).

9. Finally, calculate the average IFN concentration of the wells that fall within the linear range (usually two or three wells per sample). This value corresponds to the final concentration of type I interferon in the original sample (F).

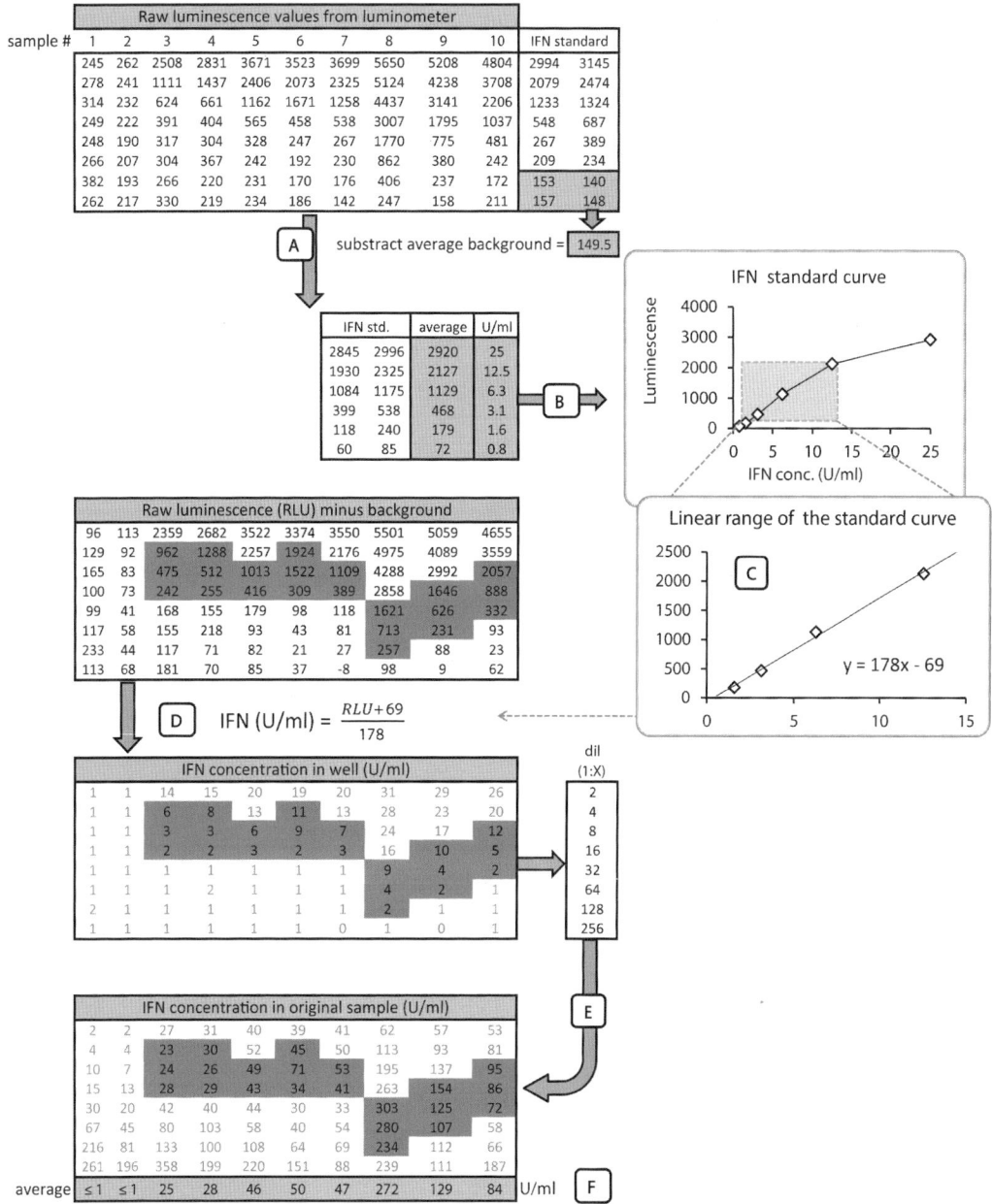

Fig. 2 Workflow on how to calculate the concentration of IFN in the original sample from the readout of the luminometer

4   Notes

1. For more detail on the construction of CEC-32 chicken-IFN reporter cells, *see* ref. 12.

2. To have enough RNA, each well of a 24-well plate should contain around 3×10^5 avian cells.

3. For normalization, a housekeeping gene such as GAPDH is generally used. It is advised to ensure that the reference gene selected is stable under the conditions of each experiment by performing stability analysis. When the mRNA level of the reference gene is not stable during the experimental procedure, such as during prolonged infection with a virus, we use an external reference gene for normalization. The external reference gene we use is luciferase, which is added as mRNA (commercially available) to the RLT lysis buffer (1 ng/sample) prior to RNA isolation and cell lysis. This guarantees that the external reference RNA and the host RNA are subject to the same treatment prior to cDNA synthesis.

4. To calculate the fold change of IBV total RNA, C_t deviation is calculated versus a fixed C_t value (e.g., $C_t = 30$), because no IBV is present in the non-infected cells that are used as control in all the experiments.

5. Interferon containing samples can be stored at 4 °C overnight. Storage at –20 °C ensures long time (>months) stability. One freeze–thaw cycle can reduce the IFN-activity of a sample by 40 %. To avoid repeated freeze-thawing of samples, avoid measuring the same sample twice. To achieve this, ensure that at least one of the dilutions of the samples falls within the linear range of the standard curve.

6. To select the appropriate dilutions it is advisable to perform a pilot experiment. One can either make an extensive twofold serial dilution series (for example, 2–1,024 times dilution), or use tenfold pre-dilutions. In our hands IFN production by avian cells rarely exceeds 5,000 U/ml, and therefore, a maximum of 1,000 times pre-dilution should suffice.

7. Medium is removed simply by emptying the 96-well plate in the waste and gently tapping it dry on a stack of tissues. Sterility is not an issue, the cells will only be incubated for another 6 h.

8. CEC-32 cells should *not be allowed to dry out*! Make sure you transfer the samples to the cells within minutes. Drying of the monolayer will decrease the luminescence and thereby negatively influence the assay. When transferring the diluted samples to 96-well plate with CEC-32 cells, start with the lowest concentration and work your way up the dilutions. In this way the same tips can be used for multiple dilutions. Do not forget to add medium to the negative controls.

Acknowledgements

We would like to thank Peter Staeheli for supplying the CEC-32 chicken type I interferon reporter cell line.

References

1. Rusinova I, Forster S, Yu S et al (2013) Interferome v2.0: an updated database of annotated interferon-regulated genes. Nucleic Acids Res 41:D1040–D1046

2. Ivashkiv LB, Donlin LT (2014) Regulation of type I interferon responses. Nat Rev Immunol 14:36–49

3. Zhong Y, Tan YW, Liu DX (2012) Recent progress in studies of arterivirus- and coronavirus-host interactions. Viruses 4:980–1010

4. Sick C, Schultz U, Staeheli P (1996) A family of genes coding for two serologically distinct chicken interferons. J Biol Chem 271:7635–7639

5. Sekellick MJ, Ferrandino AF, Hopkins DA et al (1994) Chicken interferon gene: cloning, expression, and analysis. J Interferon Res 14:71–79

6. Sick C, Schultz U, Munster U et al (1998) Promoter structures and differential responses to viral and nonviral inducers of chicken type I interferon genes. J Biol Chem 273:9749–9754

7. Sen GC, Sarkar SN (2007) The interferon-stimulated genes: targets of direct signaling by interferons, double-stranded RNA, and viruses. Curr Top Microbiol Immunol 316:233–250

8. Villanueva AI, Kulkarni RR, Sharif S (2011) Synthetic double-stranded RNA oligonucleotides are immunostimulatory for chicken spleen cells. Dev Comp Immunol 35:28–34

9. Li YP, Handberg KJ, Juul-Madsen HR et al (2007) Transcriptional profiles of chicken embryo cell cultures following infection with infectious bursal disease virus. Arch Virol 152:463–478

10. Daviet S, Van Borm S, Habyarimana A et al (2009) Induction of Mx and PKR failed to protect chickens from H5N1 infection. Viral Immunol 22:467–472

11. Forlenza M, Kaiser T, Savelkoul HF et al (2012) The use of real-time quantitative PCR for the analysis of cytokine mRNA levels. Methods Mol Biol 820:7–23

12. Schwarz H, Harlin O, Ohnemus A et al (2004) Synthesis of IFN-beta by virus-infected chicken embryo cells demonstrated with specific antisera and a new bioassay. J Interferon Cytokine Res 24:179–184

Chapter 22

Studying the Dynamics of Coronavirus Replicative Structures

Marne C. Hagemeijer and Cornelis A.M. de Haan

Abstract

Coronaviruses (CoVs) generate specialized membrane compartments, which consist of double membrane vesicles connected to convoluted membranes, the so-called replicative structures, where viral RNA synthesis takes place. These sites harbor the CoV replication–transcription complexes (RTCs): multi-protein complexes consisting of 16 nonstructural proteins (nsps), the CoV nucleocapsid protein (N) and presumably host proteins. To successfully establish functional membrane-bound RTCs all of the viral and host constituents need to be correctly spatiotemporally organized during viral infection. Few studies, however, have investigated the dynamic processes involved in the formation and functioning of the (subunits of) CoV RTCs and the replicative structures in living cells. In this chapter we describe several protocols to perform time-lapse imaging of CoV-infected cells and to study the kinetics of (subunits of) the CoV replicative structures. The approaches described are not limited to CoV-infected cells; they can also be applied to other virus-infected or non-infected cells.

Key words Coronavirus, Nonstructural proteins, Live-cell imaging, Replication–transcription complex, Dynamics, Fluorescence recovery after photobleaching, Fluorescence loss in photobleaching

1 Introduction

Coronavirus (CoV) replicative structures are impressive multicomponent assemblies that consist of no less than 16 viral replicase proteins (the nonstructural proteins (nsps) [1–11]), the nucleocapsid (N) protein [10, 12–15], an as yet unknown number of host proteins, and an elaborate network of endoplasmic reticulum (ER)-derived double membrane vesicles and convoluted membranes [2, 4, 9, 16]. These "replication organelles" have been studied in lots of detail over the last few decades, generating a wealth of exciting information with respect to their composition, morphology, and functioning during the viral life cycle. Unfortunately one of the most, and perhaps unintentionally, overlooked areas is the real-time spatiotemporal dynamic behavior of the formation and functioning of the replicative structures

Helena Jane Maier et al. (eds.), *Coronaviruses: Methods and Protocols*, Methods in Molecular Biology, vol. 1282, DOI 10.1007/978-1-4939-2438-7_22, © Springer Science+Business Media New York 2015

themselves but also the individual (membranous) associated components in living cells.

To study the dynamic processes underlying the formation and functioning of the CoV replicative structures and/or replication-associated proteins, one needs noninvasive means to visualize them in living cells. Recombinant CoVs expressing replicase proteins fused to fluorescent proteins (FPs) serve as an excellent tool to perform such real-time imaging studies in their native environment. Alternatively, these fluorescent fusion proteins may be expressed upon transfection of expression plasmids prior to infection, if desirable in combination with plasmids expressing host proteins fused to FPs. Using confocal laser scanning microscopy (CLSM) several live-cell imaging techniques can be employed to investigate the real-time kinetics of the replicative structures and its associated components. Once cells are infected with the recombinant FP-expressing CoVs, one can simply follow the fate of the replicative structures or its components over time, either under native or perturbed conditions, resulting in valuable spatiotemporal information with respect to the formation and behavior of the replicative structures and its associated components (*see* Fig. 1 for an example of a time-lapse experiment of MHV-nsp2GFP-infected LR7 cells). Alternatively, a (part of the) pool of FPs may be selectively and irreversibly photobleached and the temporal fate of the remaining non-bleached proteins may be followed [reviewed in [17–19]]. From the latter measurements parameters can be extracted, which among others reflect (the lack of) mobility

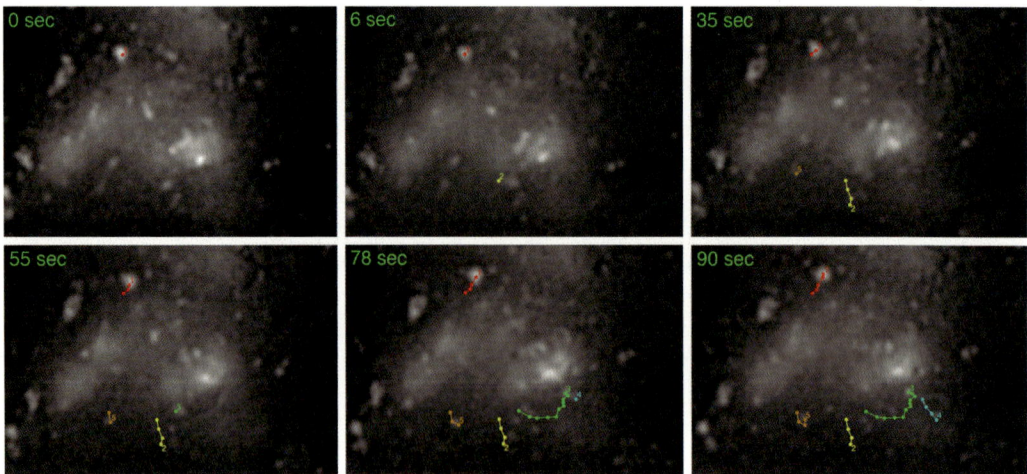

Fig. 1 Time-lapse imaging of MHV-nsp2GFP in LR7 cells. LR7 cells were infected with MHV-nsp2GFP [MHV expressing an nsp2-GFP fusion protein [21]] and at 6 h p.i. a time-lapse experiment was performed as described in Subheadings 3.1 and 3.2. Nsp2-GFP-positive foci were manually tracked over time using MtrackJ [22] and examples of the displacement of these structures have been indicated by the different *numbered* and *colored lines*

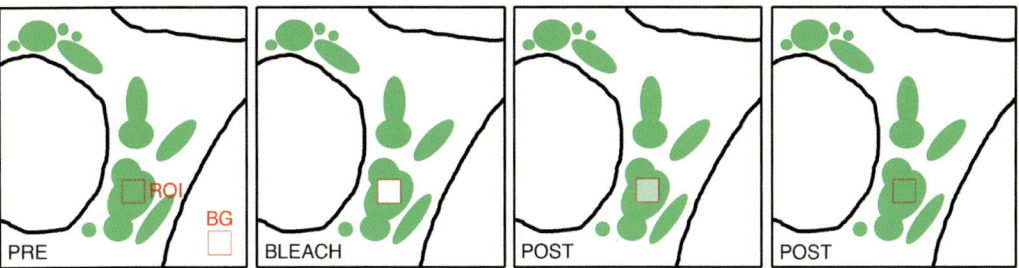

Fig. 2 Schematic overview of a FRAP experiment. A specific region of interest (ROI) targeting (part of) the FP-tagged replicative structure(s) is irreversibly photobleached and the recovery of fluorescence into the bleached area is monitored over time. In this example, the FP-tagged proteins are mobile as recovery of the fluorescence in the bleached area is observed [as has been previously observed for the MHV N protein [14]]. In the absence of recovery, the FP-tagged proteins are immobile [as has been observed for example for the MHV nsp2 protein [21]]. The *green structures* in the cell are a schematic representation of the replicative structures. *BG* background ROI for qualitative FRAP analysis

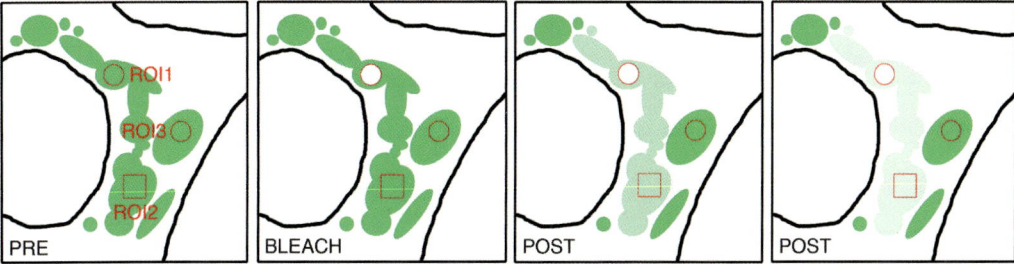

Fig. 3 Schematic overview of a FLIP experiment. The replicative structure(s) is repeatedly photobleached in ROI1 (*circle*) and the loss of fluorescence is monitored in ROI2 (*squared box*) or ROI3 (*circle*) over time. In this example the fluorescence of the FP-tagged structure is lost over time in ROI2, but not ROI3, indicating continuity between the membrane compartments of ROI1 and ROI2, but not between those of ROI1 and ROI3. Another cell in the field of view may be used to monitor/correct for photobleaching during the FLIP assay

[fluorescence recovery after photobleaching (FRAP)] or the (dis) continuity between different membrane compartments [fluorescence loss in photobleaching (FLIP)]. Figures 2 and 3 illustrate the principles of the FRAP and FLIP approaches, respectively, with typical fluorescence intensity graphs that can be obtained from these type of live-cell imaging experiments depicted in Fig. 4.

In this chapter we describe (1) a general protocol for setting up a time-lapse imaging experiment of CoV-infected cells to study the mobility of the replicative structures, followed by (2) two photobleaching approaches (FRAP and FLIP) to study the kinetics and continuity of the replicative structures and/or associated proteins, respectively and finally (3) how to analyze the quantitative kinetic data to obtain different parameters describing the dynamic behavior of the investigated structures/proteins in living cells.

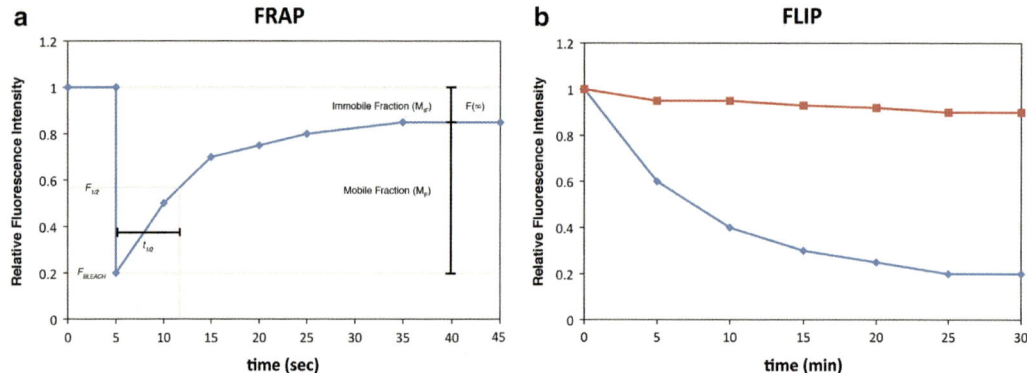

Fig. 4 Fluorescence graphs of typical FRAP and FLIP experiments. Two examples of typical fluorescence graphs when performing FRAP (*left graph*) or FLIP (*right graph*) experiments. (**a**) $F\infty$: the plateau level reached at the end of the experiment, M_f: mobile fraction of the FP-tagged replicative structures, M_{iF}: the immobile fraction of the FP-tagged proteins present at the replicative structures, $t_{1/2}$: halftime of the recovery. (**b**) The decrease of fluorescence (*diamonds/blue line*) indicates that continuity between membrane compartments exists. If no or hardly any decrease is observed, the different membrane compartments are not continuous (for example ROI3 in Fig. 3). The *squares/red line* represents the loss of fluorescence of a control cell that is not photobleached and serves as a control that the observed loss of fluorescence is due to migration of FP-tagged proteins into the bleached area and not due to general photobleaching of the cell itself

Although, the live-cell imaging protocols described in this chapter uses CoV-infected cells, these approaches can easily be applied to other virus-infected or non-infected cells.

2 Materials

1. Mouse LR7 fibroblasts.

2. Cell culture medium: Dulbecco's Modified Eagle Medium (DMEM), 10 % fetal bovine serum (FBS), 100 IU/ml penicillin, 100 µg/ml streptomycin.

3. Imaging medium: DMEM without phenol-red, 10 % FBS and 25 mM HEPES pH 7.4.

4. Recombinant MHV(s) expressing viral FP-tagged proteins or wild-type MHV-A59.

5. Confocal Laser Scanning Microscope (CLSM) set up with a heated stage or environmental chamber.

6. Live-cell imaging clusters: 2- or 4-well Lab-Tek chamber slides or 35 mm MatTek #1.5 glass bottom dishes.

7. Plasmid DNA and transfection reagent (e.g., FuGENE 6 or Lipofectamine 2000) for (co)transfection of cells with a (viral or host) reporter plasmid prior to infection.

3 Methods

All procedures have to be performed at room temperature (RT) and under sterile conditions, unless otherwise stated in the text. We assume that the investigator has sufficient knowledge on how to operate a confocal microscope when following the live-cell imaging protocols described below.

3.1 Visualizing the CoV Replicative Structures and the Replicase Subunits

1. Plate mouse LR7 cells in culture medium on either Lab-Tek chamber slides or MatTek glass bottom dishes (#1.5 thickness) in such a density that 24–48 h after plating, the cells have reached 70–80 % confluency.

2. Prior to live-cell imaging infect the cells with recombinant FP-tagged MHV at a MOI of 10 for 1 h diluted in culture medium (infection medium) at 37 °C/5 % CO_2. Alternatively, transfect the cells with expression plasmids using standard protocols (e.g., by using FuGENE 6 or Lipofectamine 2000) 16–24 h prior to infection with recombinant FP-tagged or wild-type MHV.

3. After 1 h of infection, replace the infection medium with culture medium and allow incubation to continue at 37 °C/5 % CO_2.

4. At the preferred time point post infection (p.i.), replace the culture medium with pre-warmed (37 °C) imaging medium and transfer the Lab-Tek chamber slide or MatTek glass bottom dish to a [humidified (5 % CO_2)] heated incubation chamber (37 °C) on the microscope stage of the confocal microscope (*see* **Note 1**).

3.2 Real-Time Dynamics of the CoV Replicative Structures

1. After performing all the steps in Subheading 3.1 adjust the pinhole size of the CLSM to 1 airy unit (AU) and the scan area to 512×512 pixels with a bit depth of 12-bit. Do not apply any averaging during the acquisition process (*see* **Note 2**).

2. Turn on the required lasers to detect (and bleach, *see* Subheadings 3.3 and 3.4) the FP-tagged replicase protein(s) and adjust the laser power to a low percentage to prevent photobleaching of the FPs.

3. Start scanning the live cells using continuous acquisition and adjust the gain and offset for each laser channel (if applicable) to optimize the amount of fluorescent signal that the detectors will detect but without registering any oversaturated pixels in the field of view (*see* **Note 3**).

4. If the confocal microscope has the option to use an autofocus strategy to minimize (x,y,z) drift during the acquisition of the

time-lapse images, set up the autofocus strategy according to the manufacturer's instructions at this point.

5. Empirically determine the acquisition parameters, i.e., the number of time points to be imaged in a specific time interval that results in the acquisition of the maximum number of frames but the least amount of photobleaching of the cells.

6. Set up a time series using the parameters determined in **step 5** and start imaging the real-time dynamics of the CoV replicative structures.

7. Collect at least 5–10 individual time-lapse movies per experiment (Fig. 1).

3.3 FRAP of the CoV Replicative Structure (Subunits)

1. Perform **steps 1–5** as described in Subheading 3.2 but open the pinhole completely to acquire the maximum amount of fluorescent signal emitted from the whole cell (*see* **Note 4**).

2. Specify a region of interest (ROI) that targets (part of) the replicative structure that will be irreversibly photobleached (Fig. 2).

3. Set up a time-series that consists of (1) five pre-bleach images using the laser power established in **step 1**, (2) a photobleach event using 100 % laser power, and (3) an empirically determined number of post-bleach images (*see* **Note 5**).

4. Once the optimal bleaching and temporal parameters have been established perform approximately 5–10 FRAP experiments to acquire a sufficient number of data sets to extract qualitative FRAP parameters (*see* Subheading 3.5).

3.4 Continuity Between CoV Replicative Structures: FLIP

1. Perform **steps 1–5**, as described in Subheading 3.2, and open the pinhole completely.

2. Select a field of view, which contains at least two infected and/or (co)transfected cells but preferably more (*see* **Note 6**).

3. Specify two ROIs in the targeted cell, one that will repeatedly be photobleached (Fig. 3, ROI1) and one where the loss of fluorescence intensity will be measured over time (Fig. 3, ROI2 and 3).

4. Set up an imaging protocol that includes (1) the acquisition of five pre-bleach images, (2) a 100 % laser power bleach event, and (3) the collection of ten post-bleach images. Steps (2) and (3) will have to be repeated for multiple cycles of which the number of cycles has to be determined empirically by the investigator. The cell(s) should be imaged during steps (1) and (3) using the acquisition settings as determined in **step 1**.

5. Perform approximately 5–10 FLIP experiments to acquire enough data sets to determine whether continuity between different membrane compartments exists.

**3.5 Analysis
of the Obtained
Live-Cell Imaging Data**

The majority of the commercial confocal microscope systems have excellent software packages installed on their workstations to perform (automated) analysis of the obtained data sets (e.g., softWoRx from Applied Precision or ZEN from Zeiss). If such software packages or automated approaches are not available, the open source program ImageJ [20] is a good alternative, from which qualitative parameters can be determined using the steps below as a general guideline.

1. After export from the quantitative (raw) data from the CLSM subtract the background fluorescence signal (F_{BG}) from the fluorescence signal from the bleached area (F_{ROI}) for each time point:

$$F_{ROI,BG}[t] = F_{ROI}[t] - F_{BG}[t] \tag{1}$$

2. Correct these measurements for any photobleaching that occurred over time during the image acquisition using the background-subtracted fluorescence intensity of the whole cell (F_{CELL}; *see* **Note 7**):

$$F_{ROI,CORR}[t] = F_{ROI,BG}[t] / (F_{CELL}[t] - F_{BG}[t]) \tag{2}$$

3. Normalize the obtained data to the fluorescent intensity of the first pre-bleach background-subtracted time point's fluorescent intensity (*see* **Note 7**):

$$F_{NORM}[t] = F_{ROI,CORR}[t] \times (F_{CELL}[0] - F_{BG}[0]) / F_{ROI,BG}[0] \tag{3}$$

4. Using the corrected and normalized fluorescence data, plot the fluorescence intensity in a particular region of interest (ROI) as shown in Fig. 4 (Fig. 4a, b are examples of a FRAP and FLIP graph, respectively).

5. The mobile fraction (M_f) can be determined using the generated fluorescence intensity plot(s) and the following formula:

$$M_f = (F[\infty] - F_{BLEACH}) / (1 - F_{BLEACH}) \tag{4}$$

6. The half-life of the bleached FPs ($t_{1/2}$) can be determined from the fluorescence intensity graph by determining $F_{1/2}$ (*see* **Note 8**):

$$F_{1/2} = (F[\infty] + F_{BLEACH}) / 2 \tag{5}$$

4 Notes

1. Whether or not CO_2 will be used during image acquisition depends on the microscope setup of the investigator, i.e., if an external CO_2 tank is connected to the incubation chamber. If not, HEPES-buffered culture medium can be used during

image acquisition. Both approaches have successfully been used in our laboratory.

2. Imaging at a high magnification of individual cells is preferable when performing time-lapse imaging of CoV replicative structures; the authors recommend using an oil immersion 40×/1.3NA or 63×/1.4NA objective.

3. Use the range indicator function of the confocal microscope to avoid imaging of oversaturated pixels, as data sets containing oversaturated pixels cannot be used for quantitative data analysis.

4. Opening the pinhole completely is important to ensure that the maximum amount of light emitted from the fluophores is collected and not only from a single focal plane.

5. When applying a single photobleach event of 100 % laser power, corresponding to the wavelength of the FP, does not result in a fluorescent intensity drop to background levels, the number of iterations should be increased. Once the recovery of the fluorescence in the bleached area reaches a plateau the acquisition of additional frames can be stopped.

6. During FLIP experiments the cell of interest is repeatedly photobleached at a high intensity laser power and therefore at least one control cell, i.e., a cell that is not photobleached, should be in the field of view to monitor whether loss of fluorescence is due to diffusion of the FPs and not due to photobleaching of the cell.

7. Correction for unwanted photobleaching is performed by using the fluorescent intensity (fluctuations) of the whole cell during the FRAP experiment. Such a control is not possible when performing FLIP as the loss of fluorescence is measured. Instead, the background-subtracted fluorescence of a whole non-bleached cell, which has been fitted to the equation $y(t) = \exp(-t/x)$, should be used for correction. For normalization of FLIP data use the following formula: $F_{\text{NORM}}[t] = F_{\text{ROI,CORR}}[t] \times \left(1 / F_{\text{ROI, BG}}[0]\right)$. We would like to recommend the investigator to consult reference [19] for more comprehensive in depth details on both FRAP and FLIP imaging approaches and data analysis.

8. Calculate the $F_{1/2}$ using formula (5) and determine the corresponding time point on the x-axis of the FR graph to determine $t_{1/2}$.

Acknowledgements

The authors would like to thank Xufeng Wu and Fleur de Haan for valuable suggestions and discussions.

References

1. Deming DJ, Graham RL, Denison MR et al (2007) Processing of open reading frame 1a replicase proteins nsp7 to nsp10 in murine hepatitis virus strain A59 replication. J Virol 81:10280–10291

2. Goldsmith CS, Tatti KM, Ksiazek TG et al (2004) Ultrastructural characterization of SARS coronavirus. Emerg Infect Dis 10:320–326

3. Graham RL, Sims AC, Brockway SM et al (2005) The nsp2 replicase proteins of murine hepatitis virus and severe acute respiratory syndrome coronavirus are dispensable for viral replication. J Virol 79:13399–13411

4. Knoops K, Kikkert M, Worm SH et al (2008) SARS-coronavirus replication is supported by a reticulovesicular network of modified endoplasmic reticulum. PLoS Biol 6:e226

5. Oostra M, te Lintelo EG, Deijs M et al (2007) Localization and membrane topology of coronavirus nonstructural protein 4: involvement of the early secretory pathway in replication. J Virol 81:12323–12336

6. Prentice E, McAuliffe J, Lu X et al (2004) Identification and characterization of severe acute respiratory syndrome coronavirus replicase proteins. J Virol 78:9977–9986

7. Reggiori F, Monastyrska I, Verheije MH et al (2010) Coronaviruses Hijack the LC3-I-positive EDEMosomes, ER-derived vesicles exporting short-lived ERAD regulators, for replication. Cell Host Microbe 7:500–508

8. Shi ST, Schiller JJ, Kanjanahaluethai A et al (1999) Colocalization and membrane association of murine hepatitis virus gene 1 products and De novo-synthesized viral RNA in infected cells. J Virol 73:5957–5969

9. Snijder EJ, van der Meer Y, Zevenhoven-Dobbe J et al (2006) Ultrastructure and origin of membrane vesicles associated with the severe acute respiratory syndrome coronavirus replication complex. J Virol 80:5927–5940

10. Ulasli M, Verheije MH, de Haan CA et al (2010) Qualitative and quantitative ultrastructural analysis of the membrane rearrangements induced by coronavirus. Cell Microbiol 12:844–861

11. van der Meer Y, Snijder EJ, Dobbe JC et al (1999) Localization of mouse hepatitis virus nonstructural proteins and RNA synthesis indicates a role for late endosomes in viral replication. J Virol 73:7641–7657

12. Bost AG, Carnahan RH, Lu XT et al (2000) Four proteins processed from the replicase gene polyprotein of mouse hepatitis virus colocalize in the cell periphery and adjacent to sites of virion assembly. J Virol 74:3379–3387

13. Denison MR, Spaan WJ, van der Meer Y et al (1999) The putative helicase of the coronavirus mouse hepatitis virus is processed from the replicase gene polyprotein and localizes in complexes that are active in viral RNA synthesis. J Virol 73:6862–6871

14. Verheije MH, Hagemeijer MC, Ulasli M et al (2010) The coronavirus nucleocapsid protein is dynamically associated with the replication-transcription complexes. J Virol 84:11575–11579

15. Verheije MH, Raaben M, Mari M et al (2008) Mouse hepatitis coronavirus RNA replication depends on GBF1-mediated ARF1 activation. PLoS Pathog 4:e1000088

16. Gosert R, Kanjanahaluethai A, Egger D et al (2002) RNA replication of mouse hepatitis virus takes place at double-membrane vesicles. J Virol 76:3697–3708

17. Lippincott-Schwartz J, Snapp E, Kenworthy A (2001) Studying protein dynamics in living cells. Nat Rev Mol Cell Biol 2:444–456

18. Lippincott-Schwartz J, Altan-Bonnet N, Patterson GH (2003) Photobleaching and photoactivation: following protein dynamics in living cells. Nat Cell Biol S7–14

19. Bancaud A, Huet S, Rabut G et al (2010) Fluorescence perturbation techniques to study mobility and molecular dynamics of proteins in live cells: FRAP, photoactivation, photoconversion, and FLIP. Cold Spring Harb Protoc 2010, pdb top90

20. Rasband W (1997-2014) ImageJ. U. S. National Institutes of Health. Bethesda MD

21. Hagemeijer MC, Verheije MH, Ulasli M et al (2010) Dynamics of coronavirus replication-transcription complexes. J Virol 84:2134–2149

22. Meijering E, Dzyubachyk O, Smal I (2012) Methods for cell and particle tracking. Methods Enzymol 504:183–200

Chapter 23

Preparation of Cultured Cells Using High-Pressure Freezing and Freeze Substitution for Subsequent 2D or 3D Visualization in the Transmission Electron Microscope

Philippa C. Hawes

Abstract

Transmission electron microscopy (TEM) is an invaluable technique used for imaging the ultrastructure of samples and it is particularly useful when determining virus–host interactions at a cellular level. The environment inside a TEM is not favorable for biological material (high vacuum and high energy electrons). Also biological samples have little or no intrinsic electron contrast, and rarely do they naturally exist in very thin sheets, as is required for optimum resolution in the TEM. To prepare these samples for imaging in the TEM therefore requires extensive processing which can alter the ultrastructure of the material. Here we describe a method which aims to minimize preparation artifacts by freezing the samples at high pressure to instantaneously preserve ultrastructural detail, then rapidly substituting the ice and infiltrating with resin to provide a firm matrix which can be cut into thin sections for imaging. Thicker sections of this material can also be imaged and reconstructed into 3D volumes using electron tomography.

Key words High-pressure freezing, Freeze substitution, Transmission electron microscopy, Sapphire discs, Electron tomography

---

1 Introduction

The method of preservation of samples for TEM can influence image interpretation so it is important to stabilize the sample with as little change from the in vivo state as possible. There are two main methods of stabilizing (fixing) samples for TEM: chemical fixation and cryo-fixation. There are advantages and disadvantages to both. Chemical fixation is the most common method and, although time consuming, is an easy, repeatable method that requires very little specialized equipment. Alternatively, cryo-fixation methods seek to preserve samples in as near the in vivo state as possible by stabilizing instantaneously and reducing or eliminating the use of chemicals. Cryo-fixation methods are preferable to chemical fixation methods; however, they have significant technical and financial disadvantages.

Helena Jane Maier et al. (eds.), *Coronaviruses: Methods and Protocols*, Methods in Molecular Biology, vol. 1282,
DOI 10.1007/978-1-4939-2438-7_23, © Springer Science+Business Media New York 2015

Standard chemical fixation protocols are readily available that give reproducible results (for example [1]), with minimal use of specialized equipment. These protocols are relatively quick and easy to do. The main disadvantage of using chemical fixation is that the introduction of toxic chemicals to the sample can have an unknown effect on ultrastructure. Fixative penetrates even soft biological material slowly which allows changes to occur within the sample before it is fully stabilized, for example redistribution and extraction of ions and soluble proteins [2], extraction and rearrangement of phospholipids [3], mismatch in osmotic conditions leading to organelle blooming and non-isotropic shrinkage [4]. There is no such thing as a "universal fixative" and fixatives do not preserve all structures within cells equally. Another important disadvantage of chemical fixation is that at every stage during the process antigens in the sample are destroyed. Therefore, it is not possible to carry out immunogold labelling experiments using chemically fixed and epoxy resin embedded samples.

The only viable alternative to the deleterious effects of chemical fixation and dehydration is to preserve samples by freezing the water present rapidly enough to prevent ice crystals forming. If samples are frozen quickly enough the water inside the sample is vitrified and both soluble and non-soluble structures are held in a glass-like matrix (amorphous ice), stabilizing the sample instantaneously. Amorphous ice is non-destructive, but to achieve full vitrification of cellular water very high cooling rates are required [5]. If these cooling rates are not reached, crystalline ice forms and solutes within the cell are trapped between the crystals forming a network of segregated compartments. When viewed in the microscope, this is known as ice segregation artifact and is particularly obvious in badly frozen nuclei where it appears as a "cracking" pattern (Fig. 1). Once frozen, the sample may be stored in liquid nitrogen (−196 °C) before further processing. The temperature cannot be allowed to climb above the re-vitrification point of water (around −140 °C) or ice crystals will form. The advantage of using freezing techniques to preserve samples is that they are stabilized instantaneously without the need for chemicals. However, the disadvantages are that at atmospheric pressure good freezing only occurs within a few microns of the surface at best, and the techniques involved are difficult and time-consuming requiring dedicated equipment.

There are many different cryo-fixation methods available for use but the depth of good preservation is limited to 20–40 μm at atmospheric pressure, whichever method is used [6]. An alternative is to freeze the sample at high pressure, an idea first postulated by Moor and Riehle in 1968 [7]. At higher pressures, water expands less during freezing, and hence, less heat of crystallization is produced, so adequate cryo-fixation is achieved at reduced cooling rates [8, 9].

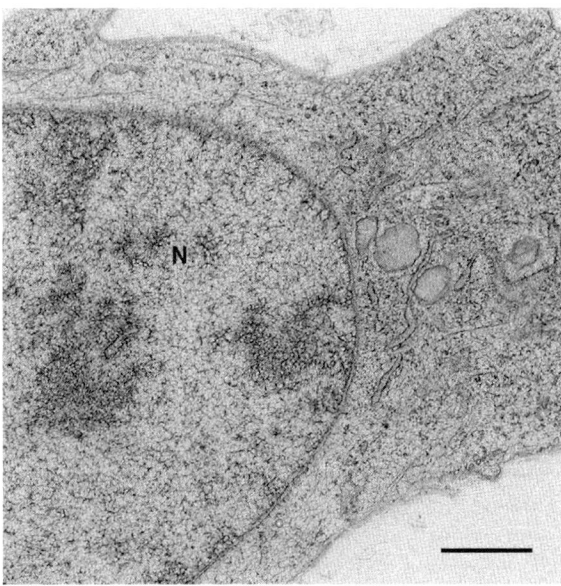

Fig. 1 Example of ice segregation artifact in a cell culture Vero cell. This appears as a "cracking" pattern and is particularly obvious in the nucleus (N) of badly frozen cells. Scale bar indicates 1 μm

At atmospheric pressure amorphous ice is produced at a freezing rate of several 10,000 °C/s; however, if samples are frozen at 210 MPa, this required freezing rate drops to several 1,000 °C/s [5]. Freezing at high pressure allows vitrification to occur to a depth of 200 μm [6, 10]. For this reason "high-pressure freezing" (HPF) can be used to prepare many different types of sample, ranging from suspensions to small pieces of solid tissue and is widely regarded as the optimal cryo-fixation method for general electron microscopy [6]. For a comprehensive review of high-pressure freezing and freeze substitution, *see* McDonald [11].

After high-pressure freezing, samples are processed for morphological or immunocytochemical studies by freeze substitution (FS). During FS amorphous ice is replaced by solvent, generally acetone [12], containing one or more chemical additives. The addition of chemicals at this stage does not affect the preservation of tissue as the sample has already been stabilized in the high-pressure freezer. After substitution, samples are infiltrated with acrylic resin, and the resin polymerized. Ultrathin sections can be cut at room temperature and examined in the transmission electron microscope.

There are many published protocols for freeze substitution which vary considerably. One of the greatest variations is the time that samples are kept in substitution medium prior to embedding. It is widely accepted that leaving samples in fixative and/or solvents during room temperature chemical fixation can rearrange cell components, especially lipids [3], and extract soluble cytoplasmic contents [2]. At low temperatures this process slows but does not stop. Therefore, there has to be a balance between sufficient time in substitution media for full replacement of water, and prolonged substitution leading to extraction of cell components. Dedicated freeze substitution units are available which finely control the temperature changes required during substitution, and in some cases can be programmed to control the addition of solvents. These units are expensive to purchase; however, recently a protocol has been developed which negates the need for dedicated freeze substitution units [13].

Here we describe a method for high-pressure freezing and freeze substitution of cells in culture that minimizes mechanical or chemical stress prior to freezing and gives consistent preservation of cellular architecture. The thermal load of the sample is reduced by the use of "naked" sapphires, and by avoiding the use of cryoprotectants or fillers. Reducing the thermal load significantly increases the quality of freezing.

2 Materials

2.1 Chemical Reagents

1. Fetal calf serum.
2. Appropriate cell culture media.
3. Pure methanol.
4. Uranyl acetate (UA) crystals.
5. Analytical grade acetone (99.9 %).
6. 20 % (w/v) solution of UA in methanol (*see* **Notes 1** and **2**).
7. Freeze substitution (FS) medium: 2 % (v/v) uranyl acetate in analytical grade acetone (*see* **Notes 1** and **2**).
8. Lowicryl HM20, made to manufacturers specification (*see* **Note 3**).
9. A plentiful supply of liquid nitrogen.
10. Epoxy resin blocks, previously polymerized in BEEM capsules for mounting samples.

2.2 Hardware

1. Dark glass screw top bottle to store 2 % uranyl acetate solution (at 4 °C).
2. Clear glass screw top bottle for mixing Lowicryl HM20 resin.
3. Ultrafine forceps (long, narrow handles).
4. Cryo forceps for transferring sample tubes to and from the liquid nitrogen dewar.
5. Liquid nitrogen dewar (1 l).
6. Polystyrene liquid nitrogen holder, shallow.

7. Adjustable, illuminated magnifying lamp.

8. High-pressure freezer (e.g., Leica HPM100, or ABRA HPM010) with all necessary associated inserts, spacers, etc.

9. Freeze substitution unit (e.g., Leica AFS2 or equivalent) with all necessary associated containers, embedding molds, etc.

10. Mini hacksaw.

11. Razor blade.

12. Transmission electron microscope, 120 or 200 kV (preferable for tomographical studies).

2.3 Consumables

1. 3 mm sapphire coverslips.

2. Appropriate cell culture plates.

3. Nickel single hole TEM grids.

4. 1.5 ml Eppendorf tubes, screw top, two holes punched below cap.

5. Filter paper.

6. Plastic Pasteur pipettes.

7. Formvar (or equivalent) coated copper 200 mesh, hexagonal, thin bar grids (for thin sections, morphological imaging).

8. Formvar (or equivalent) coated single slot grids (for thick sections, electron tomographical imaging).

3   Methods

Please make sure to follow local chemical safety procedures and ensure appropriate PPE is used.

3.1 Preparation of Cells

1. Choose sapphire discs appropriate to the sample holder associated with your high-pressure freezer (HPF). We use either 3 mm or 6 mm sapphire discs designed for the Leica HPM100 (Leica Microsystems), or 3 mm sapphire discs designed for the ABRA HPM010 (RMC Products). Discs are supplied in solvent, so rinse discs briefly in culture medium and incubate discs in fetal calf serum for 60 min at 37 °C. This provides a proteinaceous layer for cells to adhere to (*see* **Note 4**).

2. In a suitable cell culture vessel (24-, 12-, or 6-well plate) place discs on the base of each well and add the appropriate volume of cell suspension. Ensure the discs are flat on the base of each well and are not floating in the media. Incubate at 37 °C for an appropriate amount of time so that the sapphires are coated with an approximately 80 % confluent cell monolayer.

3. Infect the cells as appropriate.

3.2 Fixation of Cells by High-Pressure Freezing

1. One hour before you want to freeze your samples, start cooling down the high-pressure freezer. Make sure you have enough liquid nitrogen available for the entire process. Fill a small (approx. 1 l) dewar with liquid nitrogen for transporting your samples. Fill a small polystyrene box with liquid nitrogen to use to cool down your tools. Ensure these vessels are in close proximity to the freezer, and that you have the means to refill them as needed.

2. Once the HPF is cool and stable do a test freezing run to check it is working correctly.

3. In an MBSC, remove cell culture medium from wells containing the sapphire discs and replace with warmed, fresh medium.

4. Take the cell culture dishes containing the sapphire discs to the HPF and load the sapphires into the appropriate holder as in Fig. 2. There is no need to add fillers/cryoprotectants (*see* **Note 5**). There is no need to encase in aluminum planchettes (*see* **Note 6**).

5. Quickly load the sample holder into the HPF and freeze.

6. Remove the frozen "sapphire disc sandwich" from the sample holder under liquid nitrogen in the polystyrene box, remembering to use cooled tools to do so. An illuminated magnifying lamp is useful for this step.

7. Remove the screw top and cool a labelled Eppendorf in liquid nitrogen, making sure it has had two holes punched near the top of the tube.

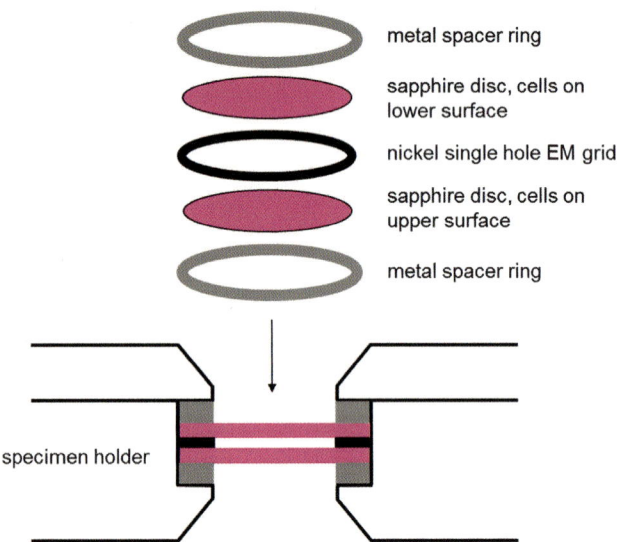

Fig. 2 Diagram illustrating the method used to load a Bal-tec HPM010 (now sold as ABRA HPM010) HPF holder. The method can be adapted for other manufacturers' models; however, the basic "sandwich" structure should remain constant. Spacer rings, sapphire discs and nickel grid all 3 mm in diameter. Sapphire discs thickness 100 μm, nickel grid thickness 50 μm. Diagram from [14]

8. When cold, place the sapphire disc sandwich inside the Eppendorf, replace the screw top, and leave to float in liquid nitrogen until all samples have been frozen and placed in labelled Eppendorf tubes. Quickly transfer these Eppendorf tubes, one by one, to the liquid nitrogen transfer dewar, replace the lid, and transfer to appropriate liquid nitrogen storage. Samples must be stored in the liquid phase, not in the gaseous phase.

3.3 Freeze Substitution

This process is started during the late afternoon and the freeze substitution (FS) unit programmed to start substitution the following morning.

1. One hour before use, cool down the FS unit with liquid nitrogen to its lowest temperature, typically –160 °C. We have an AFS1 and AFS 2 (Leica Microsystems); however, this FS method can be programmed into any other FS unit you may have.

2. Ensure the exhaust of the FS unit chamber is fed into a fume hood, and that you are working in close proximity to the fume hood. It is preferable to place any waste solvent/resin straight into the hood.

3. Place an appropriate number of aluminum cups (supplied with the FS unit) into the cold chamber (one per sample).

4. Using a small liquid nitrogen transfer dewar collect the appropriate samples from the liquid nitrogen store and place the dewar in the fume hood next to the FS unit.

5. Place an Eppendorf containing one sapphire sandwich into each of the cooled aluminum cups. The samples will still be in a small amount of liquid nitrogen. This will boil off as the Eppendorf and chamber equilibrate overnight.

6. Program the FS unit to follow a short FS/short warm up cycle (Fig. 3). For this particular type of sample this protocol was found to be superior to longer protocols (*see* **Note 7**). Here is an example of timings; adjust as necessary although keep the time intervals consistent with the protocol.

Temperature to rise from –196 °C to –160 °C overnight.
05:30 Temperature to rise from –160 °C to –90 °C at 20° per hour (3.5 h)
09:00 Temperature held at –90 °C, addition of FS medium (see below) (1 h)
10:00 Temperature to rise from –90 °C to –50 °C at 20° per hour (2 h)
12:00 Temperature held at –50 °C during resin infiltration

7. The following morning, before 09:00 ensure you have all the appropriate consumables, tools, solutions, and containers for waste within easy reach of the FS unit.

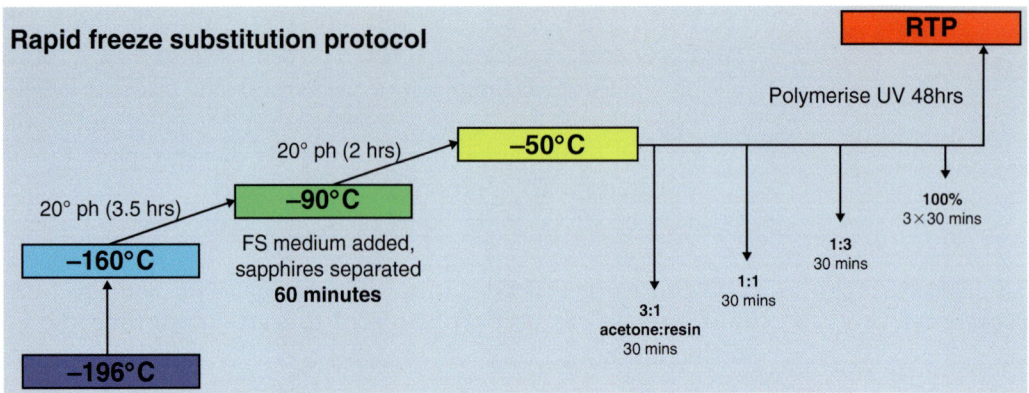

Fig. 3 Diagram summarizing short freeze substitution/short warm up cycle protocol [14]

8. Cool 1–2 ml of FS medium (*see* **Note 1**) in a clean aluminum cup by placing in FS unit chamber.

9. At 09:00, undo screw top from Eppendorf and empty sapphire sandwich into its aluminum cup.

10. Using a cooled plastic pipette gently place 0.5 ml cold FS medium in each aluminum cup so that the sapphire sandwiches are covered. Sandwiches will probably separate in the liquid; take care to note how the sapphire discs separate, so you know which side the cells cover.

11. Using cooled fine forceps ensure the discs lay in the FS medium, cells facing uppermost (*see* **Note 8**). Remove aluminum cup containing any unused FS medium and place in fume hood to get to room temperature before disposal.

12. Leave samples in FS medium until program has reached –50 °C (12:00).

13. Before 12:00, place a clean aluminum cup in the FS unit chamber containing 1 ml Lowicryl HM20 (*see* **Note 3**) and 3 ml acetone. Mix well and leave for a few minutes to cool. Adjust volumes according to the number of samples you have, but keep the ratio the same. The volumes described here assume 4 aluminum cups are present.

14. At 12:00, using a cooled plastic pipette remove the FS medium from each sample aluminum cup and place in a waste aluminum cup also in the chamber.

15. Taking care to note any sapphire movement, gently add 1 ml of cold 1:3 resin–acetone mix to each aluminum cup containing a sample, using a clean cooled pipette. Leave for 30 min.

16. Remove the aluminum cup containing any leftover 1:3 resin mix and place in fume hood.

17. Place a clean aluminum cup into the chamber containing 2 ml Lowicryl HM20 and 2 ml acetone. Mix well and allow to cool.

18. After 30 min, using a cooled plastic pipette remove the 1:3 resin mix from each sample aluminum cup and place in the waste aluminum cup also in the chamber.

19. Taking care to note any sapphire movement, gently add 1 ml of cold 1:1 resin–acetone mix using a clean cooled pipette. Leave for 30 min.

20. Remove the aluminum cup containing any leftover 1:1 resin mix and place in fume hood.

21. Place a clean aluminum cup into the chamber and mix 3 ml Lowicryl HM20 and 1 ml acetone. Allow to cool.

22. After a further 30 min, using a cooled plastic pipette remove the 1:1 resin mix from each sample aluminum cup and place in the waste aluminum cup also in the chamber.

23. Again, taking care to note any sapphire movement, gently add 1 ml of cold 3:1 resin–acetone mix using a clean cooled pipette. Leave for 30 min.

24. Remove the aluminum cup containing any leftover 3:1 resin mix and place in fume hood.

25. Place a clean aluminum cup into the chamber containing 14 ml Lowicryl HM20 resin. Allow to cool.

26. After 30 min, using a cooled plastic pipette remove the 3:1 resin mix from each sample aluminum cup and place in the waste aluminum cup also in the chamber.

27. Taking care to note any sapphire movement, gently add 1 ml of cold resin using a clean cooled pipette. Leave for 30 min.

28. After 30 min, remove the resin and place in the waste aluminum cup in the chamber.

29. Repeat **steps 27** and **28** twice.

30. Cool a flat embedding mold (supplied with FS unit) in the chamber during the last infiltration step.

31. Using cooled tools carefully place each sapphire, cells facing uppermost, into a compartment within the embedding mold, ensuring the disc sits on the metal base, and gently cover with cold Lowicryl resin (*see* **Note 9**—handling Lowicryl resin).

32. Cool a UV transparent plastic cup (supplied with FS unit) and place over the embedding mold. Seal with a small amount of Lowicryl resin so that oxygen is excluded during polymerization.

33. Place the UV light (supplied with FS unit) over the chamber, program in the polymerization protocol, and start the program:
    - 48 h at –50 °C, with UV
    - Temperature rise from –50 °C to room temperature (20 °C) at 20° per hour (3.5 h), with UV
    - 48 h at room temperature, with UV

34. Dispose of waste chemicals according to local regulations.

35. After polymerization remove the embedding mold and push out the hardened blocks. Place in to labelled boxes. Blocks can be stored indefinitely (*see* **Note 10**). When ready to cut sections first prepare some polymerized epoxy resin blocks from BEEM capsules to provide support for the Lowicryl blocks, as follows:

- Remove the pointed end of the polymerized epoxy resin blocks with a mini hacksaw to provide a flat surface.

- Carefully trim the Lowicryl resin block to the region containing the sapphire disc using a fresh razor blade.

- Using epoxy glue, stick the Lowicryl resin block containing the sapphire onto the flat surface of the epoxy resin block, with the sapphire uppermost. Allow to dry.

- With a razor blade, trim the thin layer of Lowicryl resin from around the edges of the sapphire disc.

- Immerse the block briefly in liquid nitrogen to dislodge the sapphire disc. Discard the disc, the cells will remain embedded in the Lowicryl resin.

36. With an ultramicrotome, section the Lowicryl block *en face* to produce sections containing longitudinal views of cells in the monolayer, all in the same orientation and plane (*see* **Note 11**).

37. Collect sections on Formvar-coated grids (*see* **Note 12**). When using the 2 % uranyl acetate freeze substitution medium it is not necessary to add any further contrast.

38. Collect images using a transmission electron microscope (*see* **Note 13**). See [1, 14] for examples of cells prepared using this method.

39. For immunogold labelling studies the procedure above is very similar; however, use a FS medium containing a lower concentration of uranyl acetate, for example 0.2 % [14].

4 Notes

1. FS medium: first, prepare 20 % (w/v) solution of UA in methanol. This will need to be kept on a stirrer in a fume hood for over an hour in order for all the UA crystals to dissolve. Then, prepare 2 % (v/v) uranyl acetate by adding 0.5 ml 20 % UA in methanol to 4.5 ml 99.9 % analytical grade acetone. Mix well. Store in dark glass bottle at 4 °C.

2. It is not necessary to store 99.9 % analytical grade acetone under a molecular sieve to keep water-free (as recommended in some older protocols). During storage under a molecular sieve the acetone discolors over time, so it is possible the sieve adds some form of contaminant. We have never recorded any deleterious effects of using analytical grade acetone which has not been stored under a molecular sieve.

3. Place each resin component into a screw top glass jar and mix the resin by bubbling dry nitrogen gas through the mixture, while holding the lid close to the neck of the jar. This eliminates oxygen from the jar which is important as Lowicryl HM20 does not polymerize effectively in the presence of oxygen.

4. Preliminary experiments using "naked" sapphire discs indicated that cell adhesion was a potential weakness as cells were lost during processing. Collagen IV, fetal calf serum (FCS), Matrigel Basement Membrane Matrix (BD Biosciences), and carbon were tested as substrate pretreatments to aid cell adhesion. Discs were incubated at 37 °C for 60 min in neat FCS, Matrigel or collagen IV solution (made to manufacturers specification), or discs were coated with 10 nm carbon using a high vacuum carbon coater (Agar Scientific). Most retention of cells was seen using FCS as the substrate pretreatment.

5. To remove intercellular air pockets within samples protocols in the past have included inert, non-penetrating "fillers," for example 1-hexadecene. To reduce the possibility of artifact, cryoprotectant/fillers were not used during this protocol. We did not see any detrimental effect of omitting this step, indeed it could be considered advantageous as the fillers act as a heat sink and reduce the cooling rate of the sample and prevent penetration of FS medium.

6. Many protocols protect the sapphire discs during freezing using aluminum planchettes (supplied with the high-pressure freezer). These planchettes act as a barrier between liquid nitrogen and sapphire discs and act as a heat sink, reducing the cooling rate of the sample. In our protocol liquid nitrogen jets directly onto the sapphires producing excellent freezing across the whole sapphire disc.

7. Protocols with long FS times and/or long warm up times were investigated but the short FS and short warm up protocol described here was found to produce consistently good freezing across large areas of the sapphire, provide good ultrastructural detail, allow minimal visible extraction of cytoplasmic contents and have the practical benefit of a shortened procedure.

8. It is possible to see the cells on the sapphire discs down the FS unit binocular eyepieces by picking the disc up, above the liquid surface and slightly angling it towards the light. Cells will appear rough on the glass; if cells are not present the disc will have a smooth, reflective surface.

9. Lowicryl HM20 resin can be difficult to work with at first. It has low surface tension and will creep along surfaces if drops escape from pipettes. It has a very strong odor so keep in the fume hood when possible. When filling the embedding mold,

allow resin to settle and then top up to make sure the volume has been filled correctly. Wipe away any drips from the FS unit chamber with paper towel or else the embedding mold will become stuck to the chamber during polymerization.

10. The resin should have a pink hue which demonstrates total polymerization. However, if the resin is colorless but hard, this indicates sufficient polymerization for cutting.

11. Collect all the sections (even if incomplete) as the material will be close to the surface of the block face. Cells in culture can be very thin so it is easy to cut through the whole cell sheet while waiting to collect a complete section.

12. If collecting thick sections for electron tomography, use Formvar (or equivalent) coated single slot grids. Also collect some thin sections (on Formvar, or equivalent, coated 200 mesh grids) for orientation.

13. Lowicryl HM20 is not as beam stable in the TEM as epoxy resins. The Formvar support will help with this, but avoid long beam exposures at high intensities.

References

1. Maier HJ, Hawes PC, Cottam EM et al (2013) Infectious bronchitis virus generates spherules from zippered endoplasmic reticulum membranes. MBio 4(5)

2. Zierold K (1991) Cryofixation methods for ion localisation in cells by electron probe microanalysis: a review. J Microsc 161: 357–366

3. Maneta-Peyret L, Compere P, Moreau P et al (1999) Immunocytochemistry of lipids: chemical fixatives have dramatic effects on the preservation of tissue lipids. Histochem J 31:541–547

4. Lee RMKW (1984) A critical appraisal of the effects of fixation, dehydration and embedding on cell volume. In: Revel JP, Barnard T et al (eds) The science of biological specimen preparation, Scanning Electron microscopy, Inc., AMF O'Hare, Chicago, IL, pp p61–p70

5. Studer D, Humbel BM, Chiquet M (2008) Electron microscopy of high pressure frozen samples: bridging the gap between cellular ultrastructure and atomic resolution. Histochem Cell Biol 130:877–889

6. Studer D, Michel M, Muller M (1989) High pressure freezing comes of age. Scanning Microsc Suppl 3:253–268

7. Moor H, Riehle U (1968) Snap-freezing under high pressure: a new fixation technique for freeze-etching. Proc 4th Eur Reg Conf Electron Microsc 2: 33–34

8. Studer D, Graber W, Al-Amoudi A et al (2001) A new approach for cryofixation by high-pressure freezing. J Microsc 203:285–294

9. Vanhecke D, Graber W, Studer D (2008) Close-to-native ultrastructural preservation by high pressure freezing. Methods Cell Biol 88:151–164

10. Sartori N, Richter K, Dubochet J (1993) Vitrification depth can be increased more than 10-fold by high pressure freezing. J Microsc 172:55–61.11

11. McDonald KL (2014) Out with the old and in with the new: rapid specimen preparation procedures for electron microscopy of sectioned biological material. Protoplasma 251:429–448

12. Studer D, Chiquet M, Hunzkier EB (1996) Evidence for a distinct water-rich layer surrounding collagen fibrils in articular cartilage extracellular matrix. J Struct Biol 117:81–85

13. McDonald KL, Webb RI (2011) Freeze substitution in 3 hours or less. J Microsc 243: 227–233

14. Hawes PC, Netherton CL, Mueller M et al (2007) Rapid freeze-substitution preserves membranes in high-pressure frozen tissue culture cells. J Microsc 226:182–189

INDEX

Helena Jane Maier et al. (eds.), *Coronaviruses: Methods and Protocols*, Methods in Molecular Biology, vol. 1282,
DOI 10.1007/978-1-4939-2438-7, © Springer Science+Business Media New York 2015

Printed by Printforce, the Netherlands